T0327625

Chemical Ecology of Insect Parasitoids

To Maura, Eliana, Emilio and Ludovica

Chemical Ecology of Insect Parasitoids

Edited by

Eric Wajnberg

Institut National de la Recherche Agronomique (INRA)
Sophia Antipolis Cedex
France

and

Stefano Colazza

Department of Agricultural and Forest Sciences
University of Palermo
Palermo
Italy

WILEY-BLACKWELL

A John Wiley & Sons, Ltd., Publication

Library of Congress Cataloging-in-Publication Data

Chemical ecology of insect parasitoids / edited by Eric Wajnberg and Stefano Colazza.
     pages cm
  Includes bibliographical references and index.
  ISBN 978-1-118-40952-7 (cloth)
1. Semiochemicals. 2. Plant chemical ecology. 3. Parasitoids. 4. Plant parasites. 5. Insect-plant relationships. I. Wajnberg, E. II. Colazza, Stefano.
  SB933.5.C47 2013
  632'.7–dc23

                        2012049864

A catalogue record for this book is available from the British Library.

Wiley also publishes its books in a variety of electronic formats. Some content that appears in print may not be available in electronic books.

Cover image: A parasitoid wasp (*Cotesia vestalis*) and a diamondback moth (*Plutella xylostella*) larva on a broccoli leaf. Photograph courtesy of Jarmo Holopainen. The molecule pictured on the right is 4-methylquinazoline, and the molecule on the left is (4R,5S)-5-hydroxy-4-decanolide.
Cover design by Steve Thompson

Set in 10/12.5 pt Minion by Toppan Best-set Premedia Limited

1   2013

Contents

Contributors

T. Martijn Bezemer
Department of Terrestrial Ecology
Netherlands Institute of Ecology (NIOO-KNAW)
P.O. Box 50
6700 AB Wageningen
The Netherlands

Maria Carolina Blassioli-Moraes
Embrapa
Genetic Resources and Biotechnology
Caixa Postal 02372
70770-917 Brasília-DF
Brazil

Miguel Borges
Embrapa
Genetic Resources and Biotechnology
Caixa Postal 02372
70770-917 Brasília-DF
Brazil

Stefano Colazza
Department of Agricultural and Forest Sciences
University of Palermo
Viale delle Scienze, 13
90128 Palermo
Italy

Antonino Cusumano
Department of Agricultural and Forest Sciences
University of Palermo
Viale delle Scienze, 13
90128 Palermo
Italy

Nina E. Fatouros
Wageningen University
Laboratory of Entomology
Department of Plant Sciences
PO Box 8031
6700 EH Wageningen
The Netherlands

Geoff M. Gurr
E.H. Graham Centre for Agricultural Innovation
NSW Department of Primary Industries and Charles Sturt University
P.O. Box 883
Orange, NSW, 2800
Australia

Jeffrey A. Harvey
Department of Terrestrial Ecology
Netherlands Institute of Ecology (NIOO-KNAW)
P.O. Box 50
6700 AB Wageningen
The Netherlands

Sari J. Himanen
MTT Agrifood Research Finland
Plant Production Research
Lönnrotinkatu 5
50100 Mikkeli
Finland

Jarmo K. Holopainen
Department of Environmental Science
University of Eastern Finland
P.O. Box 1627
70211 Kuopio
Finland

Martinus E. Huigens
Dutch Butterfly Conservation
PO Box 506
6700 AM Wageningen
The Netherlands

Ryoko T. Ichiki
Japan International Research Centre for Agricultural Sciences
Owashi
Tsukuba
Ibaraki 305-8686
Japan

Yooichi Kainoh
Faculty of Life and Environmental Sciences
University of Tsukuba
Tsukuba
Ibaraki 305-8572
Japan

Raul Alberto Laumann
Embrapa
Genetic Resources and Biotechnology
Caixa Postal 02372
70770-917 Brasília-DF
Brazil

Torsten Meiners
Freie Universität Berlin
Department of Applied Zoology/Animal Ecology
Haderslebener Str. 9
12163 Berlin
Germany

Satoshi Nakamura
Japan International Research Centre for Agricultural Sciences
Owashi
Tsukuba
Ibaraki 305-8686
Japan

Paul J. Ode
Colorado State University
Department of Bioagricultural Sciences and Pest Management
Fort Collins, CO 80523-1177
USA

Timothy D. Paine
University of California, Riverside
Department of Entomology
Riverside, CA 92521
USA

Ezio Peri
Department of Agricultural and Forest Sciences
University of Palermo
Viale delle Scienze, 13
90128 Palermo
Italy

Guy M. Poppy
Centre for Biological Sciences
Building 85
University of Southampton
Highfield Campus
Southampton SO17 1BJ
UK

Donna M.Y. Read
School of Agriculture and Wine Science
Charles Sturt University
P.O. Box 883
Orange, NSW, 2800
Australia

Michael Rostás
Bio-Protection Research Centre
PO Box 84
Lincoln University 7647
Canterbury
New Zealand

Joachim Ruther
Institute of Zoology
Chemical Ecology Group
University of Regensburg
Universitätsstrasse 31
93053 Regensburg
Germany

Marja Simpson
E.H. Graham Centre for Agricultural Innovation
NSW Department of Primary Industries and Charles Sturt University
P.O. Box 883
Orange, NSW, 2800
Australia

Roxina Soler
Department of Terrestrial Ecology
Netherlands Institute of Ecology (NIOO-KNAW)
P.O. Box 50
6700 AB Wageningen
The Netherlands

Eric Wajnberg
INRA
400 Route des Chappes
BP 167
06903 Sophia Antipolis Cedex
France

Nicole Wäschke
Freie Universität Berlin
Department of Applied Zoology/Animal Ecology
Haderslebener Str. 9
12163 Berlin
Germany

1

Chemical ecology of insect parasitoids: towards a new era

Stefano Colazza[1] and Eric Wajnberg[2]

[1] Department of Agricultural and Forest Sciences, University of Palermo, Italy
[2] INRA, Sophia Antipolis Cedex, France

Abstract

Over the course of evolutionary time, insect parasitoids have developed diverse strategies for using chemical compounds to communicate with various protagonists within their environment (i.e. conspecifics, their hosts, and the plants on which their hosts are living). Unravelling the evolutionary meaning of such chemical communication networks not only provides new insights into the ecology of these insects but also contributes to improving the use of parasitoids for the control of insect pests in biological control programmes. A book covering our current knowledge of the chemical ecology of insect parasitoids is therefore particularly timely and will appeal to a large number of potential readers worldwide, from university students to senior scientists. Internationally recognized specialists were invited to contribute chapters to this book, examining the main topics and exploring the most interesting issues in the field of chemical ecology of insect parasitoids. The chapters are organized so as to present the most significant knowledge and discoveries made over recent decades, and their potential uses in pest control.

1.1 Introduction

For several million years, plants, insects and their natural enemies have coevolved on the basis of information flows within food webs (Krebs & Davies 1987). As a consequence, they were – and still are – continuously exposed to selection pressures which drive evolution according to a process referred to as an 'arms race' (Dawkins & Krebs 1979). Different ecological features of interacting species can evolve in response to selection pressures, leading species and their populations to evolve and improve their reproductive success.

Chemical Ecology of Insect Parasitoids, First Edition. Eric Wajnberg and Stefano Colazza.
© 2013 John Wiley & Sons, Ltd. Published 2013 by John Wiley & Sons, Ltd.

Most of the time, such responses cause individuals to react more effectively to signals coming from their biotic and abiotic environment. Among the different types of signals that can influence these ecological interactions, chemical cues, called semiochemicals (from the Greek '*semeion*', a mark or signal), play the major role (Nordlund 1981, Vinson 1985, Vet & Dicke 1992). These compounds can be classified into two groups, named pheromones and allelochemicals.

Pheromones (from the Greek '*pherein*', to carry, and '*horman*', to excite or stimulate) are chemical signals that mediate interactions between individuals of the same species, and they are often described on the basis of their function (Wyatt 2010). Since their discovery, many pheromones have been identified and synthesized and a number of techniques have been developed to use them in Integrated Pest Management (IPM) programmes against insect pests (Ridgway *et al.* 1990). In particular, pheromones are widely used in IPM to monitor insect pest populations and to interfere with their behaviour, thus reducing or preventing agricultural damage (Witzgall *et al.* 2010). On the other hand, signals that operate interspecifically (between different species) are termed 'allelochemicals' and may be called synomones, kairomones or allomones, depending on their ecological and biological functions (Dicke & Sabelis 1988, Ruther *et al.* 2002).

The study of the ecological functions of semiochemicals is the main subject of chemical ecology (Ruther *et al.* 2002, Eisner 2003, Bergstrom 2007, Colazza *et al.* 2010, Wortman-Wunder & Vivanco 2011). In the past two decades, a plethora of studies have demonstrated the importance of chemical cues for ecological processes at the individual, population and ecosystem levels (Takken & Dicke 2006). However, although there have been considerable advances in understanding, there are still many critical questions that lack answers (Meinwald & Eisner 2008). The new possibilities offered by genomic and proteomic tools will undoubtedly result in increased understanding over the coming years (Kessler & Baldwin 2002, Vermeer *et al.* 2011).

Parasitoids represent fascinating model organisms for evolutionary and ecological studies because of their species richness, ecological impact and economic importance (Godfray 1994, Wajnberg *et al.* 2008). They belong mainly to two orders, Hymenoptera and Diptera. Within the Hymenoptera, there are about 45 families containing parasitoids. Within the Diptera, most species of parasitoids occur in the family Tachinidae.

Evidence that semiochemicals can modify the behaviour of insect natural enemies has inspired researchers to explore the possibility of using semiochemicals to conserve and/or enhance the efficacy of natural enemies in cropping systems (Pickett *et al.* 1997, Khan *et al.* 2008). However, the use of semiochemicals integrated with natural enemies in IPM is still limited, despite the fact that important research has been done in recent years to elucidate the interactions between semiochemicals and natural enemies in a multitrophic context (see Soler, Bezemer & Harvey, Chapter 4, this volume).

During the last few decades, many studies have investigated the chemically mediated foraging behaviour of parasitoids in an attempt to understand the factors that guide parasitoids to their hosts. In order to provide a quantitative overview of this work, a literature survey was performed. This consisted of interrogating the Scopus database using the term 'parasitoids'. In total, we obtained 10,463 references published within the period 1935–2011 (about 140 papers published per year). Among these, 458 (i.e. 4.38%) also used the terms 'synomone', 'kairomone' and/or 'allomone'. Figure 1.1 summarizes the cumulative number of references published over the years and the frequency of the use of the terms 'synomone', 'kairomone' and/or 'allomone'.

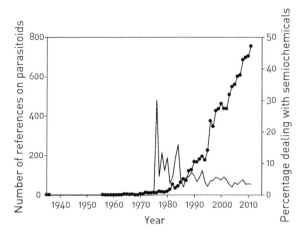

Figure 1.1 Changes in the number of references found in the Scopus abstracts and citations database for the period 1935–2011, using the term 'parasitoids' (*line with dots*) and the frequency of those that also used the terms 'synomone', 'kairomone' and/or 'allomone' (*line without dots*).

From about the mid-1970s onwards there has been an increasing interest in publishing papers dealing with parasitoids, coupled with a tendency to focus on chemical ecology. After about a decade, the frequency of papers using the words 'synomone', 'kairomone' and/ or 'allomone' stabilized at around 4.4%. The most likely reason for this is not so much a decline in interest in the chemical ecology of these model animals, but an increase in interest in other aspects of the biology and ecology of parasitoids, for example behavioural ecology (see below). In other words, there was – and still is – a constant interest in understanding the mechanisms involved in the way that insect parasitoids produce and use chemical compounds to communicate with the various different protagonists within their environment (i.e. conspecifics, their hosts, and the plants on which their hosts are living). Unravelling the evolutionary meaning of such a chemical communication network can provide new insights into the ecology of these insects, and especially on how to improve their use for the control of harmful pests in biological control programmes. Therefore, a book covering the current state of knowledge on the chemical ecology of insect parasitoids seems particularly timely and capable of appealing to a large number of potential readers worldwide.

1.2 Integrating behavioural ecology and chemical ecology in insect parasitoids

In 2008, Wajnberg *et al.* (2008) edited a book on the behavioural ecology of insect parasitoids covering cutting-edge research into decision-making processes in insect parasitoids and their implications for biological control. The goal of behavioural ecology is to understand the behavioural decisions adopted by animals to maximize their long-term reproductive success, and for this research theoretical maximization models are frequently used. We visualize this current book on the chemical ecology of insect parasitoids as a complementary volume to Wajnberg *et al.* (2008), extending our understanding and knowledge of insect parasitoids and their use for controlling pests in biological control programmes. After all, chemical ecology – and especially that of insect parasitoids – and behavioural ecology are both based on the accurate observation and analysis of animal behaviour, and the final goal of both volumes is to find ways to improve the efficacy of these insects in controlling pests and protecting crops.

Research efforts in chemical ecology have mainly been based on the development of chemical tools (i) to identify the chemical compounds involved in the way in which insect parasitoids interact with their environment, both at the intra- and interspecific levels; (ii) to understand the associated metabolic pathways; and (iii) to synthesize these compounds for use in enhancing pest control strategies through field-release applications. Such a chemical-based approach has resulted in a rapid improvement in understanding over the past decades, as can be seen in the different chapters within this book. However, we think that it has also precluded the development of theoretical research that aims to understand the optimal decision-making strategy that should be adopted by these insects in terms of releasing and/or perceiving chemical signals in their environment. The fact is that insect parasitoids have been subjected to extreme evolutionary pressure, over the course of time, to use such chemical tools and there are obviously cost/benefit ratio issues that need to be optimized in order to increase the ability of these animals to contribute genetically to the following generations.

The reason that such a theoretical approach has not yet been sufficiently developed, in our view, is most probably because the chemical ecology community mainly consists of scientists who have more expertise in chemistry than in theoretical ecology. There are, however, a handful of studies that have used optimality models to illustrate the chemical ecology of insect parasitoids. For example, Hoffmeister & Roitberg (1998) developed a theoretical model to identify the optimal persistence duration (or decay rate) of a contact pheromone used by a herbivorous insect to signal the presence of its own eggs to both the marking female and conspecifics. The point is that this pheromone is also exploited by a specialized parasitoid that attacks the herbivore's offspring, so there is an evolutionary game being played between needing to signal the presence of eggs to conspecific females while avoiding their detection by parasitoids. The approach developed by these authors effectively takes into account the physiological costs associated with the marking strategy. Examples like this are still rare, and one goal of the present volume is to foster research in this area, bridging the gap between the behavioural and chemical ecology of insect parasitoids, with the final aim of developing more efficient biological control programmes.

1.3 The use of chemical ecology to improve the efficacy of insect parasitoids in biological control programmes

In recent years, significant progress in understanding insect behaviour and advances in analytical chemistry have led to the identification and production of thousands of semi-ochemical compounds. Important research has been conducted to investigate how these compounds can be used commercially, and many of them are now contributing to established practices in Integrated Pest Management (IPM) (Suckling & Karg 2000, Witzgall *et al.* 2010). In this respect, an increasing knowledge of the influence of semiochemicals on parasitoid and predator behaviour has opened up new possibilities in pest control strategy.

Semiochemical-based manipulations normally include either 'pheromone-based tactics' or 'allelochemical-based tactics'. Pheromone-based tactics now represent one of the major strategies in ecologically based orchard pest management, leading to considerable success in both direct and indirect insect pest control. The most successful applications for the direct control of pest populations concern the release of sex pheromones to disrupt mating in the target pests (Witzgall *et al.* 2010). In contrast, allelochemical-based tactics represent a relatively new approach that mainly uses plant volatiles. The most promising application of allelochemical-based tactics involves the use of herbivore-induced plant volatiles (HIPVs) to manipulate the natural enemies of the pest species in order to attract and conserve them in the vicinity of the crops to be protected. HIPVs are semiochemicals that mediate many multitrophic interactions in both above- and below-ground plant–insect communities (Soler *et al.* 2007, Soler, Bezemer & Harvey, Chapter 4, this volume). These volatiles have received increased attention for their role in attracting natural enemies of insect pests (Ode, Chapter 2, this volume). In the last decade, the results of several field experiments have been published demonstrating that the release of HIPVs can indeed augment, conserve or enhance the efficacy of natural enemies. However, allelochemical-based tactics, especially based on the use of HIPVs, are lagging far behind the development of applications of pheromones. In this respect, it has to be noted that genetically modified plants have recently been shown to provide new opportunities for semiochemical applications. For example, plants can be engineered to produce (E)-β-farnesene to mimic the natural aphid alarm response in order to increase foraging by aphid predators and parasitoids (Yu *et al.* 2012).

1.4 Overview

Remarkable advances in our understanding of the chemical ecology of insect parasitoids have occurred in recent years. In this book, we have assembled papers written by internationally recognized experts who are at the forefront of their field. The chapters are organized in order to present the most important knowledge and discoveries made over the past few decades, and on their potential use in pest control strategy. In addition to this introductory chapter, the book contains 12 chapters, organized into two parts. The first part addresses the basic aspects of parasitoid behaviour, and the second focuses on possible strategies for manipulating the behaviour of insect parasitoids to increase their pest control ability by means of chemical cues in different ecosystems and under different agricultural practices. Specific relevant case studies are also presented.

The first part of the book starts with a chapter focusing on plant defence responses to a diversity of pests and abiotic stressors, and their effects on insect parasitoids (Chapter 2). Plants are indeed an important component of the foraging environment of insect parasitoids, orchestrating the presence of complex signalling networks available to these insects. This is the topic addressed by the following chapter, which considers in detail the role of volatile and non-volatile compounds, coupled with biotic and abiotic factors, in shaping the variability of these chemicals in both time and space (Chapter 3).

Chemical signals are not only relevant at the above-ground level. There are also signals at the below-ground level that are involved in structuring the ecology of plant–insect interactions. Understanding the role played by these different signals requires a multitrophic approach, and that is what is discussed in the following chapter (Chapter 4). Chemical signals are also used by phoretic insect parasitoids that 'hitch-hike' on hosts in order to increase access to potential hosts to attack. Several important and fascinating studies have been carried out on this topic, which is addressed in Chapter 5. In the past decade, astonishing progress has been made to further our understanding of pheromone-mediated communication in parasitic wasps, especially for mate finding and recognition, aggregation, or host-marking behaviour, and this is presented in Chapter 6. The two remaining chapters of the first part of the book address the particular case of dipteran tachinid species that demonstrate oviposition strategies which differ from those adopted by hymenopteran wasps (Chapter 7), and the potential consequences of climate change on the chemical ecology of insect parasitoids in general (Chapter 8).

The second part of this book focuses on applications for biological control. The first chapter (Chapter 9) starts by providing a detailed overview of how semiochemicals can be used to manipulate the foraging behaviour of insect parasitoids in order to increase their impact on pest populations. This can be done either through facilitating their ability to locate and attack their hosts, or by increasing their recruitment within agroecosystems. The following chapters then address the application of chemical cues for enhancing the pest control efficacy of parasitic wasps in arable crops (Chapter 10), orchards and vineyards (Chapter 11), organic cropping systems (Chapter 12) and forest trees (Chapter 13).

1.5 Conclusions

This book is intended for anyone interested in understanding how insects, and parasitic wasps in particular, use chemical compounds to communicate with others and to discover resources to exploit. The book will be of interest to research scientists and their students working in the academic world (research centres and universities) and also to teachers in graduate schools and universities that teach insect chemical ecology. Furthermore, the background information gathered together in this book could be used to encourage high-school students and stimulate research in the field of chemical ecology of insect parasitoids. Biological control practitioners will also find the technical information needed to improve pest control efficacy through the release of insect parasitoids in the field.

Throughout the book, critical research questions are explicitly identified, acknowledging the gaps in current knowledge. Ultimately, the goal of the book is to foster synergistic research that will eventually lead to a better understanding of the fields of chemical and behavioural ecology of parasitic wasps.

Acknowledgements

We thank Antonino Cusumano for assistance in interrogating the Scopus database, and Helen Roy and Ezio Peri for critical comments on this chapter. We also wish to thank the referees who read and commented critically on one or more chapters. They include Miguel Borges, Stefano Colazza, Jeff Harvey, Jarmo Holopainen, Martinus Huigens, Yooichi Kainoh, Jocelyn Millar, Satoshi Nakamura, Paul Ode, Tim Paine, Ezio Peri, Guy Poppy, Michael Rostás, Helen Roy, Joachim Ruther, Roxina Soler and Eric Wajnberg. Finally, we express our sincere thanks to the staff at Wiley-Blackwell for their excellent help and support during the production of this book.

Although much editing work has been done, the information provided within each chapter remains the sole responsibility of the individual authors.

References

Bergstrom, G. (2007) Chemical ecology = chemistry + ecology! *Pure and Applied Chemistry* **79**: 2305–23.

Colazza, S., Peri, E., Salerno, G. and Conti, E. (2010) Host searching by egg parasitoids: exploitation of host chemical cues. In: Cônsoli, F.L., Parra, J.R.P. and Zucchi, R.A. (eds) *Egg Parasitoids in Agroecosystems with Emphasis on Trichogramma*. Springer, Dordrecht, pp. 97–147.

Dawkins, R. and Krebs, J. (1979) Arms races between and within species. *Proceedings of the Royal Society of London B* **205**: 489–511.

Dicke, M. and Sabelis, M.W. (1988) Infochemical terminology: based on cost–benefit analysis rather than origin of compounds. *Functional Ecology* **2**: 131–9.

Eisner, T. (2003) Chemical ecology: can it survive without natural products chemistry? *Proceedings of the National Academy of Sciences USA* **100**: 14517–8.

Godfray, H.C.J. (1994) *Parasitoids: Behavioural and Evolutionary Ecology*. Princeton University Press, Princeton, NJ.

Hoffmeister, T.S. and Roitberg, B.D. (1998) Evolution of signal persistence under predator exploitation. *Ecoscience* **5**: 312–20.

Kessler, A. and Baldwin, I.T. (2002) Plant responses to insect herbivory: The emerging molecular analysis. *Annual Review of Plant Biology* **53**: 299–328.

Khan, Z.R., James, D.G., Midega, C.A.O. and Pickett, C.H. (2008) Chemical ecology and conservation biological control. *Biological Control* **45**: 210–24.

Krebs, J.R. and Davies, N.B. (1987) *An Introduction to Behavioural Ecology*. Blackwell Publishing Ltd., Oxford.

Meinwald, J. and Eisner, T. (2008) Chemical ecology in retrospect and prospect. *Proceedings of the National Academy of Sciences USA* **105**: 4539–40.

Nordlund, D.A. (1981) Semiochemicals: a review of the terminology. In: Nordlund, D.A., Jones, R.J. and Lewis, W.J. (eds) *Semiochemicals: Their Role in Pest Control*. John Wiley & Sons, New York, pp. 13–28.

Pickett, J.A., Wadhams, L.J. and Woodcock, C.M. (1997) Developing sustainable pest control from chemical ecology. *Agriculture, Ecosystems and Environment* **64**: 149–56.

Ridgway, R.L., Silverstein, R.M. and Inscoe, M.N. (1990) *Behavior-Modifying Chemicals for Insect Management: Applications of Pheromones and other Attractants*. Marcel Dekker, New York.

Ruther, J., Meiners, T. and Steidle, J.L.M. (2002) Rich in phenomena – lacking in terms: a classification of kairomones. *Chemoecology* **12**: 161–7.

Soler, R., Harvey, J.A. and Bezemer, T.M. (2007) Foraging efficiency of a parasitoid of a leaf herbivore is influenced by root herbivory on neighbouring plants. *Functional Ecology* **21**: 969–74.

Suckling, D.M. and Karg, G. (2000) Pheromones and semiochemicals. In: Rechcigl, J. and Rechcigl, N. (eds) *Biological and Biotechnical Control of Insect Pests*. CRC Press, Boca Raton, FL, pp. 63–99.

Takken, W. and Dicke, M. (2006) Chemical ecology: a multidisciplinary approach. In: Dicke, M. and Takken, W. (eds) *PhD Spring School Chemical Communication: From Gene to Ecosystem*. Wageningen, 19–23 March 2005. Springer, Dordrecht, pp. 1–8.

Vermeer, K.M.C.A., Dicke, M. and de Jong, P.W. (2011) The potential of a population genomics approach to analyse geographic mosaics of plant–insect coevolution. *Evolutionary Ecology* **25**: 977–92.

Vet, L.E.M. and Dicke, M. (1992) Ecology of infochemical use by natural enemies in a tritrophic context. *Annual Review of Entomology* **37**: 141–72.

Vinson, S.B. (1985) The behavior of parasitoids. In: Kerkut, G.A. and Gilbert, L.I. (eds) *Comprehensive Insect Physiology, Biochemistry, and Pharmacology, Vol. 9*. Pergamon Press, Elmsford, NY, pp. 417–69.

Wajnberg, E., Bernstein, C. and van Alphen, J. (2008) *Behavioral Ecology of Insect Parasitoids: From Theoretical Approaches to Field Applications*. Blackwell Publishing Ltd., Oxford.

Witzgall, P., Kirsch, P. and Cork, A. (2010) Sex pheromones and their impact on pest management. *Journal of Chemical Ecology* **36**: 80–100.

Wortman-Wunder, E. and Vivanco, J.M. (2011) Chemical ecology: definition and famous examples. In: Vivanco, J.M. and Weir, T. (eds) *Chemical Biology of the Tropics: An Interdisciplinary Approach*. Springer-Verlag, Berlin, Heidelberg, pp. 15–26.

Wyatt, T.D. (2010) Pheromones and signature mixtures: defining species-wide signals and variable cues for individuality in both invertebrates and vertebrates. *Journal of Comparative Physiology A* **196**: 685–700.

Yu, X.D., Pickett, J., Ma, Y.Z., Bruce, T., Napier, J., Jones, H.D. and Xia, L.Q. (2012) Metabolic engineering of plant-derived (*E*)-β-farnesene synthase genes for a novel type of aphid-resistant genetically modified crop plants. *Journal of Integrative Plant Biology* **54**: 282–99.

Part 1
Basic concepts

2

Plant defences and parasitoid chemical ecology

Paul J. Ode

Department of Bioagricultural Sciences and Pest Management, Colorado State University, Fort Collins, USA

Abstract

Plants play a central role in the chemical ecology of most insect parasitoids. While parasitoid interactions with other organisms can be loosely categorized as those involving semiochemical communication with other insects (e.g. pheromones, kairomones, synomones) and those involving plant defensive chemicals, there is a tremendous amount of overlap between these rather artificial categories. Although certainly not all aspects of parasitoid chemical ecology can be viewed through the lens of plant defensive chemistry, a surprising amount can; and this viewpoint is an excellent way of examining the complexity of some of the issues involved. Parasitoid chemical-based communication has been well studied, particularly in terms of the use of plant- and/or herbivore-associated volatiles, which are released following herbivore damage, as kairomones to locate their insect hosts. Likewise, plant defence responses link parasitoids with the effects of an impressive array of stressors (including herbivores, plant pathogens, and abiotic factors associated with climate change). Plant defensive toxins often have strong negative effects on insect herbivores, which in turn may reduce host quality and ultimately parasitoid fitness. In other cases, herbivorous hosts may actively sequester plant defensive chemistry as a defence against parasitoids and predators. In still other cases, plant defensive chemistry may compromise the ability of herbivorous insects to mount a successful immune response against parasitoids, resulting in increased fitness of these parasitoids. While the vast majority of parasitoid chemical ecology studies to date have focused on relatively simple pairwise aspects of these relationships, parasitoid chemical ecology is influenced by multiple trophic levels spanning several kingdoms as well as a wide array of abiotic factors. In this chapter, I argue that an

Chemical Ecology of Insect Parasitoids, First Edition. Eric Wajnberg and Stefano Colazza.
© 2013 John Wiley & Sons, Ltd. Published 2013 by John Wiley & Sons, Ltd.

understanding of plant defence responses to a diversity of attackers and abiotic stressors is important to understanding the chemical ecology of many insect parasitoids. A more holistic approach to parasitoid chemical ecology can yield novel insights into not only how parasitoids relate to their environment, but also how multitrophic community relationships are structured and maintained.

2.1 Introduction

Like all insects, parasitoids interact with other organisms in ways that are largely chemically mediated. Parasitoids attract and locate mates through the use of sex pheromones. Ovipositing females use a wide variety of host-derived and herbivore-induced plant volatiles (HIPVs) to locate suitable hosts, sometimes even using host sex pheromones or epideictic pheromones as kairomones to locate host eggs. Other chemically mediated interactions are decidedly negative, notably those involving plant chemical defences against herbivores that also decrease parasitoid fitness. Not surprisingly, given the vast number of parasitoids whose hosts are herbivorous insects, plant defensive chemistry plays a prominent role in the chemical ecology of parasitoids. Plant defensive chemistry is well documented to have negative effects on parasitoid fitness by reducing the quality of their herbivorous hosts. In some cases, plant defensive chemistry has been shown to exhibit negative effects on the third trophic level when parasitoids are directly exposed to plant toxins. Plant defence signalling pathways link responses to plant pathogens – and even to abiotic factors such as drought and greenhouse gases – with insect herbivores and, by extension, their parasitoids. While I am not suggesting that all facets of parasitoid chemical ecology can be explained by focusing on plant defensive chemistry, a large proportion of insect parasitoids are directly or indirectly influenced by plant chemistry, and such a focus does go a long way in advancing our understanding of how parasitoids are integral members of the wide variety of communities and ecosystems in which they exist.

The vast majority of studies involving parasitoid chemical ecology have taken bitrophic or tritrophic approaches focusing on host–parasitoid relationships and, to a lesser extent, plant–herbivore–parasitoid relationships (Price et al. 1980, Ode 2006). Nearly all of these studies have considered systems with only one species per trophic level. However, parasitoids obviously live in a much more complex world consisting of multiple species per trophic level and interact with far more than just the trophic level or two below. Chemical signals that parasitoids receive are embedded in an incredibly complex array of other signals and noise, out of which parasitoids have to decipher a reliable message (Hilker & McNeil 2008; see also Wäschke, Meiners & Rostás, Chapter 3, this volume). It is increasingly recognized that above-ground trophic interactions influence and are influenced by trophic interactions that occur below ground and that these above-ground and below-ground interactions are mediated through changes in plant defensive chemistry (see Soler, Bezemer & Harvey, Chapter 4, this volume). Few studies have thus far examined plant defences against more than one species of simultaneous attackers and their effects on parasitoids. Those few studies that have focused on defence against multiple simultaneous attackers demonstrate complex and, often non-additive, effects on herbivore and parasitoid preference and performance patterns (e.g. Rodriguez-Saona et al. 2005). Furthermore, plant defences against chewing herbivores often influence and are influenced by changes in plant defences against bacterial, viral and fungal pathogens (Stout *et al.* 2006). Such multi-kingdom interactions are probably ubiquitous. As ecologists increasingly make use of the genomic tools available, we are making increasing strides in our understanding of how parasitoids interact with the biotic world around them. Finally, we are increasingly aware of how climate change (e.g. changes in temperature and rainfall) is affecting trophic relationships that involve parasitoids (Stireman *et al.* 2005, see also Holopainen, Himanen & Poppy, Chapter 8, this volume). In this chapter, I provide a brief overview of these areas, highlighting how a holistic, systems-oriented view of parasitoid chemical ecology will advance our understanding of the importance of this widespread and species-rich group of organisms.

2.2 Plant defences against a diversity of attackers

Most parasitoid species attack either insects that feed on plants or the natural enemies of insect herbivores. As autotrophs, plants are key components of the majority of food webs on Earth. Not only are plants the source of food for an amazing variety of herbivores (both invertebrate and vertebrate), they are also sources of nutrition for a wide array of micro-organisms (e.g. viruses, bacteria, fungi). Not surprisingly, plants employ diverse, often specialized, and integrated defences against their various attackers. Plant defences can be broadly categorized as being constitutive (expressed in the same pattern regardless of whether the plant is attacked by herbivores and/or pathogens) or induced (expressed at markedly elevated levels in response to herbivory and/or pathogen attacks) (Agrawal 2007). Examples abound of how both types of chemical defences can negatively affect the parasitoids of insect herbivores either through compromised host quality or through direct exposure to unmetabolized plant toxins encountered in the haemolymph of their hosts (Harvey 2005, Ode 2006). Likewise, HIPVs, such as green leaf volatiles (GLVs) and volatile terpenes, released by plants upon herbivore attack and tissue damage, are known to attract parasitoids and predators of insect herbivores (Heil 2008, Gols *et al.* 2011). Attack by plant pathogens may induce defences that not only affect the likelihood of further attack by other plant pathogens but may also influence the expression of defences against a wide range of insect herbivores (e.g. Hatcher 1995, Stout *et al.* 2006, Gange *et al.* 2012). Attack by root herbivores may induce defences that interact with defences against above-ground herbivores, with possible consequences for their natural enemies (van Dam & Heil 2011, Schausberger *et al.* 2012, Soler *et al.* 2012a). In order to understand how plant attackers that are not hosts of the parasitoid can affect plant defensive chemistry that influences host herbivores and their parasitoids, it is useful to consider how plant defence signalling pathways are activated and expressed.

2.2.1 Plant defence signalling pathways

Whereas induced resistance to plant pathogens has been acknowledged for much of the past century, the realization that plants may also respond to herbivore attack by subsequently increasing their investment in defence has only emerged since the early 1970s (Green & Ryan 1972, Karban & Baldwin 1997). Most of our understanding of plant responses to herbivore damage has come from studies involving chewing insects that cause extensive damage to plant tissue (Karban & Baldwin 1997, Smith *et al.* 2009). These studies have focused primarily on the Solanaceae (e.g. tomato and tobacco) and the Brassicaceae (notably, *Arabidopsis thaliana*). More recently, studies have examined how plants respond to piercing/sucking insects that feed in one location and typically cause minimal physical damage (Walling 2000, 2008). Plant responses to damage begin with signals (elicitors) from the damage source that activate a cascade of gene expression, resulting in toxin and/or volatile production that serves as induced resistance against future attacks. Different attackers are known to elicit different combinations of plant defence pathways (Walling 2000). Three phytohormones (salicylic acid (SA), jasmonic acid (JA), and ethylene (ET)) are well studied in terms of their regulation of plant defence pathways (Howe 2004, van Loon *et al.* 2006, von Dahl & Baldwin 2007, Koornneef & Pieterse 2008), although other phytohormones (e.g. abscisic acid (ABA)) are also important regulators of plant defence pathways (Robert-Seilaniantz *et al.* 2011) and novel pathways are likely to be involved in at least some systems (Walling 2000). While there are many important exceptions (see review by Stout *et al.* 2006), plants are typically protected from biotrophic plant pathogens (which must feed on living plant tissue) and many piercing/sucking insects through the action of SA-dependent defence pathways, whereas plants are typically protected from chewing insects and necrotrophic pathogens (which kill plant tissues before consuming them) by JA-/ET-dependent defence pathways (Kessler & Baldwin 2002, Koornneef & Pieterse 2008).

Piercing/sucking insects and biotrophic plant pathogens
Many piercing/sucking insects are similar to biotrophic plant pathogens in that their feeding results in limited, localized damage to plant tissues. Feeding occurs in one location for extended periods of time, often days or weeks. Therefore, it is not surprising that piercing/sucking insects that feed on phloem (e.g. aphids, whiteflies, leafhoppers) initiate plant defence pathways that are similar to those activated by biotrophic plant pathogens (Walling 2000). Elicitors from plant pathogens (e.g. lipids, polysaccharides, peptides) bind with plant receptors, releasing reactive oxygen species (ROS; see Fig. 2.1) that induce a hypersensitive response (HR; see Fig. 2.1), killing nearby plant cells and producing antibiotics that limit the spread of pathogens (Walling 2000, Kessler & Baldwin 2002, Stout *et al.* 2006, Smith *et al.* 2009). This defence strategy can be very effective against biotrophic plant pathogens because they require living plant tissue on which to survive. Reactive oxygen species also stimulate the production of SA, which results in systemic acquired resistance (SAR) that induces resistance to a broad array of plant pathogens as well as piercing/sucking insects such as whiteflies and aphids. Elicitors from the salivary secretions of phloem-feeding (e.g. aphids; see Lapitan *et al.* 2007) and cell-content-feeding insects (e.g. thrips) typically induce similar increases in the expression of pathogen resistance genes in plants, as do plant pathogens (Walling 2000, 2008, Smith *et al.* 2009). Other plant pathogens induce JA- and ET-regulated defence pathways, resulting in induced systemic resistance to such pathogens as well as many other piercing/sucking insects (Walling 2000). SA suppresses the JA-/ET-dependent pathway as well as JA-regulated wound responses (see 'Wound responses and chewing insects' below), and JA-regulated wound responses suppress SA-regulated expression of SAR. These suppressive interactions (i.e. 'cross-talk'; see Section 2.2.4) between different defence pathways are thought to be critical in modulating plant defence responses to these attackers. Furthermore, elicitors from the salivary secretions of some whiteflies induce a novel defence pathway that involves neither JA nor SA (Walling 2000).

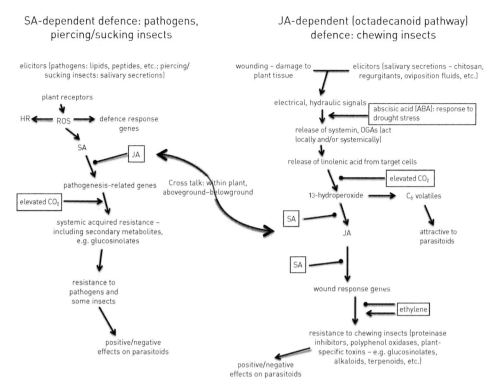

SA-dependent defence: pathogens, piercing/sucking insects

JA-dependent (octadecanoid pathway) defence: chewing insects

Figure 2.1 SA- and JA-dependent plant defence pathways (compiled from several sources, especially Walling 2000, Kessler & Baldwin 2002, Skibbe *et al.* 2008) and the effects of the major plant hormones and environmental stressors (outlined in *boxes*) on these pathways. Effects with an *arrow* represent positive effects on the pathway. Effects ending with a *closed circle* represent inhibitory effects on the defence pathway. See text for details. SA: salicylic acid, JA: jasmonic acid, ROS: reactive oxygen species, HR: hypersensitive response, OGA: oligogalacturonides.

Compared with the production of volatiles in response to damage caused by chewing insects, relatively little is known about the pathways involved in volatile production in response to damage by phloem-feeding insects. Plants attacked by aphids are known to produce blends of terpenoids and C_6 volatiles that are attractive to parasitoids (Du *et al.* 1998, Páre & Tumlinson 1999, Williams *et al.* 2005).

Wound responses and chewing insects
Unlike piercing/sucking insect herbivores, chewing insects cause extensive tissue damage as they continually move throughout the plant and sometimes between plants to consume new plant tissues. Physical damage caused by chewing insect herbivores results in gene expression in plants similar, but not identical, to that resulting from mechanical wounding (Walling 2000). Responses to chewing insects are often highly species-specific and include local and systemic production of deterrents, digestion inhibitors, toxins, as well as volatiles that both repel would-be herbivores and sometimes attract natural enemies of these herbivores. JA, a key part of the octadecanoid (C_{18} fatty acids; oxylipin) pathway, largely coordinates these responses, which are best studied in solanaceous plants such as tomato

(for more extensive details, see Walling 2000, Kessler & Baldwin 2002). Initial plant responses to wounding include electrical, hydraulic or chemical signals, which may act locally or be transported systemically through the phloem and/or xylem, to initiate the octadecanoid pathway. One of the best-studied signals is systemin, an oligopeptide that is the primary wound signal transported in the phloem of several species of Solanaceae. Systemin-binding proteins in the plasma membrane of target cells trigger the release of linolenic acid, a key substrate for the octadecanoid pathway. Herbivore-specific elicitors from salivary and regurgitant secretions are thought to modify defence responses at the systemin wound cascade. The enzyme lipoxygenase (LOX) converts linolenic acid into 13-hydroperoxide. One pathway for 13-hydroperoxide involves hydrolysis by hydroperoxide lyase (HPL) to produce traumatin (thought to be involved in wound healing) and C_6 volatiles (e.g. green leaf volatiles, many of which serve to deter feeding by insect herbivores as well as attracting their natural enemies; see Heil 2008). The other major pathway for 13-hydroperoxide involves conversion to JA through a series of enzymatic reactions. JA and its conjugates activate a suite of wound response genes. Expression of wound response genes results in the production of defensive compounds toxic to attacking herbivores, volatiles that are repellent or toxic to herbivores and often attractive to their natural enemies, and resistance to future attack by herbivores and pathogens (Karban & Baldwin 1997, Bostock 1999, Walling 2000, Kessler & Baldwin 2002). Anti-herbivore compounds resulting from the octadecanoid pathway include anti-digestive proteins (proteinase inhibitors), anti-nutritive enzymes (polyphenol oxidases), and a suite of plant-specific toxins (e.g. alkaloids, glucosinolates, furanocoumarins) (Memelink *et al.* 2001, Kessler & Baldwin 2002, De Geyter *et al.* 2012). The plant hormone abscisic acid (ABA) is known to activate the octadecanoid pathway (Fig. 2.1), whereas the hormone ethylene may have antagonistic or enhancing effects on wound response gene expression, depending on the plant–herbivore system involved (Fig. 2.1).

While plant responses to wounding or to damage done by chewing insects are similar in many respects, it is important to recognize that these responses are not equivalent. Compared to mechanical wounding, herbivore damage typically results in increased volatile production and the production of herbivore-specific blends of volatiles (Dicke 1999, Páre & Tumlinson 1999). Herbivore-specific elicitors from the regurgitant and oral/salivary secretions of feeding insect herbivores result in elevated production of JA and greater expression of wound response genes (e.g. via the WRKY superfamily of transcription factors and other transcription factors that appear to function upstream of JA) than that observed from wounding alone (Walling 2000, Skibbe *et al.* 2008). WRKY transcription factors are proteins produced by plants in response to a wide variety of biotic and abiotic stressors that regulate gene expression. Similarly, oviposition fluids of some bruchid beetles act as herbivore-specific elicitors that result in elevated defences against beetle larvae (Doss *et al.* 2000). In some cases, the application of regurgitant in the absence of wounding is sufficient to initiate the octadecanoid defence pathway (Felton & Tumlinson 2008, Hilker & Meiners 2010). Furthermore, elicitors from some insect regurgitants express defence genes that appear to be independent of the JA/octadecanoid pathway. Such elicitors enable plants to tailor their defensive responses against specific herbivores.

2.2.2 Plant volatiles and parasitoids

Plant volatiles released in response to chewing herbivore damage may serve as direct defences when they are interpreted by herbivores as repellents or indicators of crowded feeding conditions (e.g. Kessler & Baldwin 2001). Frequently, volatiles are thought to act as indirect plant defences when they serve as reliable signals of the presence of suitable hosts and prey for parasitoids and predators (Vet & Dicke 1992, De Moraes *et al.* 1998, Kessler & Baldwin 2001, Turlings & Wäckers 2004). Plant volatiles play a key role in the host location and acceptance decisions made by parasitoids (Vet & Dicke 1992, Dicke & Vet 1999, Turlings *et al.* 2002). Volatiles include C_6 compounds ('green leaf volatiles') derived from linolenic acid via the octadecanoid pathway (see above) and terpenoids from the isoprenoid pathway, among others. Field studies using transformed lines of the wild tobacco *Nicotiana attenuata* – where three genes coding enzymes (lipoxygenase, hydroperoxide lyase and allene oxide) in the octadecanoid pathway involved in volatile production had been silenced – were more vulnerable to herbivory by the specialist *Manduca quinquemaculata* and produced lower levels of the volatiles that potentially attract natural enemies of this herbivore (Kessler *et al.* 2004). Oral secretions of *Manduca sexta* elicit the production of volatile terpenoids by field populations of *N. attenuata*. WRKY-silenced plants produced significantly lower levels of JA and volatile terpenes (e.g. *cis*-α-bergamotene) and were less attractive to predators of *M. sexta* (Skibbe *et al.* 2008). Exogenously applied JA restored volatile terpene production by these plants. It remains unclear whether herbivore-induced plant volatiles (HIPVs) represent an active form of plant defence (e.g. a 'cry for help'; see Price *et al.* 1980, Dicke & Baldwin 2010) or are simply adopted by parasitoids as reliable signals of the presence and status of their hosts. If HIPVs are selected because they attract natural enemies of herbivores, then signalling theory suggests that HIPVs should be costly for the plant to produce (Godfray 1995, Vet & Godfray 2008). Parasitoids that reduce herbivore load on plants are expected to have a positive effect on plant fitness, although many koinobionts (parasitoids that permit their hosts to continue feeding) may induce their hosts to feed more, resulting in even more damage to the plant (Godfray 1994, Coleman *et al.* 1999). No studies in natural systems to date have measured the effects of volatile production and attractiveness to herbivore natural enemies on plant fitness, preventing an assessment of the ecological significance of herbivore-induced volatiles as indirect plant defences (van der Meijden & Klinkhamer 2000, Dicke & Baldwin 2010, Hare 2011). Insect parasitoids, on the other hand, clearly benefit from the production of herbivore-induced plant volatiles. Many parasitoids rely on distinct volatile blends released by herbivore-damaged plants to locate appropriate host species at the appropriate developmental stage. As is true of non-volatile induced defences, feeding by herbivores often results in different volatile profiles than wounding alone, and these are more attractive to parasitic wasps (e.g. Turlings *et al.* 1990, Dicke 1999, Páre & Tumlinson 1999). Feeding damage and pre-digestive salivary secretions from chewing herbivores frequently induce the production and release of herbivore-specific sets of volatiles that attract their parasitoids (Dicke 1999, Páre & Tumlinson 1999, Heil 2008). In some cases, the volatile organic compounds released by herbivore-attacked plants can be highly specific to particular plant–herbivore combinations. Parasitoids can use these specific volatiles to locate specific host species and even hosts at an appropriate developmental stage (Turlings & Benrey 1998). An elegant study of the braconid *Cardiochiles nigriceps* shows just how highly specific such interactions involving plant volatiles and parasitoids can be (De Moraes *et al.* 1998). *C. nigriceps* females are attracted strongly to the volatile profiles emitted by tobacco plants attacked by its preferred host *Heliothis virescens*, but not to the volatile profiles emitted when plants are attacked by the closely related *Helicoverpa zea*. This preference is due to differences in the volatile profiles released by tobacco when attacked by these two herbivores. Furthermore, these volatile differences and wasp preferences persist even when damaged plant tissues and herbivores are removed, indicating that volatile production is systemically induced. Interestingly, similar patterns were found with the same parasitoid and herbivores on cotton and corn, suggesting that highly specialized plant–host herbivore–parasitoid interactions based on herbivore-specific induced plant volatiles are widespread (De Moraes *et al.* 1998). Not only can plant defence responses differ as a function of herbivore species and even life stage, but they can also differ as a function of which parasitoid species has attacked a given herbivore. In a study of cabbage, *Brassica oleracea*, Poelman *et al.* (2011) found that plant defence responses to two species of herbivores, *Pieris brassicae* and *P. rapae*, were strongly dependent on which species of parasitoid had attacked these herbivores. *Pieris* regurgitant quality was influenced by the species of parasitoid by which they were attacked, apparently resulting in differential gene expression related to JA signalling. This, in turn, differentially influenced the subsequent oviposition preference of the diamondback moth, *Plutella xylostella* (Poelman *et al.* 2011). Another herbivore-specific elicitor of plant defences involves the cabbage white butterfly (*Pieris brassicae*) and the production of benzyl cyanide, an anti-aphrodisiac transferred from male butterflies to females during mating. Benzyl cyanide is transferred to the plant surface during oviposition, where it induces plant gene expression that promotes intensive foraging by *Trichogramma brassicae*, an egg parasitoid of *P. brassicae* (Fatouros *et al.* 2008). Other examples of oviposition-induced plant defences exist (Hilker & Meiners 2006).

2.2.3 Plant toxins and parasitoids

Plant toxins are known to provide defence against insect herbivores. Plant defensive toxins may be produced constitutively and/or they may be induced by herbivore feeding damage through the plant defence signalling pathways discussed above (Memelink *et al.* 2001, De Geyter *et al.* 2012). Unlike the herbivore-induced plant volatiles discussed above, plant defensive chemistry often has adverse effects on parasitoid fitness when their herbivorous insect hosts feed on plants containing these toxins (Ode 2006). Negative effects of plant toxins on parasitoid fitness arise either because host herbivore quality is decreased or because developing parasitoids directly encounter unmetabolized toxins in the tissues of their herbivorous hosts. Whereas the interactions between HIPVs and parasitoid foraging and acceptance behaviours are relatively well characterized ecologically, chemically, and even in terms of the signalling pathways involved, much less is known about how parasitoids interact with plant defensive toxins consumed by their herbivorous insect hosts. Many studies have documented host plant species or cultivar differences in their effects on parasitoid fitness measures, and it is likely that differences in plant defensive chemistry profiles explain such patterns (see reviews by Price *et al.* 1980, Hare 1992, 2002, Hunter 2003, Harvey 2005, Ode 2006). For instance, differences in the glucosinolate profiles of wild cabbage populations are correlated strongly with fitness effects on herbivores and their larval and pupal parasitoids (Gols *et al.* 2009, Harvey & Gols 2011). Far fewer studies have clearly documented the tritrophic effects of specific plant defensive chemicals on parasitoid fitness. Those that have, typically make use of artificial diets for herbivore hosts to which defined quantities of plant toxins have been added. A classic example of this approach involved the parasitoid *Hyposoter exiguae*, which suffered increased mortality and morphological deformities when developing in tomato fruitworm (*Heliothis zea*) hosts fed an artificial diet containing high concentrations of the glycoalkaloid α-tomatine (Campbell & Duffey 1979). This approach is limited to those host species for which artificial diets have been developed. Another approach that shows great promise in establishing the tri-trophic effects of plant defence chemistry is the use of experiments that manipulate the JA signalling pathway, either via exogenous applications of JA-mimics or through gene-silencing techniques, to alter the production of defensive toxins. However, to the extent that these approaches have included insect parasitoids, such studies have generally focused on changes in HIPVs attractive to parasitoids and predators (so-called 'indirect defences'). An exception is a study where JA exogenously applied to tomato plants positively affected parasitism rates (by 37%) of the beet armyworm *Heliothis zea* by the parasitoid *Hyposoter exiguae*, but also negatively affected parasitoid performance (reduced pupal weight and increased development time) (Thaler 1999). Such manipulations can help to establish the causal connections between plant chemistry and parasitoid fitness.

A handful of studies suggest that generalist herbivores and their parasitoids are more susceptible to plant toxins than specialist host–parasitoid associations are, implying that generalist herbivores are less well equipped to detoxify plant toxins than specialists and that these differences have consequences at higher trophic levels (e.g. Lampert *et al.* 2011a). For instance, the generalist noctuid moth *Spodoptera frugiperda* and its generalist parasi-toid, *Hyposoter annulipes* were more negatively affected by concentrations of nicotine in their artificial diet than were the specialist herbivore *Manduca sexta* and its specialist para-sitoid *Cotesia congregata* (Barbosa *et al.* 1986, 1991). Another artificial diet study involving this system documented the negative effects of nicotine across all three trophic levels: *M. sexta*, *C. congregata* and its hyperparasitoid *Lysibia nana* (Harvey *et al.* 2007). Similar experiments using artificially selected lines of the ribwort plantain *Plantago lanceolata*, which differ in their iridoid glycoside concentrations, suggested that these compounds have negative effects on generalist, but not specialist, herbivores and their parasitoids (Harvey *et al.* 2005).

The degree to which plant defensive chemistry negatively affects parasitoids depends in large part on whether their hosts sequester plant toxins as defences against their natural enemies as well as on the detoxification abilities of their hosts. Several herbivorous insects are known to sequester toxins from plants that they feed upon and use these as defences against their parasitoids and predators (e.g. Nishida 2002, Ode 2006, Lampert *et al.* 2011b) and at least one study has shown that parasitoids and their hyperparasitoids may seques-ter plant toxins that accumulate in their herbivorous host (van Nouhuys *et al.* 2012). Studies differentiating the negative effects of plant defensive chemistry on host quality from the effects of direct exposure of plant toxins on developing parasitoids are rare. An exception to this is presented in a series of studies involving a wild parsnip furanocou-marin (xanthotoxin), the association between the specialist parsnip webworm (*Depres-saria pastinacella*) and its specialist parasitoid (*Copidosoma sosares*), and the association between the generalist cabbage looper (*Trichoplusia ni*) and its oligophagous parasitoid (*Copidosoma floridanum*). Xanthotoxin incorporated into an artificial diet had a substan-tially more negative effect on *C. floridanum* survival and clutch size than it did on *C. sosares* survival and clutch size (Lampert *et al.* 2011a). The negative effect of xanthotoxin on *C. floridanum* was partly explained by a reduction in host quality. Compared with *D. pastinacella*, *T. ni* pupal weight and survival were reduced when reared on artificial diets containing xanthotoxin. Because *T. ni* is much less efficient at metabolizing xanthotoxin than is *D. pastinacella*, more xanthotoxin passes unmetabolized into the haemolymph of *T. ni*. Therefore, developing *C. floridanum* embryos and larvae are exposed to much higher levels of xanthotoxin than developing *C. sosares* embryos and larvae, despite their hosts having fed on diets containing the same concentrations of xanthotoxin (Lampert *et al.* 2011a). Xanthotoxin does have negative effects on *D. pastinacella* and *C. sosares*, but only at much higher concentrations than are lethal to *T. ni* and *C. floridanum* (Lampert *et al.* 2008, 2011a).

Little is known about the ability of parasitoids to detoxify the plant toxins that they encounter during their larval development. In large part, this is because few studies have examined whether parasitoids directly encounter and consume plant toxins. One example is the parasitoid *Hyposoter exiguae*, which encounters both rutin and α-tomatine in the haemolymph of its host, *Heliothis zea* (Campbell & Duffey 1981, Bloem & Duffey 1990). Studies of *C. sosares* and *C. floridanum* indicate that while embryos and larvae do encounter unmetabolized xanthotoxin in their respective hosts, they are unable to metabolize it. Instead, the effects of this plant defensive chemical on these two parasitoids are mediated by their respective hosts' metabolic capabilities (McGovern *et al.* 2006, Lampert *et al.* 2008, 2011a). A few studies have shown that parasitoids possess functional cytochrome P450 systems and are able to metabolize xenobiotics such as pesticides (Ode 2006, Oakeshott *et al.* 2010). Furthermore, variation in field populations of some parasitoids and the ability of some species to respond to artificial selection for increased resistance indicates genetic variation for the ability to metabolize toxins (Ode 2006). Clearly, much more work is needed to determine the extent to which plant chemistry acts as a selective force on parasitoids and, in turn, the extent to which parasitoids can respond to variation in plant defences and function as indirect defences against insect herbivores.

On the other hand, plant toxins consumed by insect herbivores may have positive effects on parasitoids if the ingested plant toxins compromise the immune system of the host herbivore. One of the primary immune responses of herbivore larvae against parasitoids is encapsulation, a process whereby host haematocytes form a melanized, multiple-layer capsule around a parasitoid egg or first instar, resulting in asphyxiation of the parasitoid (Strand & Pech 1995, González-Santoyo & Córdoba-Aguilar 2012). In one such study, small cabbage white butterfly larvae (*Pieris rapae*) were less able to encapsulate eggs of the parasitoid *Cotesia glomerata* when the butterfly larvae fed on wild cabbage plants that contained higher glucosinolate concentrations (Bukovinszky *et al.* 2009). Host plant quality (presumably due to differences in plant chemistry) has been shown to influence the number of haematocytes produced as well as the activity of phenoloxidase, a key enzyme in the melanization process, in the cabbage looper *Trichoplusia ni* (Shikano *et al.* 2010). Similarly, larvae of the buckeye butterfly *Junonia coenia* are less able to mount a successful encapsulation response to injected glass beads when fed a diet with high concentrations of iridoid glycosides (Smilanich *et al.* 2009). On the other hand, some insect hosts are known to 'self-medicate', whereby they preferentially consume plants higher in toxins when they are parasitized than they would if not parasitized. In these cases, the benefits of getting rid of the parasites appear to outweigh the costs of ingesting plant material that is toxic. The generalist arctiid moth *Grammia incorrupta* preferentially feeds on plants higher in toxic pyrrolizidine alkaloids when parasitized by tachinid flies (Singer *et al.* 2009, Smilanich *et al.* 2011). Finally, larval *Drosophila melanogaster* parasitized by the eucoilid wasp *Leptopilina heterotoma* preferentially feed on diets higher in alcohol, which is lethal to the parasitoid, making it more likely that the fly will successfully survive parasitism (Milan *et al.* 2012).

Whether parasitoids have a positive selective impact on plant fitness such that plants invest less in chemically based defences remains an intriguing, yet unresolved, question. If parasitoids are effective as agents of indirect defence and if plant toxins have negative effects on parasitoids, selection for a reduced plant investment in chemical defences should mitigate the negative effects on parasitoids. Furthermore, if plant defensive chemistry is costly, then the effective reduction in herbivore pressure by parasitoids is expected to select for a reduction in investment in plant defences. Given the widespread successful use of parasitoids (and predators) in insect biological control programmes, there is good reason to believe that parasitoids do have a positive effect on plant fitness. Furthermore, there are several studies that show increased plant reproductive success when parasitoids are present (see review in Ode 2006). To date, no studies have conclusively shown that parasitoids exert sufficient selection on plant fitness through the reduction of herbivore pressure such that plants are able to invest less in costly chemical-based defences (but see Ode *et al.* 2004 for suggestive, circumstantial evidence of this in the wild parsnip–parsnip webworm–*Copidosoma sosares* tritrophic system).

2.2.4 Cross-talk between plant defence pathways

As discussed in Section 2.2.1, induced defences against phloem-feeding insects and biotrophic pathogens are typically associated with SA-based plant defences, whereas induced defences against chewing insect herbivores, some phloem-feeding insects, and necrotrophic pathogens are associated with JA-/ET-based plant defences (Koornneef & Pieterse 2008, Smith *et al.* 2009, Thaler *et al.* 2012). While there are certainly many important exceptions (Thaler *et al.* 2004, Stout *et al.* 2006), these generalizations about associations between classes of attackers and the types of plant defence pathways they induce do hold up in a wide range of systems (Kessler & Baldwin 2002, Koornneef & Pieterse 2008, Smith *et al.* 2009). Individual plants often experience sequential or even simultaneous attack from multiple herbivores and plant pathogens over the course of their lifespan, inducing a diversity of plant defence pathways within a plant. Induced plant defences exact significant fitness costs, and one means by which plants are thought to regulate these costs is through the coordinated activation of different defence pathways (Thaler *et al.* 2012).

One such regulatory mechanism is 'cross-talk' between defence pathways such that activation of one defence pathway modulates the expression of another pathway. The strength and direction of cross-talk is highly context-dependent. SA, when produced in sufficient quantities, is known to suppress production of JA as well as interfere with JA-mediated defence gene expression. Similarly, JA interferes with the expression of pathogenesis-related protein genes that are part of the SA defence pathway (Walling 2000). In general, higher levels of SA or JA result in antagonistic, suppressive interactions with the other defence pathway. Lower levels of JA or SA either have no effect or may even have synergistic effects on the other defence pathways (Smith *et al.* 2009). Whether suppression of JA defence pathways by SA (and vice versa) occurs depends on the timing of the expression of the different pathways (Thaler *et al.* 2002). If too much time has elapsed between different classes of attackers, there is little likelihood of cross-talk.

Transcriptional factors such as WRKY are also known to mediate cross-talk between the JA and SA defence pathways (Li *et al.* 2004). Furthermore, several additional phytohormones (e.g. ABA, auxins, brassinosteroids, cytokinins, gibberellic acid) are known to function as positive or negative regulators of these defence pathways (Robert-Seilaniantz *et al.* 2011). Cross-talk is generally viewed as a mechanism by which plants are able to exhibit flexibility and adaptability in their responses to a diversity of attackers (Koornneef & Pieterse 2008), although this idea needs formal testing (Thaler *et al.* 2012). Negative cross-talk between JA- and SA-mediated defence pathways frequently results in enhanced defence against one class of attackers at the expense of increased susceptibility to another class of attackers. For instance, tobacco (*Nicotiana tabacum*) plants inoculated with tobacco mosaic virus systemically induce SA, which – in addition to slowing the spread of this virus – inhibits the production of JA. Consequently, this increases susceptibility to damage by the tobacco hornworm *Manduca sexta* (Orozco-Cardenas *et al.* 1993, Preston *et al.* 1999). Conversely, oral secretions from the generalist beet armyworm (*Spodoptera exigua*) induce ET, which in turn suppresses SA in wild tobacco (*N. attenuata*), allowing JA-based defences to work against this herbivore (Diezel *et al.* 2009). In general though, while JA may inhibit the SA defence pathway, the suppressive effect is often not as strong as the reverse (Thaler *et al.* 2002). ET and the cross-talk regulatory gene NPR1 (non-expressor of pathogenesis-related protein gene) appear to be important in adjusting this balance in favour of JA-based defences directed against chewing insect herbivores (Thaler *et al.* 2012). Few molecular studies of cross-talk have included a bioassay that examines the effects of cross-talk on insect herbivores (e.g. Cui *et al.* 2005) and very few studies have examined JA–SA cross-talk in the field (e.g. Thaler *et al.* 1999, Rayapuram & Baldwin 2007). Although JA–SA cross-talk probably has fitness consequences for the plant, studies documenting these costs have not yet been carried out (Thaler *et al.* 2012).

While cross-talk is an important strategy for plants to manage a suite of defence responses against a multitude of attackers, this strategy is often co-opted by these same attackers. The bacterial pathogen *Pseudomonas syringae* produces coronatine, a JA-mimic, which activates the JA pathway and suppresses the SA pathway, allowing it to thrive in *Arabidopsis thaliana* (Brooks *et al.* 2005, Cui *et al.* 2005). Oviposition by the specialist *Pieris brassicae* and the generalist *Spodoptera littoralis* trigger cellular changes in *Arabidopsis* that induce the SA pathway and suppress the JA defence pathway. It is interesting that larval *S. littoralis* (but not larval *P. brassicae*, presumably because this species is adapted to glucosinolate defences through a nitrile-specifier protein in its midgut; Wittstock *et al.* 2004) growth rates were higher on plants receiving egg extract, suggesting that this herbivore has manipulated the plant's defence cross-talk mechanisms to its own benefit (Bruessow *et al.* 2010). That *P. brassicae* has no shared evolutionary history with *A. thaliana* is interesting and perhaps suggests that the ability to manipulate the plant's cross-talk mechanisms is phylogenetically conserved. Similarly, salivary secretions from the beet armyworm *S. exigua* have been shown to induce the systemic acquired resistance (SAR) pathway, thereby inhibiting the production of glucosinolate-based defences in *Arabidopsis* (Weech *et al.* 2008). Collectively, these studies suggest that various plant attackers are able to disable plant defence pathways directed against themselves through the induction of another plant defence pathway and subsequent cross-talk.

Cross-talk and parasitoids

As discussed above, several studies, mostly involving plants in the Brassicaceae or Solanaceae, have examined the effects of exogenously applied SA or JA (or their mimics) on various insect herbivores and their parasitoids. Given the relationships between the SA and JA pathways described above, it may be reasonable to infer that cross-talk between plant defence pathways affects parasitoids. However, few studies have explicitly examined the consequences of cross-talk on insect parasitoids. Plant pathogens may stimulate the SA-dependent defence pathway at the expense of compromising the expression of the JA-dependent defence pathway. As a consequence, chewing herbivores and their parasitoids may be the beneficiaries of cross-talk. In one recent field study, a lepidopteran leafminer (*Tischeria ekebladella*) developed more quickly and suffered increased parasitism rates when it fed on oak trees (*Quercus robur*) infected with a powdery mildew (*Erysiphe alphitoides*) (Tack *et al.* 2012). Because the leafminer and the powdery mildew do not have direct contact with one another, these interactions are likely to be mediated by cross-talk between SA- and JA-based defences, although the activity of these pathways was not measured in this study. In another field study of a plant pathogen (*Podosphaera plantaginis*), a plant (*Plantago lanceolata*), a herbivore (*Melitaea cinxia*) and a parasitoid (*Cotesia melitaearum*), *C. melitaearum* showed mixed fitness responses when they attacked *M. cinxia* butterfly larvae that fed on plants infected with the fungus. Parasitoids were smaller but their broods tended to be female-biased (van Nouhuys & Laine 2008). In a greenhouse study, peanuts (*Arachis hypogaea*) infected with white mould fungus (*Sclerotium rolfsii*) emitted plant volatiles that were more attractive to the beet armyworm (*Spodoptera exigua*) and its parasitoid, *Cotesia marginiventris* (Cardoza *et al.* 2003). Not all plant pathogen–plant–herbivore–parasitoid interactions are positive. Brood sizes of the parasitoid *Copidosoma bakeri* are smaller when their noctuid hosts (*Agrotis ipsilon*) feed on perennial ryegrass (*Lolium perenne*) infected with the endophytic fungus *Neotyphodium lolii*, which produces toxic ergot alkaloids (Bixby-Brosi & Potter 2012). Two studies have examined the interactions of phloem-feeding insects and chewing insects and their parasitoids. One study involves two herbivores, the phloem-feeding aphid *Macrosiphum euphorbiae* and the chewing herbivore *Spodoptera exigua*, which induce the SA- and JA-regulated defence pathways, respectively, in tomato (Rodriguez-Saona *et al.* 2005). Tomato plants attacked by aphids are preferred host plants for *S. exigua*, although prior aphid attack had no effect on the parasitoid *Cotesia marginiventris*. The most explicit examination to date of the effects of cross-talk between plant defence responses involves the chewing herbivore *Pieris brassicae* and its parasitoid *Cotesia glomerata*, the phloem-feeding *Brevicoryne brassicae* and its parasitoid *Diaeretiella rapae*, and cabbage (*Brassica oleraceae*) (Soler *et al.* 2012b). *P. brassicae* caterpillars and their parasitoids developed faster and were larger when plants were also attacked by aphids, suggesting that aphid attack induced SA-based defence, which suppressed JA-based defences. Supporting this, the authors found that transcript levels of two genes coding for JA biosynthesis were suppressed in aphid-infested plants (Soler *et al.* 2012b). Caterpillar suppression of SA-based defences was much weaker than aphid suppression of JA-based defences. More work explicitly focusing on the effects of cross-talk on insect parasitoids is warranted, especially as parasitoids (and other natural enemies) are widely acknowledged as indirect plant defences. Parasitoids are integral components of most, if not all, plant–insect interactions. Ignoring the role of parasitoids severely compromises our understanding of how plant–plant pathogen–herbivore communities function. Parasitoids probably influence herbivory pressures on plants and thereby the selective impact of herbivores on plant defences. Not only may cross-talk between SA- and JA pathways explain complex interactions between different classes of herbivores and their parasitoids, they may also explain interactions between plant pathogens, insect herbivores and their parasitoids, as well as aboveground–below-ground interactions (see below).

2.3 Above-ground–below-ground interactions and parasitoids

Plants are hosts to a diverse community of above-ground and below-ground herbivores and their natural enemies, pollinators, pathogens, and mutualist fungal/bacterial associates, each with their unique relationships. Historically, the interactions of below-ground and above-ground communities of herbivores and pathogens with roots and shoots have been studied independently, in part because these communities are physically separated. However, it is increasingly recognized that above-ground and below-ground herbivores and plant pathogens have potentially strong effects on one another (see reviews by Bezemer & van Dam 2005, Kaplan *et al.* 2008, van Dam 2009, van Dam & Heil 2011; see also Soler, Bezemer & Harvey, Chapter 4 , this volume). Above-ground and below-ground interactions are necessarily plant-mediated and potentially involve a wide array of herbivores and pathogens. These interactions may be positive, negative or neutral, depending on the identity and number of species involved. Such interactions (especially those involving plant pathogens and chewing herbivores) between above-ground and below-ground communities are frequently mediated by cross-talk between SA- and JA-regulated induced defence pathways. Furthermore, the effects of below-ground herbivores on above-ground herbivores are often stronger than the reverse pattern, possibly reflecting source–sink relationships between roots and shoots (Bezemer *et al.* 2003, Bezemer & van Dam 2005, Erb *et al.* 2008, Kaplan *et al.* 2008; but see Soler *et al.* 2007, Pierre *et al.* 2011). Many of these studies focus on aphids (phloem feeders) as the above-ground herbivores. On the other hand, feeding by chewing herbivores below-ground may have adverse effects on above-ground chewing herbivores (Bezemer & van Dam 2005). For instance, feeding damage by cabbage root fly larvae (*Delia radicum*) induced higher glucosinolate levels in the leaves of *Brassica nigra*, which was correlated with reduced above-ground herbivory by *Phyllotreta* spp. flea beetles (Soler *et al.* 2009). Induced defence signals triggered by feeding on the roots are systemically transported to the shoots and leaves. Similarly, root pathogens may induce systemic (systemic acquired resistance; SA-signalling pathway) defences that are expressed in the shoots, providing protection against foliar pathogens (Pieterse *et al.* 2002). These results are broadly suggestive of cross-talk between chewing herbivores that induce the JA-dependent signalling pathway and phloem feeders that respond to the SA-dependent signalling pathway. In turn, above-ground–below-ground herbivore interactions alter volatile profiles and attractiveness to natural enemies (e.g. Rasmann & Turlings 2007). Further complicating the relationship between below-ground and above-ground herbivory is the suggestion that below-ground herbivory results in water stress and ABA responses to drought that influence above-ground herbivores (Erb *et al.* 2011). Separating such effects is not a trivial undertaking.

Nearly all studies of the effects of below-ground–above-ground herbivore interactions on parasitoids have focused on the production of volatiles and their role in attracting parasitoids. In one study, root herbivory by scarab beetle and weevil larvae on the marsh thistle *Cirsium palustre* increased parasitism of a tephritid seed predator by *Pteromalus elevatus* and *Torymus chloromerus* (Masters *et al.* 2001). Other studies have demonstrated negative effects of below-ground herbivory on the host searching and acceptance behaviour of parasitoids (van Dam 2009). Below-ground mutualistic associations with mycorrhizal fungi may affect plant interactions with above-ground herbivores. In several cases, mycorrhizal associates increase above-ground plant growth, which was of benefit to above-ground insect herbivores (van Dam & Heil 2011). A study of seven herbaceous dicots and grasses demonstrated that below-ground mycorrhizal associates increased both above-ground biomass and above-ground herbivore performance (Kempel *et al.* 2010). However, the increase in plant biomass and herbivore performance disappeared if these plants had previously been attacked by herbivores, which presumably induced plant defences against chewing herbivores. The arbuscular mycorrhizal fungus *Glomus mosseae* alters the HIPV emissions of bean plants (*Phaseolus vulgaris*) when attacked by the spider mite *Tetranychus urticae*. Levels of two terpenoids synthesized *de novo* upon attack by the spider mite, β-ocimene and β-caryophyllene, are increased when plants are infected with the fungus. These changes to HIPV profiles make them more attractive to the predatory mite *Phytoseiulus persimilis* (Schausberger *et al.* 2012). In another study involving *G. mosseae*, tomatoes whose roots were colonized by this fungus were more attractive to foraging *Aphidius ervi*, an important parasitoid of the potato aphid *Macrosiphum euphorbiae* (Guerrieri *et al.* 2004). However, not all studies of below-ground fungal associates show an increase in above-ground HIPVs (e.g. Fontana *et al.* 2009). While most studies have focused on the effects of above-ground–below-ground interactions on volatile production and attractiveness to parasitoids, at least one study has demonstrated positive effects of below-ground organisms on the fitness traits, rather than attraction, of above-ground parasitoids. Below-ground communities of grassland microorganisms and nematodes have a positive effect on the survival and body size of *Aphidius colemani*, a parasitoid of the bird cherry-oat aphid, *Rhopalosiphum padi* (Bezemer *et al.* 2005).

Considering the vast array of interactions involving plants, herbivores and their natural enemies, plant pathogens, mutualisms with fungal and bacterial associates, the bewildering and often unpredictable diversity of outcomes of these interactions is not surprising. When confronted with multiple players, each with their own set of selective pressures, general statements regarding the outcome of below-ground and above-ground interactions have been difficult to make. This is certainly true when trying to make sense of above-ground–below-ground interactions involving parasitoids. One approach to understanding how above-ground and below-ground components interact with one another is to identify keystone species that are important to how these communities function versus 'passenger' species that, while they may be strongly affected by their interactions with other species, have only a minor effect on other species in the community (van Dam & Heil 2011). Far too often, parasitoids are treated as passenger species, but increasing evidence suggests that they may in fact function as keystone species in that they may reduce herbivore pressure and increase plant fitness.

2.4 Climate change and parasitoid chemical ecology

Increased periods and frequency of drought events and elevated CO_2 are widespread consequences of global climate change. Both drought events and elevated CO_2 are known to influence plant resistance to herbivory (e.g. Koricheva *et al.* 1998, Stiling & Cornelissen 2007, Robinson *et al.* 2012). However our understanding of the effects of climate change related events on plant chemistry is limited. Furthermore, our understanding of how climate change affects parasitoids is even less well understood (see Holopainen, Himanen & Poppy, Chapter 8, this volume). Abiotic stressors such as ozone, light and temperature all result in higher levels of reactive oxygen species, which cause damage to plant lipids and proteins thereby eliciting the expression of plant defence pathways (Holopainen &

Gershenzon 2010). The plant response pathways that are initiated by elevated CO_2 and other greenhouse gases are further modulated by the timing and severity of attacks by plant pathogens and different types of insect herbivores.

Drought

Studies of drought effects on herbivory have often been designed as tests of either the plant stress hypothesis (White 1984), which predicts that defences should decline and available nitrogen should increase under water stress conditions, or the plant vigour hypothesis (Price 1991), which argues that herbivore performance should be positively correlated with overall plant growth. Increased drought has been correlated frequently with decreased plant defensive chemistry. In a study of garlic mustard, *Alliaria petiolata* (Brassicaceae), experimentally applied drought stress significantly decreased glucosinolate concentrations (Gutbrodt *et al.* 2011). The specialist herbivore *Pieris brassicae* preferred to feed less, but developed faster and attained greater body mass when feeding on garlic mustard experiencing drought stress. The generalist *Spodoptera littoralis* preferred to feed on drought-stressed plants, presumably because these plants produce reduced levels of glucosinolates (Gutbrodt *et al.* 2011). The authors suggest that the difference between feeding preference and larval performance in *P. brassicae* may be explained as a means of avoiding attack by parasitoids. Similarly, a field study showed that experimentally induced drought conditions reduced plant defensive chemistry of the grass *Holcus lanatus* and resulted in increased herbivore (*S. littoralis*) performance (Walter *et al.* 2012). However, the effects of drought on plant direct defences are not universally negative. Drought increased the expression of direct defences in tomato, and the generalist herbivore *S. exigua* performed more poorly on drought-stressed plants (English-Loeb *et al.* 1997). The apparent lack of consistency in terms of the relationship between drought, available nitrogen, and plant direct defences against herbivores arises from the fact that plant investment in different types of direct defences against herbivores does not follow simple trade-offs based on the amount of nitrogen allocated to growth versus defence (e.g. Hamilton *et al.* 2001).

Few studies have explicitly made the link between drought, plant defences, and resistance against herbivores. ABA is a stress response hormone produced in response to a wide range of environmental stressors including drought (Hirayama & Shinozaki 2010). The ABA pathway is primarily antagonistic with the JA/ET defence pathway (see Fig. 2.1) and it is thought that its expression takes precedence over the JA pathway under conditions of drought stress (Fujita *et al.* 2006, Ton *et al.* 2009). In a study of several field populations of *Boechera stricta*, a close relative of *Arabidopsis thaliana*, Siemens *et al.* (2012) found a negative relationship between drought stress and glucosinolate production, suggesting that drought conditions induce the ABA stress response pathway in *Boechera stricta* at the expense of defence against herbivores. At least one study has examined the effect of drought stress on parasitoids. Larvae of the generalist moth *Spodoptera exigua* performed better on drought-stressed cotton, presumably because these plants produced reduced levels of direct defences (Olson *et al.* 2009). Furthermore, drought-stressed plants increased their production of volatile terpenoids, which resulted in higher parasitism rates of *S. exigua* by the parasitoid *Microplitis croceipes*.

Elevated CO_2

Two recent meta-analyses have examined the relationship between elevated CO_2, plant traits including defensive chemistry, and herbivore responses (Stiling & Cornelissen 2007,

Robinson *et al.* 2012). Elevated CO_2 is often correlated with reduced herbivore abundance and, in some cases, elevated CO_2 results in increased plant biomass and increased concentrations of defensive compounds (Stiling & Cornelissen 2007). Total plant nitrogen, which is often correlated with herbivore performance (Mattson 1980), tended to be negatively correlated with elevated CO_2, whereas total carbohydrates were positively correlated with elevated CO_2 (Robinson *et al.* 2012). Correspondingly, herbivore responses to the effects of elevated CO_2 differed between phloem feeders (which responded positively to elevated CO_2) and foliage feeders (which tended to respond negatively to elevated CO_2). However, many individual studies show widely variable relationships between CO_2, plant growth and defence, and herbivory (Robinson *et al.* 2012). Concentrations of so-called nitrogen-based defences (e.g. alkaloids) were found to decrease, whereas carbon-based defences (e.g. phenolics, tannins) increased under elevated CO_2 (Robinson *et al.* 2012). However, another recent meta-analysis (Ryan *et al.* 2010, cited in Robinson *et al.* 2012) found that nitrogen-based and carbon-based plant defences were as likely to increase as they were to decrease in response to elevated CO_2. While it is tempting to view the relationship between atmospheric CO_2 levels and plant chemical defences against herbivores as being determined by a simple ratio of $C:N$ available for investment in growth versus investment in a broad class of chemical defences (e.g. alkaloids vs. phenolics vs. tannins, or nitrogen-based defences vs. carbon-based defences), this is a deceptively simplistic view when one considers the wide range of selective factors that determine specific patterns of plant chemical defences against herbivores (Hamilton *et al.* 2001, Nitao *et al.* 2002).

Despite our understanding of the relationships between drought, plant chemistry, herbivory and parasitoids, few studies have attempted to examine the relationship between CO_2 levels and parasitoids as mediated by changes in plant chemistry. Several studies have shown that elevated CO_2 increases resistance to pathogens (Matros *et al.* 2006, Ye *et al.* 2010) and that this is probably due to an increase in the expression of SA-mediated defence pathways (Huang *et al.* 2012; see Fig. 2.1). Elevated CO_2 has been shown to interfere with JA-mediated plant defences against chewing herbivores (Fig. 2.1). Soybean plants (*Glycine max*) grown under field conditions with elevated CO_2 were more susceptible to damage by the Japanese beetle (*Popillia japonica*) and the western corn rootworm (*Diabrotica virgifera virgifera*). Elevated CO_2 suppressed the expression of lipoxygenase genes involved in the JA signalling pathway, resulting in a decrease in the production of proteinase inhibitors (Zavala *et al.* 2008).

While it may be reasonable to expect such relationships, no studies to date have examined the effects of elevated CO_2 on JA–SA cross-talk and their consequences for parasitoids. A handful of studies have demonstrated that elevated CO_2 changes plant defensive chemistry and have examined these effects on herbivores and their parasitoids. For example, elevated atmospheric CO_2 resulted in increased populations of the aphid *Myzus persicae*, but parasitism rates by *Aphidius matricariae* were unchanged (Bezemer *et al.* 1998). Herbivory by diamondback moth larvae (*Plutella xylostella*) induces higher glucosinolate concentrations in *Arabidopsis thaliana* under elevated, but not ambient, concentrations of atmospheric CO_2 (Bidart-Bouzat *et al.* 2005). Such effects of CO_2 on the inducibility of glucosinolates probably have consequences at higher trophic levels. Elevated CO_2 can have variable (positive, neutral or negative) effects on volatile terpenoid production and on the subsequent attraction of insect predators and parasitoids (Vuorinen *et al.* 2004, Bidart-Bouzat & Imeh-Nathaniel 2008 and references therein, Holopainen & Gershenzon 2010). Under elevated CO_2, the parasitoid *Cotesia plutellae* (=*C. vestalis*) is unable to detect the volatiles released by cabbage plants fed upon by its diamondback moth host, *Plutella xylostella* (Vuorinen *et al.* 2004). Much more remains to be done to further explore the consequences of increased CO_2 and other greenhouse gases on herbivores and their parasitoids.

2.5 Conclusions

Despite long-standing calls to include higher trophic levels in studies of plant–herbivore interactions (Price *et al.* 1980, Ode 2006), there remain many tantalizing areas where parasitoids and predators probably play an important role but where we have little evidence to support our understanding of such relationships. A few studies (some that are discussed in this chapter and others in succeeding chapters) have explicitly studied the complex links that exist, for example, between parasitoids and above-ground–below-ground herbivory, between parasitoids and climate change effects on herbivores, and between parasitoids and plant pathogen–insect herbivore interactions. In many cases, we are left trying to piece together likely relationships from the fragmented study of different parts of these complex relationships but, as this chapter has hopefully conveyed, the outcomes of such relationships are difficult to predict. In particular, more holistic studies that focus on the fitness impacts that parasitoids have on other trophic levels are clearly needed. Much remains to be done in terms of exploring how parasitoids interact with other trophic levels. Too often, plant–insect studies ignore higher trophic levels (or treat them simply as interesting but unimportant phenomena that do not have a bearing on the topics being investigated). Likewise, plant–pathogen interaction studies tend to ignore interactions with other trophic levels. However, parasitoids may have a major role in regulating herbivore populations and, in doing so, may act as selective agents on plant investment in defence. Incorporation of increased trophic complexity in plant–insect interaction studies will undoubtedly yield useful insights and may well help to resolve the myriad (often conflicting) hypotheses regarding plant–insect interactions. In addition, incorporation of the third trophic level as well as interactions with plant pathogens and climate change will dramatically improve our understanding of factors involved in plant pest invasions and how to effectively manage them.

Acknowledgements

I am grateful to Eric Wajnberg and Stefano Colazza for inviting me to write this chapter. I thank Jeff Harvey for insightful comments on an earlier draft of this chapter.

References

Agrawal, A.A. (2007) Macroevolution of plant defense strategies. *Trends in Ecology and Evolution* 22: 103–9.
Barbosa, P., Gross, P. and Kemper, J. (1991) Influence of plant allelochemicals on the tobacco hornworm and its parasitoid, *Cotesia congregata. Ecology* 72: 1567–75.

Barbosa, P., Saunders, J.A., Kemper, J., Trumbule, R., Olechno, J. and Martinat, P. (1986) Plant allelo-chemicals and insect parasitoids: effects of nicotine on *Cotesia congregata* (Say) (Hymenoptera: Braconidae) and *Hyposoter annulipes* (Cresson) (Hymenoptera: Ichneumonidae). *Journal of Chemical Ecology* **12**: 1319–28.

Bezemer, T.M. and van Dam, N.M. (2005) Linking aboveground and belowground interactions via induced plant defenses. *Trends in Ecology and Evolution* **20**: 617–24.

Bezemer, T.M., de Deyn, G.B., Bossinga, T.M., van Dam, N.M., Harvey, J.A. and van der Putten, W.H. (2005) Soil community composition drives aboveground plant–herbivore–parasitoid interactions. *Ecology Letters* **8**: 652–61.

Bezemer, T.M., Jones, T.H. and Knight, K.J. (1998) Long-term effects of elevated CO_2 and temperature on populations of the peach potato aphid *Myzus persicae* and its parasitoid *Aphidius matricariae*. *Oecologia* **116**: 128–35.

Bezemer, T.M., Wagenaar, R., van Dam, N.M. and Wäckers, F.L. (2003) Interactions between above- and belowground insect herbivores as mediated by the plant defense system. *Oikos* **101**: 555–62.

Bidart-Bouzat, M.G. and Imeh-Nathaniel, A. (2008) Global change effects on plant chemical defenses against insect herbivores. *Journal of Integrative Plant Biology* **50**: 1339–54.

Bidart-Bouzat, M.G., Mithen, R. and Berenbaum, M.R. (2005) Elevated CO_2 influences herbivory-induced defense responses of *Arabidopsis thaliana*. *Oecologia* **145**: 415–524.

Bixby-Brosi, A.J. and Potter, D.A. (2012) Endophyte-mediated tritrophic interactions between a grass-feeding caterpillar and two parasitoid species with different life histories. *Arthropod–Plant Interactions* **6**: 27–34.

Bloem, K.A. and Duffey, S.S. (1990) Interactive effect of protein and rutin on larval *Heliothis zea* and the endoparasitoids *Hyposoter exiguae*. *Entomologia Experimentalis et Applicata* **54**: 149–60.

Bostock, R.M. (1999) Signal conflicts and synergies in induced resistance to multiple attackers. *Physiological and Molecular Plant Pathology* **55**: 99–109.

Brooks, D.M., Bender, C.L. and Kunkel, B.N. (2005) The *Pseudomonas syringae* phytotoxin coronatine promotes virulence by overcoming salicylic acid-dependent defences in *Arabidopsis thaliana*. *Molecular Plant Pathology* **6**: 629–39.

Bruessow, F., Gouhier-Darimont, C., Buchala, A., Metraux, J.-P. and Reymond, P. (2010) Insect egg suppress plant defence against chewing herbivores. *Plant Journal* **62**: 876–85.

Bukovinszky, T., Poelman, E.H., Gols, R., Prekatsakis, G., Vet, L.E.M., Harvey, J.A. and Dicke, M. (2009) Consequences of constitutive and induced variation in plant nutritional quality for immune defence of a herbivore against parasitism. *Oecologia* **160**: 299–308.

Campbell, B.C. and Duffey, S.S. (1979) Tomatine and parasitic wasps: potential incompatibility of plant antibiosis with biological control. *Science* **205**: 700–2.

Campbell, B.C. and Duffey, S.S. (1981) Alleviation of α-tomatine-induced toxicity to the parasitoid, *Hyposoter exiguae*, by phytosterols in the diet of the host, *Heliothis zea*. *Journal of Chemical Ecology* **7**: 927–46.

Cardoza, Y.J., Teal, P.E.A. and Tumlinson, J.H. (2003) Effect of peanut plant fungal infection on oviposition preference by *Spodoptera exigua* and on host-searching behavior by *Cotesia marginiventris*. *Environmental Entomology* **32**: 970–6.

Coleman, R.A., Barker, A.M. and Fenner, M. (1999) Parasitism of the herbivore *Pieris brassicae* L. (Lep., Pieridae) by *Cotesia glomerata* L. (Hym., Braconidae) does not benefit the host plant by reduction of herbivory. *Journal of Applied Entomology* **123**: 171–7.

Cui, J., Bahrami, A.K., Pringle, E.G., Hernandez-Guzman, G., Bender, C.L., Pierce, N.E. and Ausubel, F.M. (2005) *Pseudomonas syringae* manipulates systemic plant defenses against pathogens and herbivores. *Proceedings of the National Academy of Sciences USA* **102**: 1791–6.

De Geyter, N., Gholami, A., Goormachtig, S. and Goossens, A. (2012) Transcriptional machineries in jasmonate-elicited plant secondary metabolism. *Trends in Plant Science* **17**: 349–59.

De Moraes, C.M., Lewis, W.J., Paré, P.W., Alborn, H.T. and Tumlinson, J.H. (1998) Herbivore-infested plants selectively attract parasitoids. *Nature* **393**: 570–3.

Dicke, M. (1999) Are herbivore-induced plant volatiles reliable indicators of herbivore identity to foraging carnivorous arthropods? *Entomologia Experimentalis et Applicata* **91**: 131–42.

Dicke, M. and Baldwin, I.T. (2010) The evolutionary context for herbivore-induced plant volatiles: beyond the 'cry for help'. *Trends in Plant Science* **15**: 167–75.

Dicke, M. and Vet, L.E.M. (1999) Plant–carnivore interactions: evolutionary and ecological consequences for plant, herbivore and carnivore. In: Olff, H., Brown, V.K. and Drent, R.H. (eds) *Herbivores: Between Plants and Predators*. Blackwell, Oxford, pp. 483–520.

Diezel, C., von Dahl, C.C., Gaquerel, E. and Baldwin, I.T. (2009) Different lepidopteran elicitors account for cross-talk in herbivory-induced phytohormone signaling. *Plant Physiology* **150**: 1576–86.

Doss, R.P., Oliver, J.E., Proebsting, W.M., Potter, S.W., Kuy, S.R., Clement, S.L., Williamson, R.T., Carney, J.R. and Devilbiss, E.D. (2000) Bruchins: insect-derived plant regulators that stimulate neoplasm formation. *Proceedings of the National Academy of Sciences USA* **97**: 237–49.

Du, Y., Poppy, G.M., Powell, W., Pickett, J.A., Wadhams, L.J. and Woodcock, C.M. (1998) Identification of semiochemicals released during aphid feeding that attract the parasitoid *Aphidius ervi*. *Journal of Chemical Ecology* **24**: 1355–68.

English-Loeb, G., Stout, M.J. and Duffey, S.S. (1997) Drought stress in tomatoes: changes in plant chemistry and potential nonlinear consequences for insect herbivores. *Oikos* **79**: 456–68.

Erb, M., Köllner, T.G., Degenhardt, J., Zwahlen, C., Hibbard, B.E. and Turlings, T.C.J. (2011) The role of abscisic acid and water stress in root herbivore-induced leaf resistance. *New Phytologist* **189**: 308–20.

Erb, M., Ton, J., Degenhardt, J. and Turlings, T.C.J. (2008) Interactions between arthropod-induced aboveground and belowground defenses in plants. *Plant Physiology* **146**: 867–74.

Fatouros, N.E., Broekgaarden, C., Bukovinszkine'Kiss, G., van Loon, J.J.A., Mumm, R., Huigens, M.E., Dicke, M. and Hilker, M. (2008) Male-derived butterfly anti-aphrodisiac mediates induced indirect plant defense. *Proceedings of the National Academy of Science USA* **105**: 10033–8.

Felton, G.W. and Tumlinson, J.H. (2008) Plant–insect dialogs: complex interactions at the plant–insect interface. *Current Opinion in Plant Biology* **11**: 457–63.

Fontana, A., Reichelt, M., Hempel, S., Gershenzon, J. and Unsicker, S.B. (2009) The effects of arbuscular mycorrhizal fungi on direct and indirect defense metabolites of *Plantago lanceolata* L. *Journal of Chemical Ecology* **35**: 833–43.

Fujita, M., Fujita, Y., Noutoshi, Y., Takahashi, F., Narusaka, Y., Yamaguchi-Shinozaki, K. and Shinozaki, K. (2006) Cross-talk between abiotic and biotic stress responses: a current view from the points of convergence in the stress signaling networks. *Current Opinions in Plant Biology* **9**: 436–42.

Gange, A.C., Eschen, R., Wearn, J.A., Thawer, A. and Sutton, B.C. (2012) Differential effects of foliar endophytic fungi on insect herbivores attacking a herbaceous plant. *Oecologia* **168**: 1023–31.

Godfray, H.C.J. (1994) *Parasitoids: Behavioral and Evolutionary Ecology*. Princeton University Press, Princeton, NJ.

Godfray, H.C.J. (1995) Communication between the first and the third trophic levels: an analysis using biological signaling theory. *Oikos* **72**: 367–74.

Gols, R., Bullock, J.M., Dicke, M., Bukovinszky, T. and Harvey, J.A. (2011) Smelling the wood from the trees: non-linear parasitoid responses to volatile attractants produced by wild and cultivated cabbage. *Journal of Chemical Ecology* **37**: 795–807.

Gols, R., van Dam, N.M., Raaijmakers, C.E., Dicke, M. and Harvey, J.A. (2009) Are population differences in plant quality reflected in the preference and performance of two endoparasitoids wasps? *Oikos* **118**: 733–43.

González-Santoyo, I. and Córdoba-Aguilar, A. (2012) Phenoloxidase: a key component of the insect immune system. *Entomologia Experimentalis et Applicata* **142**: 1–16.

Green, T.R. and Ryan, C.A. (1972) Wound-induced proteinase inhibitor in plant leaves: a possible defense mechanism against insects. *Science* **175**: 776–7.

Guerrieri, E., Lingua, G., Digilio, M.C., Massa, N. and Berta, G. (2004) Do interactions between plant roots and the rhizosphere affect parasitoid behaviour? *Ecological Entomology* **29**: 753–6.

Gutbrodt, B., Mody, K. and Dorn, S. (2011) Drought changes plant chemistry and causes contrasting responses in lepidopteran herbivores. *Oikos* **120**: 1732–40.

Hamilton, J.G., Zangerl, A.R., DeLucia, E.H. and Berenbaum, M.R. (2001) The carbon-nutrient balance hypothesis: its rise and fall. *Ecology Letters* **4**: 86–95.

Hare, J.D. (1992) Effects of plant variation on herbivore–natural enemy interactions. In: Fritz, R.S. and Simms, E.L. (eds) *Plant Resistance to Herbivores and Pathogens: Ecology, Evolution, and Genetics.* University of Chicago Press, Chicago, pp. 278–98.

Hare, J.D. (2002) Plant genetic variation in tritrophic interactions. In: Tscharntke, T. and Hawkins, B.A. (eds) *Multitrophic Level Interactions.* Cambridge University Press, Cambridge, pp. 8–43.

Hare, J.D. (2011) Ecological role of volatiles produced by plants in response to damage by herbivorous insects. *Annual Review of Entomology* **56**: 161–80.

Harvey, J.A. (2005) Factors affecting the evolution of development strategies in parasitoid wasps: the importance of functional constraints and incorporating complexity. *Entomologia Experimentalis et Applicata* **117**: 1–13.

Harvey, J.A. and Gols, R. (2011) Population-related variation in plant defense more strongly affects survival of an herbivore than its solitary parasitoid wasp. *Journal of Chemical Ecology* **37**: 1081–90.

Harvey, J.A., van Dam, N.M., Witjes, L.M.A., Soler, R. and Gols, R. (2007) Effects of dietary nicotine on the development of an insect herbivore, its parasitoid and secondary hyperparasitoid over four trophic levels. *Ecological Entomology* **32**: 15–23.

Harvey, J.A., van Nouhuys, S. and Biere, A. (2005) Effects of quantitative variation in allelochemicals in *Plantago lanceolata* on development of a generalist and a specialist herbivore and their endoparasitoids. *Journal of Chemical Ecology* **31**: 287–302.

Hatcher, P.E. (1995) Three-way interaction between plant pathogenic fungi, herbivorous insects and their host plants. *Biological Reviews* **70**: 639–94.

Heil, M. (2008) Indirect defence via tritrophic interactions. *New Phytologist* **178**: 41–61.

Hilker, M. and McNeil, J. (2008) Chemical and behavioral ecology in insect parasitoids: how to behave optimally in a complex odorous environment? In: Wajnberg, E., Bernstein, C. and van Alphen, J.J.M. (eds) *Behavioral Ecology of Insect Parasitoids: From Theoretical Approaches to Field Applications.* Blackwell Publishing Ltd., Oxford, pp. 92–112.

Hilker, M. and Meiners, T. (2006) Early herbivore alert: insect eggs induce plant defense. *Journal of Chemical Ecology* **32**: 1379–97.

Hilker, M. and Meiners, T. (2010) How do plants "notice" attack by herbivorous arthropods? *Biological Reviews* **85**: 267–80.

Hirayama, T. and Shinozaki, K. (2010) Research on plant abiotic stress responses in the post-genome era: past, present and future. *Plant Journal* **61**: 1041–52.

Holopainen, J.K. and Gershenzon, J. (2010) Multiple stress factors and the emission of plant VOCs. *Trends in Plant Science* **15**: 176–84.

Howe, G.A. (2004) Jasmonates as signals in the wound response. *Journal of Plant Growth Regulators* **23**: 223–37.

Huang, L., Ren, Q., Sun, Y., Ye, L., Cao, H. and Ge, F. (2012) Lower incidence and severity of tomato virus in elevated CO_2 is accompanied by modulated plant induced defence in tomato. *Plant Biology* **14**: 905–13.

Hunter, M.D. (2003) Effects of plant quality on the population ecology of parasitoids. *Agricultural and Forest Entomology* **5**: 1–8.

Kaplan, I., Halitschke, R., Kessler, A., Sardanelli, S. and Denno, R.F. (2008) Constitutive and induced defenses to herbivory in above- and belowground plant tissues. *Ecology* **89**: 392–406.

Karban, R. and Baldwin, I.T. (1997) *Induced Responses to Herbivory.* University of Chicago Press, Chicago.

Kempel, A., Schmidt, A.K., Brandl, R. and Schädler, M. (2010) Support from the underground: induced plant resistance depends on arbuscular mycorrhizal fungi. *Functional Ecology* **24**: 293–300.

Kessler, A. and Baldwin, I.T. (2001) Defensive function of herbivore-induced plant volatile emissions in nature. *Science* **291**: 2141–4.

Kessler, A. and Baldwin, I.T. (2002) Plant responses to insect herbivory: the emerging molecular analysis. *Annual Review of Plant Biology* **53**: 299–328.

Kessler, A., Halitschke, R. and Baldwin, I.T. (2004) Silencing the jasmonate cascade: induced plant defenses and insect populations. *Science* **305**: 665–8.

Koornneef, A. and Pieterse, C.M.J. (2008) Cross talk in defense signaling. *Plant Physiology* **146**: 839–44.

Koricheva, J., Larsson, S. and Haukioja, E. (1998) Insect performance on experimentally stressed woody plants: a meta-analysis. *Annual Review of Entomology* **43**: 195–216.

Lampert, E.C., Dyer, L.A. and Bowers, M.D. (2011b) Chemical defense across three trophic levels: *Catalpa bignonioides*, the caterpillar *Ceratomia catalpae*, and its endoparasitoids *Cotesia congregata*. *Journal of Chemical Ecology* **37**: 1063–70.

Lampert, E.C., Zangerl, A.R., Berenbaum, M.R. and Ode, P.J. (2008) Tritrophic effects of xanthotoxin on the polyembryonic parasitoid *Copidosoma sosares* (Hymenoptera: Encyrtidae). *Journal of Chemical Ecology* **34**: 783–90.

Lampert, E.C., Zangerl, A.R., Berenbaum, M.R. and Ode, P.J. (2011a) Generalist and specialist host–parasitoid associations respond differently to wild parsnip (*Pastinaca sativa*) defensive chemistry. *Ecological Entomology* **36**: 52–61.

Lapitan, N.L.V., Li, Y.C., Peng, J.H. and Botha, A.M. (2007) Fractionated extracts of Russian wheat aphid eliciting defense responses in wheat. *Journal of Economic Entomology* **100**: 990–9.

Li, J., Brader, G. and Palva, E.T. (2004) The WRKY70 transcription factor: a node of convergence for jasmonate-mediated and salicylate-mediated signals in plant defense. *Plant Cell* **16**: 319–31.

Masters, G.J., Jones, T.H. and Rogers, M. (2001) Host-plant mediated effects of root herbivory on insect seed predators and their parasitoids. *Oecologia* **127**: 246–50.

Matros, A., Amme, S., Kettig, B., Buck-Sorlin, G.H., Sonnewald, U. and Mock, H.P. (2006) Growth at elevated CO_2 concentrations leads to modified profiles of secondary metabolites in tobacco cv. Samsun NN and to increased resistance against infection with potato virus Y. *Plant, Cell and Environment* **29**: 126–37.

Mattson, W.J. (1980) Herbivory in relation to plant nitrogen content. *Annual Review of Ecology and Systematics* **11**: 119–61.

McGovern, J.L., Zangerl, A.R., Ode, P.J. and Berenbaum, M.R. (2006) Furanocoumarins and their detoxification in a tri-trophic interaction. *Chemoecology* **16**: 45–50.

Memelink, J., Verpoorte, R. and Kijne, J.W. (2001) ORCAnization of jasmonate-responsive gene expression in alkaloid metabolism. *Trends in Plant Science* **6**: 212–19.

Milan, N.F., Kacsoh, B.Z. and Schlenke, T.A. (2012) Alcohol consumption as self-medication against blood-borne parasites in the fruit fly. *Current Biology* **22**: 488–93.

Nishida, R. (2002) Sequestration of defensive substances from plants by Lepidoptera. *Annual Review of Entomology* **47**: 57–92.

Nitao, J.K., Zangerl, A.R. and Berenbaum, M.R. (2002) CNB: requiescat in pace? *Oikos* **98**: 540–6.

Oakeshott, J.G., Johnson, R.M., Berenbaum, M.R., Ranson, H., Cristino, A.S. and Claudianos, C. (2010) Metabolic enzymes associated with xenobiotic and chemosensory responses in *Nasonia vitripennis*. *Insect Molecular Biology* **19**: 147–63.

Ode, P.J. (2006) Plant chemistry and natural enemy fitness: effects on herbivore and natural enemy interactions. *Annual Review of Entomology* **51**: 163–85.

Ode, P.J., Berenbaum, M.R., Zangerl, A.R. and Hardy, I.C.W. (2004) Host plant, host plant chemistry and the polyembryonic parasitoid *Copidosoma sosares*: indirect effects in a tritrophic interaction. *Oikos* **104**: 388–400.

Olson, D.M., Cortesero, A.M., Rains, G.C., Potter, T. and Lewis, W.J. (2009) Nitrogen and water affect direct and indirect plant systemic induced defense in cotton. *Biological Control* **49**: 239–44.

Orozco-Cardenas, M., McGurl, B. and Ryan, C.A. (1993) Expression of an anti-sense prosystemin gene in tomato plants reduces resistance toward *Manduca sexta* larvae. *Proceedings of the National Academy of Sciences USA* **90**: 8273–6.

Páre, P.W. and Tumlinson, J.H. (1999) Plant volatiles as a defense against insect herbivores. *Plant Physiology* **121**: 325–31.

Pierre, P.S., Dugravot, S., Ferry, A., Soler, R., van Dam, N.M. and Cortesero, A.-M. (2011) Aboveground herbivory affects indirect defences of brassicaceous plants against the root feeder *Delia radicum* Linnaeus: laboratory and field evidence. *Ecological Entomology* **36**: 326–34.

Pieterse, C.M.J., van Wees, S.C.M., Ton, J., van Pelt, J.A. and van Loon, L.C. (2002) Signalling in rhizobacteria-induced systemic resistance in *Arabidopsis thaliana*. *Plant Biology* **4**: 535–44.

Poelman, E.H., Zheng, S.-J., Zhang, Z., Heemskerk, N.M., Cortesero, A.-M. and Dicke, M. (2011) Parasitoid-specific induction of plant responses to parasitized herbivores affects colonization by subsequent herbivores. *Proceedings of the National Academy of Sciences USA* **108**: 19647–52.

Preston, C.A., Lewandowski, C., Enyedi, A.J. and Baldwin, I.T. (1999) Tobacco mosaic virus inoculation inhibits wound-induced jasmonic acid-mediated responses within but not between plants. *Planta* **209**: 87–95.

Price, P.W. (1991) The plant vigor hypothesis and herbivore attack. *Oikos* **62**: 244–51.

Price, P.W., Bouton, C.E., Gross, P., McPheron, B.A., Thompson, J.N. and Weis, A.E. (1980) Interactions among three trophic levels: influence of plants on interactions between insect herbivores and natural enemies. *Annual Review of Ecology and Systematics* **11**: 41–65.

Rasmann, S. and Turlings, T.C.J. (2007) Simultaneous feeding by aboveground and belowground herbivores attenuates plant-mediated attraction of their respective natural enemies. *Ecology Letters* **10**: 926–36.

Rayapuram, C. and Baldwin, I.T. (2007) Increased SA in NPR1-silenced plants antagonizes JA and JA-dependent direct and indirect defenses in herbivore-attacked *Nicotiana attenuata* in nature. *Plant Journal* **52**: 700–15.

Robert-Seilaniantz, A., Grant, M. and Jones, J.D.G. (2011) Hormone crosstalk in plant disease and defense: more than just jasmonate–salicylate antagonism. *Annual Review of Phytopathology* **49**: 317–43.

Robinson, E.A., Ryan, G.D. and Newman, J.A. (2012) A meta-analytical review of the effects of elevated CO_2 on plant–arthropod interactions highlights the importance of interacting environmental and biological variables. *New Phytologist* **194**: 321–36.

Rodriguez-Saona, C., Chalmers, J.A., Raj, S. and Thaler, J.S. (2005) Induced plant responses to multiple damagers: differential effects on an herbivore and its parasitoid. *Oecologia* **143**: 566–77.

Schausberger, P., Peneder, S., Jürschik, S. and Hoffmann, D. (2012) Mycorrhiza changes plant volatiles to attract spider mite enemies. *Functional Ecology* **26**: 441–9.

Shikano, I., Ericsson, J.D., Cory, J.S. and Myers, J.H. (2010) Indirect plant-mediated effects on insect immunity and disease resistance in a tritrophic system. *Basic and Applied Ecology* **11**: 15–22.

Siemens, D.H., Duvall-Jisha, J., Jacobs, J., Manthey, J., Haugen, R. and Matzner, S. (2012) Water deficiency induces evolutionary tradeoff between stress tolerance and chemical defense allocation that may help explain range limits in plants. *Oikos* **121**: 790–800.

Singer, M.S., Mace, K.C. and Bernays, E.A. (2009) Self-medication as adaptive plasticity: increased ingestion of plant toxins by parasitized caterpillars. *PLoS ONE* **4**: e4796.

Skibbe, M., Qu, N., Galis, I. and Baldwin, I.T. (2008) Induced plant defenses in the natural environment: *Nicotiana attenuata* WRKY3 and WRKY6 coordinate responses to herbivory. *Plant Cell* **20**: 1984–2000.

Smilanich, A.M., Dyer, L.A., Chambers, J.Q. and Bowers, M.D. (2009) Immunological cost of chemical defence and the evolution of herbivore diet breadth. *Ecology Letters* **12**: 612–21.

Smilanich, A.M., Mason, P.A., Sprung, L., Chase, T.R. and Singer, M.S. (2011) Complex effects of parasitoids on pharmacophagy and diet choice of a polyphagous caterpillar. *Oecologia* **165**: 995–1005.

Smith, J.L., De Moraes, C.M. and Mescher, M.C. (2009) Jasmonate- and salicylate-mediated plant defense responses to insect herbivores, pathogens and parasitic plants. *Pest Management Science* **65**: 497–503.

Soler, R., Badenes-Pérez, F.R., Broekgaarden, C., Zheng, S.-J., David, A., Boland, W. and Dicke, M. (2012b) Plant-mediated facilitation between a leaf-feeding and a phloem-feeding insect in a brassicaceous plant: from insect performance to gene transcription. *Functional Ecology* **26**: 156–66.

Soler, R., Bezemer, T.M., Cortesero, A.M., van der Putten, W.H., Vet, L.E.M. and Harvey, J.A. (2007) Impact of foliar herbivory on the development of a root-feeding insect and its parasitoid. *Oecologia* **152**: 257–64.

Soler, R., Schaper, S.V., Bezemer, T.M., Cortesero, A.M., Hoffmeister, T.S., van der Putten, W.H., Vet, L.E.M. and Harvey, J.A. (2009) Influence of presence and spatial arrangement of belowground insects on host-plant selection of aboveground insects: a field study. *Ecological Entomology* **34**: 339–45.

Soler, R., van der Putten, W.H., Harvey, J.A., Vet, L.E.M., Dicke, M. and Bezemer, T.M. (2012a) Root herbivore effects on aboveground multitrophic interactions: patterns, processes and mechanisms. *Journal of Chemical Ecology* **38**: 755–67.

Stiling, P. and Cornelissen, T. (2007) How does elevated carbon dioxide (CO_2) affect plant–herbivore interactions? A field experiment and meta-analysis of CO_2-mediated changes on plant chemistry and herbivore performance. *Global Change Biology* **13**: 1823–42.

Stireman, J.O., Dyer, L.A., Janzen, D.H., Singer, M.S., Lill, J.T., Marquis, R.J., Ricklefs, R.E., Gentry, G.L., Hallwachs, W., Coley, P.D., Barone, J.A., Greeney, H.F., Connahs, H., Barbosa, P., Morais, H.C. and Diniz, I.R. (2005) Climatic unpredictability and parasitism of caterpillars: implications of global warming. *Proceedings of the National Academy of Sciences USA* **101**: 17384–7.

Stout, M.J., Thaler, J.S. and Thomma, B.P.H.J. (2006) Plant-mediated interactions between pathogenic microorganisms and herbivorous arthropods. *Annual Review of Entomology* **51**: 663–89.

Strand, M.R and Pech, L.L. (1995) Immunological compatibility in parasitoid–host relationships. *Annual Review of Entomology* **40**: 31–56.

Tack, A.J.M., Gripenberg, S. and Roslin, T. (2012) Cross-kingdom interactions matter: fungal-mediated interactions structure an insect community on oak. *Ecology Letters* **15**: 177–85.

Thaler, J.S. (1999) Jasmonate-inducible plant defences cause increased parasitism of herbivores. *Nature* **399**: 686–8.

Thaler, J.S., Fidantsef, A.L., Duffey, S.S. and Bostock, R.M. (1999) Trade-offs in plant defense against pathogens and herbivores: a field demonstration of chemical elicitors of induced resistance. *Journal of Chemical Ecology* **25**: 1597–609.

Thaler, J.S., Fidantsef, A.L. and Bostock, R.M. (2002) Antagonism between jasmonate- and salicylate-mediated induced plant resistance: effects of concentration and timing of elicitation on defense-related proteins, herbivore, and pathogen performance in tomato. *Journal of Chemical Ecology* **28**: 1131–59.

Thaler, J.S., Humphrey, P.T. and Whiteman, N.K. (2012) Evolution of jasmonate and salicylate signal crosstalk. *Trends in Plant Science* **17**: 260–70.

Thaler, J.S., Owen, B. and Higgins, V.J. (2004) The role of jasmonate response in plant susceptibility to diverse pathogens with a range of lifestyles. *Plant Physiology* **135**: 530–8.

Ton, J., Flors, V. and Mauch-Mani, B. (2009) The multifaceted role of ABA in disease resistance. *Trends in Plant Science* **14**: 310–17.

Turlings, T.C.J. and Benrey, B. (1998) Effects of plant metabolites on the behavior and development of parasitic wasps. *ÉcoScience* **5**: 321–33.

Turlings, T.C.J. and Wäckers, F. (2004) Recruitment of predators and parasitoids by herbivore-injured plants. In: Cardé, R.T. and Millar, J.G. (eds) *Advances in Insect Chemical Ecology*. Cambridge University Press, Cambridge, pp. 21–75.

Turlings, T.C.J., Gouinguené, S., Degen, T. and Fritzsche-Hoballah, M.E. (2002) The chemical ecology of plant–caterpillar–parasitoid interactions. In: Tscharntke, T. and Hawkins, B.A. (eds) *Multitrophic Level Interactions*. Cambridge University Press, Cambridge, pp. 148–173.

Turlings, T.C.J, Tumlinson, J.H. and Lewis, W.J. (1990) Exploitation of herbivore-induced plant odors by host-seeking parasitic wasps. *Science* **250**: 1251–3.

van Dam, N.M. (2009) Belowground herbivory and plant defenses. *Annual Review of Ecology, Evolution, and Systematics* **40**: 373–91.

van Dam, N.M. and Heil, M. (2011) Multitrophic interactions below and above ground: en route to the next level. *Journal of Ecology* **99**: 77–88.

van der Meijden, E. and Klinkhamer, P.G.L. (2000) Conflicting interests of plants and the natural enemies of herbivores. *Oikos* **89**: 202–8.

van Loon, L.C., Geraats, B.P.J. and Linthorst, H.J.M. (2006) Ethylene as a modulator of disease resistance in plants. *Trends in Plant Science* **11**: 184–91.

van Nouhuys, S. and Laine, A.-L. (2008) Population dynamics and sex ratio of a parasitoid altered by fungal-infected diet of host butterfly. *Proceedings of the Royal Society of London B* **275**: 787–95.

van Nouhuys, S., Reudler, J.H., Biere, A. and Harvey, J.A. (2012) Performance of secondary parasitoids on chemically defended and undefended hosts. *Basic and Applied Ecology* **13**: 241–9.

Vet, L.E.M. and Dicke, M. (1992) Ecology of infochemical use by natural enemies in a tritrophic context. *Annual Review of Entomology* **37**: 141–72.

Vet, L.E.M. and Godfray, H.C.J. (2008) Multitrophic interactions and parasitoid behavioral ecology. In: Wajnberg, E., Bernstein, C. and van Alphen, J.J.M. (eds) *Behavioral Ecology of Insect Parasitoids: From Theoretical Approaches to Field Applications*. Blackwell Publishing Ltd., Oxford, pp. 231–252.

von Dahl, C.C. and Baldwin, I.T. (2007) Deciphering the role of ethylene in plant–herbivore interactions. *Journal of Plant Growth Regulators* **26**: 201–9.

Vuorinen, T., Nerg, A.-M., Ibrahim, M.A., Reddy, G.V.P. and Holopainen, J.K. (2004) Emission of *Plutella xylostella*-induced compounds from cabbages grown at elevated CO_2 and orientation behavior of the natural enemies. *Plant Physiology* **135**: 1984–92.

Walling, L.L. (2000) The myriad of plant responses to herbivores. *Journal of Plant Growth Regulators* **19**: 195–216.

Walling, L.L. (2008) Avoiding effective defenses: strategies employed by phloem-feeding insects. *Plant Physiology* **146**: 859–66.

Walter, J., Hein, R., Auge, H., Beierkuhnlein, C., Löffler, S., Reifenrath, K., Schädler, M., Weber, M. and Jentsch, A. (2012) How do extreme drought and plant community composition affect host plant metabolites and herbivore performance? *Arthropod–Plant Interactions* **6**: 15–25.

Weech, M.-H., Chapleau, M., Pan, L., Ide, C. and Bede, J.C. (2008) Caterpillar saliva interferes with induced *Arabidopsis thaliana* defence responses via the systemic acquired resistance pathway. *Journal of Experimental Botany* **59**: 2437–48.

White, T.C. (1984) The abundance of invertebrate herbivores in relation to the availability of nitrogen in stressed food plants. *Oecologia* **63**: 90–105.

Williams, L., Rodriguez-Saona, C., Pare, P.W. and Crafts-Brandner, S.J. (2005) The piercing-sucking herbivores *Lygus hesperus* and *Nezara viridula* induce volatile emissions in plants. *Archives of Insect Biochemistry and Physiology* **58**: 84–96.

Wittstock, U., Agerbirk, N., Stauber, E.J., Olsen, C.E., Hippler, M., Mitchell-Olds, T., Gershenzon, J. and Vogel, H. (2004) Successful herbivore attack due to metabolic diversion of a plant chemical defense. *Proceedings of the National Academy of Sciences USA* **101**: 4859–64.

Ye, L.F., Fu, X. and Ge, F. (2010) Elevated CO_2 alleviates damage from potato virus Y infection in tobacco plants. *Plant Science* **179**: 219–24.

Zavala, J.A., Casteel, C.L., DeLucia, E.H. and Berenbaum, M.R. (2008) Anthropogenic increase in carbon dioxide compromises plant defense against invasive insects. *Proceedings of the National Academy of Sciences USA* **105**: 5129–33.

3

Foraging strategies of parasitoids in complex chemical environments

Nicole Wäschke[1], Torsten Meiners[1] and Michael Rostás[2]

[1] Department of Applied Zoology/Animal Ecology, Freie Universität Berlin, Germany
[2] Bio-Protection Research Centre, Lincoln University, Canterbury, New Zealand

Abstract

Parasitoids of herbivorous insects need to locate their hosts in complex chemical environments. Such complexity arises from a plethora of volatile and non-volatile chemical compounds produced by the hosts, by their host plants, and by the surrounding vegetation. To complicate the issue, biotic and abiotic factors that affect hosts and habitat can lead to further variability of chemicals in both space and time. In this chapter, we investigate the chemical complexity encountered by parasitoids while searching for suitable hosts, and their foraging strategies to cope with this complexity. First we review parasitoid responses to vegetation and vegetation odour diversity at a large scale. At a small scale, we then focus on the chemical complexity of the host plants' variability and its influence on parasitoid foraging. At the level of parasitoid host recognition and acceptance, the variability of chemicals and complex interactions between host plants and host cues are explored. Finally, we review how parasitoids use their behavioural, sensory and neurophysiological adaptations to successfully locate their hosts in heterogeneous environments, and how their foraging strategies are shaped by different life history traits.

3.1 Introduction

Insect parasitoids need to find suitable hosts for reproduction, otherwise their genes will not be passed on to future generations. This crucial process of locating a host can be divided into different steps known as host habitat finding, host location, and host acceptance (Vinson 1976). Parasitoids, in their effort to track down herbivorous hosts, depend on a variety of cues from their host insect, the food plant on which the host feeds, and the vegetation that surrounds this food plant; in other words, the habitat (Godfray 1994; see also Meiners & Peri, Chapter 9, this volume).

It is well known that parasitoids use the volatiles that plants emit in response to herbivore activity in order to locate the food plants of their hosts. In nature, host plants grow in habitats where they are surrounded by other plants that also emit a plethora of volatiles. These will vary in quantity and composition and may be constitutive or induced, vegetative or floral, depending on the plant species and their physiological state (Mumm & Dicke 2010). Parasitoids in search of a host are exposed to the diversity of volatiles emitted by all the non-host plants that form the surrounding vegetation, and it is conceivable that high plant diversity, leading to enhanced volatile diversity, may directly affect the olfactory orientation of parasitoids (Randlkofer *et al.* 2010b). Furthermore, indirect effects are also possible, as the community of plant species surrounding a host plant can have profound effects on a nearby plant's metabolome (Scherling *et al.* 2010), which includes volatiles (Fig. 3.1). Parasitoids need to cope with these high levels of chemical complexity since they must identify the right cues, which are embedded in a background of other cue and non-cue compounds (Schröder & Hilker 2008).

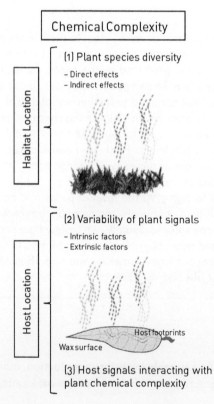

Figure 3.1 Chemical complexity encountered by parasitoids during host search. Plant diversity affects parasitoid orientation during the habitat location process. Plant odour variability and host signals interacting with plant surface chemical complexity affect parasitoids during host location at close distance.

The odour blends induced by herbivorous hosts are variable and can convey specific information enabling parasitoids to distinguish, for example, between herbivore-damaged and mechanically damaged plants (Drost *et al.* 1986, Turlings *et al.* 1990, Hilker *et al.* 2002), suitable or unsuitable herbivore species (de Moraes *et al.* 1998, Paré & Tumlinson 1999) or different host instars (Paré & Tumlinson 1999, Blassioli Moraes *et al.* 2005). Also, plant genotype effects on the volatile bouquet can be recognized by parasitoids (Hoballah *et al.* 2002). Foraging parasitoids might even be able to access information on below- and above-ground biotic and abiotic factors that affect a plant (see Soler, Bezemer & Harvey, Chapter 4, this volume) by assessing the plant's odour blend and integrating this information into their response. However, intrinsic and extrinsic factors that cause high variability in plant volatiles, and thus contribute to a chemically complex environment, do not always result in behavioural changes in the parasitoids (Hare 2011), and the mechanisms of decision making involved are not well understood.

In addition to plant volatiles, host-derived chemicals are used for identifying and recognizing hosts, in particular at close range, once the host plant has been singled out. These cues are often non-volatile and released from the faeces, silk or honeydew of herbivores (Quicke 1997). In recent years, hydrocarbon compounds from the surface of host insects have been confirmed to remain as 'footprints' on the surface of plants. Parasitoids exploit these footprints as kairomones to track down their hosts (e.g. Klomp 1981). They are confronted with chemical complexity at this small scale in the form of the composition of epicuticular plant waxes that vary according to plant genotype and environmental factors and which influence the recognition of footprint cues. We discuss the origins of plant chemical complexity and its effects on parasitoid foraging behaviour in the first part of this chapter, and in the second part we explore the principal behaviours that parasitoids have evolved to cope with this complexity (Fig. 3.2).

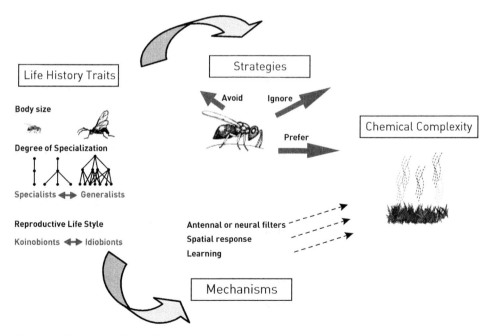

Figure 3.2 Parasitoid life history traits acting on their foraging strategies and on the mechanisms used to orientate in complex chemical environments.

3.2 Chemical complexity

3.2.1 Plant species diversity and habitat location

Parasitoids need to cope with a habitat's chemical complexity while attempting to locate the particular plant on which their insect hosts feed. Here we focus on how plant species and plant genotype diversity can contribute to chemically complex habitats. Such vegetation–host plant interactions can mediate parasitoid foraging directly or indirectly. High plant species diversity may impact parasitoid orientation directly as a result of enhanced vegetation odour complexity. Indirectly, neighbouring plants can influence the metabolic profile of the host plant (e.g. Broz *et al.* 2010) and this effect may cascade through to the third trophic level. Both aspects are discussed below.

Direct effects of plant species diversity on parasitoid orientation
Studies of diversity effects on arthropod abundance have mainly been conducted in agro-ecosystems, although tritrophic interactions in natural systems should better explain the evolution of tritrophic selective interactions. The use of chemical pesticides or artificial selection for increased crop yields results in interaction patterns of organisms from different trophic levels that do not necessarily reflect the outcome of selective pressures that exist between plants, herbivores and their parasitoids in natural systems. Furthermore, plant distributions are often more dense and homogeneous in agricultural systems. Consequently, one has to be cautious when trying to develop generalizations on diversity effects on parasitoids derived from agroecosystem studies. In the remainder of the chapter we thus specify whether findings emanate from naturally evolved interactions or from agroecosystems.

In the field, the effects of plant diversity on parasitoid abundance and parasitization rates are variable. While some parasitoid species are more abundant and show higher rates of parasitization in diversified habitats (Vanbergen *et al.* 2006, Fraser *et al.* 2007), for others higher parasitization rates were observed in monocropped habitats (Langer 1996). This difference may be due to different strategies, depending on the life history traits of the specific parasitoid species, and this is discussed in Section 3.3.3. Andow (1991) reviewed the reasons why parasitoids should be more abundant in diverse habitats. For example, host diversity is higher in polycultures, so more food and nectar resources are available to adult parasitoids. Polycultures provide more niches and refuges and can stabilize host–parasitoid interactions.

Recent studies have attempted to elucidate the effects of vegetation diversification on parasitoid orientation by following an experimental approach. Accordingly, an orientation experiment in the field showed that the time to reach the host plant took longer for parasitoids in an intercropped field than in a monoculture (Bukovinszky *et al.* 2007). Only a few studies have investigated natural, coevolved tritrophic interactions. Bezemer *et al.* (2010), studying a naturally coevolved tritrophic system, observed higher numbers of parasitoids on host plants in more simple than in complex habitats. In a biodiversity experiment, Petermann *et al.* (2010) showed a decrease in parasitoid emergence with increasing plant diversity on experimental plots. In contrast with the positive relationship between plant diversity and parasitoid abundance suggested above, these studies show that diversification may have a disruptive effect on parasitoid host location. The semiochemical diversity hypothesis formulated by Zhang & Schlyter (2004) tries to explain this disruptive effect of non-host plant species on the orientation of insects. It assumes that with higher plant species diversity, semiochemical diversity is also enhanced, and suggests that this interferes with olfactory orientation in insects. Although these ideas are based on findings pertaining to plant–herbivore interactions, they should be equally valid for parasitoids searching for herbivorous hosts.

The idea of plant species diversity causing disorientation in insects by means of enhanced odour diversity resulted in various hypotheses about possible mechanisms (reviewed by Finch & Collier 2000). In the case of parasitoids, non-host plants and diverse odorous surroundings can have different effects on their orientation behaviour (Schröder & Hilker 2008, Randlkofer *et al.* 2010b). So far, only a few laboratory studies have considered the effects of non-host plants on a parasitoid's discrimination ability. Parasitoids can be distracted by non-host plant odours rendering the target odour less detectable. This was shown in an agricultural system for the parasitoid *Dentichasmias busseolae* Heinrich (Diptera: Tachinidae), which is attracted to uninfested sorghum and maize and even more to herbivore-damaged plants. If the attractive plants are combined with a deterrent odour source (molasses grass) the parasitoid's response is weakened and the flies are only attracted to the infested host plants (Gohole *et al.* 2003b). Non-host plant odours can also completely mask the target odour, as shown for the specialist egg parasitoid *Oomyzus galerucivorus* Hedqvist (Hymenoptera: Eulophidae) occurring in calcareous grasslands. The wasps are attracted by the host plant, yarrow (*Achillea millefolium* L., Asteraceae) but fail to respond when yarrow is present in combination with the non-host plant, thyme (*Thymus vulgaris* L., Lamiaceae) (Randlkofer *et al.* 2007). However, non-host plants can also elicit a stronger attraction of parasitoids to the target odour. The wasp *Cotesia rubecula* (Marshall) (Hymenoptera: Braconidae) is more efficient when the host plant *Brassica oleraceae* L. (Brassicaceae) is offered in conjunction with non-host potato plants (Perfecto & Vet 2003). The presence of non-host plants also does not hinder the close-range foraging activities of the parasitoids *D. busseolae* and *Cotesia sesamiae* Cameron (Hymenoptera: Braconidae) (Gohole *et al.* 2005).

At the subspecies level, genotypic diversity of plants may also lead to bottom-up effects at the third trophic level (Johnson 2008). It has been proposed, however, that natural enemies are weakly affected in natural ecosystems, while heterogeneity of genotypes should have a stronger impact in agricultural monocrops, where the homogeneous background makes differences more obvious (Glinwood *et al.* 2011). To date, our knowledge of such interactions, especially when involving parasitoids, is very limited. In a field experiment with *Plutella xylostella* (L.) (Lepidoptera: Plutellidae) feeding on different *B. oleraceae* genotypes, parasitism rates by two *Diadegma* species did not correlate with plant genetic diversity (Hambäck *et al.* 2010). However, the predatory ladybird *Coccinella septempunctata* L. (Coleoptera: Coccinellidae) was more attracted to a mix of barley cultivars than to plots with single genotypes in the field. Laboratory experiments confirmed that the predator's preference was based on plant volatiles, but the mechanisms behind this behaviour are not yet understood (Ninkovic *et al.* 2011).

Indirect effects of plant species diversity on parasitoid orientation
Plant species diversity, as well as plant neighbour identity, can affect the physiology of a host plant (Broz *et al.* 2010) by changing its metabolic profile (Scherling *et al.* 2010) or at least some of its primary and secondary metabolites (Mraja *et al.* 2011). The vegetation surrounding an individual plant, and also soil factors, can therefore alter the plant's quality

as a food source for herbivores and can influence higher trophic levels (Sarfraz *et al.* 2009). Herbivores that sequester plant secondary metabolites for their own defence (Opitz & Müller 2009), for example, could become more or less suitable for parasitoids (Gols & Harvey 2009). According to the optimal foraging theory (MacArthur & Pianka 1966, Godfray 1994, Mills & Wajnberg 2008), it might be advantageous for the parasitoid to evaluate the host and the host plant quality during olfactory orientation. Parasitoids have evolved different reproductive strategies (see also Section 3.3.3) such as the idiobiont and the koinobiont lifestyle. Idiobionts retard the host's development and parasitoid larvae exploit an immobilized and static host. Hosts of koinobiont parasitoids, on the other hand, still feed and grow after being parasitized (Mills 2009) and could be affected by host plant quality, which may change over time. Especially for koinobiont parasitoids, it could be essential to assess host plant chemistry, as this can be a predictor of host performance, which is often linked to the performance of parasitoid larvae (Harvey 2005). According to the 'mother knows best' principle, the koinobiont parasitoids *Cotesia glomerata* L. (Hymenoptera: Braconidae) and *Diadegma semiclausum* Hellén (Hymenoptera: Ichneumonidae), parasitizing *Brassica*-specialized caterpillars, prefer those infested plants on which their offspring perform better. Gols *et al.* (2009) conclude that parasitoids are able to distinguish even slight differences in host plant quality from a distance by olfaction.

The link between host plant quality and long-distance olfactory evaluation by insects has not been well assessed so far, and we suggest that future studies on the orientation ability of parasitoids should also explore the role of plant species diversity and plant species identity on host plant quality.

3.2.2 Variability in host plant traits and their effects on parasitoid host location

When approaching the immediate vicinity of the host, the diversity of plant species and their odours may become less important for a parasitoid. At this stage, it is the variability of an individual host plant's chemistry that may complicate successful host finding. Chemical variability can be found in different plant traits on which parasitoids depend. For example, intrinsic (genotype, ontogeny) and extrinsic (biotic, abiotic) factors are known to cause variability in the volatiles that individual plants emit. In a similar way, these factors can also influence the composition of epicuticular plant wax, which is the surface that retains chemical cues from the host insect.

Plant signals
Intrinsic factors
Habitats differ in the selection pressures they exert on plant species, favouring specific traits that lead to the evolution of different genotypes. Phenotypically such genotypes can be expressed as chemotypes if plants differ in their secondary metabolite compositions (Langenheim 1994, Theis & Lerdau 2003, Macel & Klinkhamer 2010). Local habitat conditions have been found to generate highly divergent chemotypes in various aromatic and non-aromatic species belonging to a range of plant families (e.g. Lamiaceae: Theis & Lerdau 2003; Asteraceae: Macel & Klinkhamer 2010; Myrtaceae: Keszei *et al.* 2010; Pinaceae: Thoss *et al.* 2007; Cupressaceae: Lei *et al.* 2010; Myoporaceae: Smith *et al.* 2010). Likewise, selective breeding has generated countless genotypes of crop plants that differ in their secondary metabolic profile. Taking this into account, it is not surprising to find that variability in herbivore-induced volatiles also depends on the genotype in wild and crop plants

(e.g. Halitschke *et al.* 2000, Degen *et al.* 2004, Tamiru *et al.* 2011). Different volatile profiles may result in the differential attraction of parasitoids, as shown for *Trichogramma bournieri* Pintureau & Babault (Hymenoptera: Trichogrammatidae) and *C. sesamiae* in maize landraces infested with eggs of *Chilo partellus* Swinhoe (Lepidoptera: Crambidae) (Tamiru *et al.* 2011). However, genotype effects did not lead to changes in parasitism rates in the cotton aphid parasitoid *Lysiphlebia japonica* (Ashmead) (Hymenoptera: Braconidae) (Sun *et al.* 2011).

Furthermore, intrinsic variability is the result of ontogenetic changes. Soyabeans (*Glycine max* (L.), Fabaceae), for instance, emitted ten times more herbivore-induced volatiles while in their vegetative stage compared with individuals that already carried pods (Rostás & Eggert 2008). Likewise, higher emissions were found from younger than from older leaves of soyabean, Lima bean (*Phaseolus lunatus* L., Fabaceae) and castor (*Ricinus communis* L., Euphorbiaceae) (Radhika *et al.* 2008, Rostás & Eggert 2008). These studies showed that the distribution of volatiles acting as chemical cues follows the predictions of the optimal defence hypothesis, whereby higher levels of compounds are emitted from the more vulnerable structures of a plant and at its more vulnerable growth stages.

In the context of herbivore–parasitoid interactions, most experiments have been conducted with plants at the vegetative stage, and changes in floral volatile emissions due to herbivores feeding on vegetative tissues were neglected until recently (Lucas-Barbosa *et al.* 2011, Pareja *et al.* 2012). Pareja *et al.* (2012) showed that flowers of *Sinapis alba* L. (Brassicaeae) plants damaged by *Lipaphis erysimi* (Kaltenbach) (Hemiptera: Aphidinae) emitted lower amounts of volatiles than did the flowers of undamaged plants. Nevertheless, the parasitoid *Diaeretiella rapae* (McIntosh) (Hymenoptera: Braconidae) preferred the floral volatiles of *S. alba* plants infested with the specialist aphid *L. erysimi* more than volatiles from undamaged flowering *S. alba* plants, but was also attracted to the latter. The parasitoid's preference for lower amounts of floral volatiles released by *L. erysimi*-infested plants may be due to changes in the ratio of floral components that may indicate the presence of aphids (Pareja *et al.* 2012). These examples show that different plant parts vary in their volatile emissions during ontogenesis, but parasitoids might be able to handle this kind of variability.

Extrinsic factors
The release of plant volatiles into the environment is known to be affected by a range of abiotic factors (reviewed by Peñuelas & Llusià 2001, Hilker & McNeil 2008, Yuan *et al.* 2009, Holopainen & Gershenzon 2010, Loreto & Schnitzler 2010, Mumm & Dicke 2010). Most research so far has focused on plant volatile emission in the absence of herbivory, showing that light, temperature, CO_2, nutrients, drought or other environmental variables modulate release rates in concentration- or intensity-dependent ways. Abiotic conditions increase or decrease the emission rates of specific compounds and thus alter their ratios within a volatile blend; a pattern that is also shown for herbivore-induced odours. Fewer studies have related the effects of abiotic factors to parasitoid foraging behaviour. In general, plants exposed to abiotic stresses often show increased emission rates of various volatile compounds that are known to be involved in host habitat finding. However, this is quite variable even for a given abiotic factor. The effects of nitrogen availability, as an example, have been investigated in several plant–herbivore associations in agricultural systems. Low nitrogen availability resulted in higher herbivore-induced volatile emission in *Zea mays* L. (Poaceae), *Gossypium hirsutum* L. (Malvaceae) and *G. max* (Schmelz *et al.* 2003, Chen *et al.* 2008,

Winter & Rostás 2010). However, no change in release rates was found in *Nicotiana attenuata* Torr. ex Wats. (Solanaceae) (Lou & Baldwin 2004), while another study on *G. hirsutum* reported lower emission rates in plants that received very high or very low levels of nitrogen, and correlated this with reduced attraction on the part of the braconid *Microplitis croceipes* (Cresson) (Hymenoptera: Braconidae) (Olson *et al.* 2009). Females of *Cotesia marginiventris* (Cresson) (Hymenoptera: Braconidae), on the other hand, were equally attracted to *G. hirsutum* or *G. max* (Chen *et al.* 2008, Winter & Rostás 2010). Similar outcomes are known from studies that investigated other abiotic factors, demonstrating that although stressed plants may emit a qualitatively different volatile blend, parasitoids are not impaired in their ability to locate the host's food plant (Mumm & Dicke 2010).

The many biotic interactions between plants and insects, plants and plants, and plants and microbes also contribute to the complexity of chemical signals. Multiple attacks by herbivores, for instance, can lead to non-additive effects on the emitted volatile blend and to unpredictable outcomes in the searching behaviour of parasitoids. In nature, indeed, a plant might not be attacked by one herbivore species only but mostly by several species simultaneously (Dicke *et al.* 2009). In cabbage plants, combined infestation by the caterpillars of *P. xylostella* and *Pieris rapae* L. (Lepidoptera: Pieridae) resulted in a different odour blend compared with infestation by a single herbivore species. This change in induced odour bouquet correlated with higher attraction of *C. glomerata* parasitizing *P. rapae* but lower attraction of *Cotesia plutellae* (L.) (Hymenoptera: Braconidae) parasitizing *P. xylostella* (Shiojiri *et al.* 2001). In a similar experiment with *C. glomerata* and a different non-host species, parasitoids were more attracted to mixed infestations. However, parasitism rates were lower on mixed infestation plants compared with those infested by only the host species, probably as a result of lower host densities (Bukovinszky *et al.* 2012). Where herbivores induce different molecular plant defence pathways due to different feeding modes (e.g. sucking versus chewing), altered volatile blends may be the result of antagonistic cross-talk between jasmonic acid- and salicylic acid-mediated pathways (reviewed by Mumm & Dicke 2010; see also Ode, Chapter 2, this volume). Reduced volatile emission was shown for cotton simultaneously infested by the phloem feeding whitefly *Bemisia tabaci* (Gennadius) (Hemiptera: Aleyrodidae) and the leaf-chewing caterpillar *Spodoptera exigua* (Hübner) (Lepidoptera: Noctuidae) (Rodriguez-Saona *et al.* 2003).

Below-ground herbivory can affect above-ground plant parts and thus alter the chemical visibility of the host plant (see Soler, Bezemer & Harvey, Chapter 4, this volume). Root herbivory in maize seedlings has been observed to reduce volatile emissions induced by foliar herbivory. Infestation by *Spodoptera littoralis* Boisduval (Lepidoptera: Noctuidae) alone induced more volatiles than infestation both by the foliar herbivore *S. littoralis* and the root herbivore *Diabrotica virgifera virgifera* LeConte (Coleoptera: Chrysomelidae) (Rasmann & Turlings 2007). Naive parasitoids of *C. marginiventris* were not attracted to these doubly infested plants. However, they were able to learn the volatile blend, and experienced parasitoids were attracted to the less complex odour bouquet of doubly infested plants (Rasmann & Turlings 2007). The parasitoid *C. glomerata* avoided root-infested host plants but searched more efficiently in habitats with root-infested neighbouring plants. Root herbivory has an influence on the parasitoid's preferences and may alter plant quality in such a way that the parasitoid is no longer attracted by the plant volatiles (Soler *et al.* 2007). This effect was observable over four trophic levels (Soler *et al.* 2005).

Plant–plant interaction is an additional factor that can contribute to increased complexity in volatile signalling. Herbivore-damaged plants emit volatiles that can be perceived by neighbouring plants and, as described below, can prime them for enhanced volatile emission (Heil & Karban 2010). The primed plants release larger amounts of volatile signals when attacked by insects and thus become more attractive to parasitic wasps. This has been shown for the responses of *C. marginiventris* and *C. glomerata* to primed maize and cabbage, respectively (Ton *et al.* 2007, Peng *et al.* 2011). Alternatively, plants can simply absorb and re-release volatiles from other plant species that grow in their vicinity. The addition of re-released compounds to a plant's original blend alters its overall volatile pattern and can lead to reduced host plant recognition by herbivores, as has been shown for birch trees neighbouring *Rhododendron tomentosum* (Noe *et al.* 2008, Himanen *et al.* 2010). It seems likely that foraging parasitoids could also be affected, but this still needs to be verified.

Microorganisms can also influence the volatile bouquet of plants either by directly inducing the emission of compounds such as methyl salicylate, β-caryophyllene and others (Attaran *et al.* 2008, Mayer *et al.* 2008, Mann *et al.* 2012) or by modulating herbivore-induced blends (Cardoza *et al.* 2002, Rostás *et al.* 2006). Maize plants, for instance, emitted lower amounts of herbivore-induced volatiles when infested by a pathogenic fungus. However, the release of green leaf volatiles was not affected and neither was attraction of the parasitoids *C. marginiventris* and *Microplitis rufiventris* Kok. (Hymenoptera: Braconidae) in olfactometer bioassays (Rostás *et al.* 2006). In contrast, peanut plants (*Arachis hypogaea* L., Fabaceae) doubly challenged by the fungus *Sclerotium rolfsii* and caterpillars of *S. exigua* emitted methyl salicylate in addition to the normal herbivore-induced volatile bouquet (Cardoza *et al.* 2002). In this case, fungal infection enhanced the attraction of *C. marginiventris* towards herbivore-infested plants (Cardoza *et al.* 2003).

Host signals interacting with plant chemical complexity

Once a female parasitoid has found its host's food plant, it will search the habitat for chemical cues that are indicative of host presence and host density (Wajnberg 2006). The signals in this foraging phase are generally host-derived and thus more reliable than plant-derived cues. It can be expected that they are also less volatile and less variable, as the chemical complexity of the environment should decrease with increasing proximity to the host. One factor that can contribute to chemical complexity, which has been little studied so far, is the role of plant surface chemistry. This outermost layer of a plant's cuticle is covered by epicuticular waxes which fulfil several important physiological roles for the plant (Jetter *et al.* 2006). However, the waxy layer is also the interface of many plant–insect interactions and it is well known that components of the plant cuticle – and secondary metabolites therein – serve as host recognition cues for herbivorous insects (Müller & Riederer 2005, Badenes-Perez *et al.* 2011). Furthermore, for foraging parasitoids, wax surfaces play a somewhat indirect role as a substrate onto which insect-derived chemical signals can bind. Whenever insects move over a plant cuticle, they leave traces that can be detected by other insects. These chemical footprints have been shown to regulate intraspecific interactions. Bumblebees, for example, refrain from visiting flowers if they detect footprints from previous visits by other workers (Saleh *et al.* 2007, Wilms & Eltz 2008). But footprints also mediate interspecific interactions among carnivorous insects and between carnivores and their prey or hosts. Aphid parasitoids, for instance, avoid intraguild predation by recognizing residues from coccinelid beetles that can feed on parasitized aphids (Nakashima *et al.* 2004, 2006). Furthermore, a number of larval and adult parasitoids have been found to exploit chemical footprints to detect the presence of hosts in a host patch (Klomp 1981,

Rogers & Potter 2002, Collatz & Steidle 2008, Rostás & Wölfling 2009, Gonzalez *et al.* 2011). In egg parasitoids, where the host stage is inconspicuous, footprints of the mobile and more noticeable adult function as kairomones in host finding. This strategy of using cues from a non-appropriate host life stage to find the host stage that can be parasitized is known as 'infochemical detour' (Vet & Dicke 1992). Remarkably, for some parasitoid species, chemical footprints convey even more information than simply host presence. Females of *Trissolcus basalis* (Woix.) and *T. brochymenae* (Ashmead) (Hymenoptera: Scelionidae), which parasitize the eggs of pentatomid bugs, differentiate between the tracks of male and female bugs, and display prolonged host location behaviour on encountering the latter. The footprints of male bugs can be recognized by *T. basalis* due to the presence of *n*-nonadecane as an additional compound that is lacking in footprints from females (Colazza *et al.* 2007). In *T. brochymenae*, parasitoids have also been found to prefer the footprints of gravid bugs over virgins or female hosts that had already laid eggs, thus showing a high level of specificity (Salerno *et al.* 2009).

Footprints may be the result of active fluid secretion from the tarsal pads of adult and immature insects when in contact with the plant surface (Federle *et al.* 2002, Voigt & Gorb 2012). The excreted fluid increases the surface tension between the insect's foot and the plant surface, which enables the insect to walk on the smooth cuticles of leaves and stems. In several species, analysis of footprint chemistry showed that composition closely resembled insect cuticle coverage and mainly consisted of long-chain (C_{20} to C_{40}) *n*-alkanes, methyl-branched alkanes and alkenes, but also fatty acids and alcohols (Kosaki & Yamaoka 1996, Attygalle *et al.* 2000, Rostás & Wölfling 2009, Geiselhardt *et al.* 2011). Moreover, in some insects, tarsal pads have been reported to excrete a complex two-phasic fluid comprising lipid nanodroplets in an aqueous liquid that contains carbohydrates and amino acids (Federle *et al.* 2002, Votsch *et al.* 2002). It has been hypothesized that two-phasic footprints may be widespread among insects but have generally been overlooked. Little attention has been paid to the identification of polar compounds because this requires different analytical techniques.

To date, few studies have attempted to identify which of the footprint chemicals are used by parasitoids for host finding and recognition. In some species, *n*-alkanes of the cuticular hydrocarbon fraction of footprints elicit typical host searching behaviour (Colazza *et al.* 2007), but it has also been demonstrated that these compounds do not contain the complete information and that important minor compounds must be part of the full signal (Rostás & Wölfling 2009). The role of chemicals from the hydrophilic fraction has not yet been established, although they can be excluded in at least one host–parasitoid interaction where polar extracts of host footprints triggered no response in *C. marginiventris* (Rostás & Wölfling 2009).

How can plant chemical complexity affect the detection of host footprints? Plant wax composition varies tremendously both between plant species and within a single plant. Intrinsic factors influencing wax quantity and composition include ontogenetic state and organ-specific regulation, as can be seen in differences between upper and lower leaf surfaces in some species (Müller & Riederer 2005, Jetter *et al.* 2006). Plant cuticular waxes consist of a wide range of compounds that basically fulfil the same primary functions as insect waxes, such as protection against water loss or excessive absorption, UV radiation and pathogen defence. Alkanes, primary alcohols, aldehydes, ketones and fatty acids in the range of 20–40 carbon atoms are ubiquitous constituents. Aliphatics with secondary functional groups or triterpenoids can also occur but are more taxon-specific (Jetter *et al.* 2006).

Taking this into consideration, it is not surprising that some overlap between compounds from insect footprints and plant cuticles occurs (Rostás *et al.* 2008). Wax compounds may thus alter the parasitoid's ability to recognize host footprints, depending on the chemical composition of the plant surface. Such compounds could mask host kairomones or, if present in both plants and insects, change their proportions. So far, little knowledge is available on how plant surface chemistry influences the foraging behaviour of natural enemies. Most studies on this aspect have focused on the physical properties of epicuticular waxes, which are the result of their chemical composition. Different wax blooms can greatly affect parasitoid locomotion on plants and alter herbivore density as a consequence (Chang & Eigenbrode 2004). Evidence for a role of plant wax chemistry in modulating kairomone detection comes from only three tritrophic systems. Colazza *et al.* (2009) compared the responses of *T. basalis* females to footprints of its host *Nezara viridula* L. (Hemiptera: Pentatomidae) that had been deposited on the epicuticular or intracuticular wax layer of *Vicia faba* L. (Fabaceae) leaves. These two surfaces with altered chemical and physical properties were created by removing the epicuticular wax layer with gum arabic. While parasitoids displayed typical host finding behaviour on encountering footprints on epicuticular wax, this was not the case when the host had walked on intracuticular wax. Other approaches have used mutant plants with defects in wax biosynthesis. *Pisum sativum* L. (Fabaceae) with the mutation *wel* ('wax eliminator') shows reduced wax expression, altered wax composition, and few epicuticular crystals (Eigenbrode *et al.* 1998). These plants have been used to test the avoidance of intraspecific competition. Predatory larvae of the beetle *Hippodamia convergens* Guérom-Méneville (Coleoptera: Coccinellidae) avoided visiting leaves of wild-type *P. sativum* if these had previously been walked upon by conspecific larvae. In contrast, they did not distinguish between walked upon and control leaves when *wel*-mutants were used (Rutledge *et al.* 2008). Significant differences in the response to chemical footprints have also been found in the larval parasitoid *C. marginiventris* (Rostás *et al.* 2008). Here, leaves of two different *eceriferum* (*cer*) mutants and the corresponding wild-type background of *Hordeum vulgare* L. (Poaceae) were used as substrates for footprints of *S. exigua* larvae. The wax of the mutant leaves differed in quantity and composition, with higher ester (*cer*-za 126) or aldehyde (*cer*-yp 949) fractions compared with wild-type epicuticular wax. When given the choice, parasitoids showed more pronounced interest, in the form of antennal drumming, on wild-type and *cer*-za 126 leaves than on *cer*-yp 949 leaves. Total amounts of leaf waxes were similarly reduced in both mutants, and therefore the differential responses seem to be the result of altered chemical composition rather than the quantity of wax.

An interesting case is represented by herbivores that match their cuticular hydrocarbon profiles with the wax profile of their host plant. Such plant-induced chemical crypsis has been observed in several caterpillar species (Espelie & Brown 1990, Akino *et al.* 2004, Portugal & Trigo 2005, Piskorski *et al.* 2010). Larvae of the moth *Biston robustum* Wehrli (Lepidoptera: Geometridae) mimic the surface of their host plant and thus remain undetected by predatory ants unless they are placed on a different plant species, where they are readily attacked. If the caterpillars are allowed to feed on the new host, their hydrocarbon profile will again resemble the surface chemicals of this new plant after the next moult (Akino *et al.* 2004). It can be assumed that the footprints of these adapted herbivores should also be less easily detectable by parasitoids, as the chemical compounds that comprise the insect's footprints are generally of plant cuticular origin. However, this still needs to be proven. Examples like these suggest that chemical complexity is not only a function of plant wax variability but that host insects play an active part in order to remain hidden from natural enemies and there might be an arms race between the plant and the herbivore in natural systems. In agricultural systems, it might be an option to manipulate the chemical complexity of the wax layer through plant breeding so that hosts are no longer chemically hidden from their parasitoids.

As well as low-volatile hydrocarbon footprints, which can be active for several days (Rostás & Wölfling 2009), pheromones are another kind of host-derived cues that interact with plant waxes (Karg *et al.* 1994). In the moth *Mamestra brassicae* (L.) (Lepidoptera: Noctuidae), the sex pheromone released by the calling female during the night adsorbs to cabbage leaves for at least 24 hours and serves as a bridge in time for diurnally foraging *Trichogramma* egg parasitoids (Noldus *et al.* 1991). Whether the detection of adsorbed pheromones is influenced by variation in the epicuticular wax composition is still unknown.

3.3 Foraging strategies of parasitoids in chemically complex environments

Responses of parasitoids to enhanced environmental chemical complexity can be seen at both the behavioural and the sensory and neurophysiological level. A parasitoid can adopt different foraging strategies that might depend on its life history traits, its body size, its degree of host and host plant specialization, and its reproductive lifestyle. In the following sections, we outline these strategies and discuss the factors that might shape them.

3.3.1 Behavioural responses to chemical complexity

Depending on the benefits that parasitoids receive in chemically complex environments and on the costs they have to pay for their increased orientation efforts, parasitoids might either ignore, avoid or prefer chemically complex environments (Fig. 3.2).

Ignoring complexity
Parasitoids might ignore plant chemical complexity and use visual, vibrational and/or chemical host cues for host location instead. This change in sensory modality could take place when parasitoids change from long-range to close-range orientation after they have reached the host plant, when volatiles from surrounding plants or variable host plant cues might distract the parasitoids from the host. Although the disregarding of complex chemical cues by parasitoids has not yet been studied, it is known that parasitoids are not exclusively dependent on chemical cues during all the different phases of host location. The generalist larval parasitoid *Drino inconspicua* Meig. (Diptera: Tachinidae) was not attracted by volatiles of spruce twigs with feeding sawfly larvae, whereas movements of the host larvae significantly elicited a positive response at close range (Dippel & Hilker 1998). The pupal parasitoid *Brachymeria intermedia* (Nees) (Hymenoptera: Chalcididae) can learn visual cues from trees during habitat location, although it also uses olfactory cues (Drost & Cardé 1992).

Avoiding complexity
Parasitoids have the option to avoid areas with enhanced odour complexity, especially when their hosts choose food plants in environments of high complexity. When volatiles from neighbouring plants are dominant, parasitoids might be repelled or not attracted and do

not enter patches that may contain potential hosts. Examples of non-host plant odours enhancing odour complexity and repelling parasitoids are *D. busseolae*, which is repelled by the odour of molasses grass, which might prevent it from finding its host on sorghum (Gohole *et al.* 2003b) and *T. chilonis*, which is repelled by the odour of the non-host plant pigeonpea (*Cajanus cajan* (L.), Fabaceae) (Romeis *et al.* 1997). Loivamäki *et al.* (2008) proposed that non-host plants emitting high levels of isoprene interfere with the attraction of some parasitoids. Adding external isoprene to herbaceous plants had a repellent effect on *D. semiclausum* but not on *C. rubecula*, which does not perceive isoprene (Loivamäki *et al.* 2008). Parasitoids might compensate for enhanced odour complexity and difficulty in detecting host plant stimuli by increasing their walking or flight activity. As more complex physical vegetation structures can enhance the walking activity of parasitoids (Randlkofer *et al.* 2010a), parasitoids might also be induced to move more actively when entering areas of high odour complexity. This increased activity might also result in the parasitoid leaving an area with a complex odour composition in favour of a less complex one.

Preferring complexity

Some parasitoids may actively seek out enhanced chemical odour complexity. This can represent flower volatiles, since flowers produce the most diverse and the highest amounts of volatile compounds (Dudareva *et al.* 2004), and indicate that there is a greater probability of the presence of nectar. Naive *Trybliographa rapae* Westwood (Hymenoptera: Figitidae) females were significantly attracted to flower odours from *Fagopyrum esculentum* Moench. (Polygonaceae) and repelled by *Coriandrum sativum* L. (Apiaceae) and *Borago officinalis* L. (Boraginaceae). Significantly more *Delia radicum* L. (Diptera: Anthomyiidae) larvae were parasitized in semi-field cages where a floral resource was present (Nilsson *et al.* 2011). Parasitoids might also prefer areas of greater complexity when the presence of their host in those areas is more likely. In the field, the tansy leaf beetle *Galeruca tanaceti* L. (Coleoptera: Chrysomelidae) prefers to lay eggs in areas with enhanced plant species diversity (Randlkofer *et al.* 2007) and its specialized egg parasitoid *O. galerucivorus* might benefit from seeking its host in areas of enhanced odour complexity, although an enhanced odour complexity repels the parasitoid in this case (Randlkofer *et al.* 2007).

Spatial responses to complexity

Our present knowledge suggests that both the quality and the quantity of plant compounds present as background odour, as well as their mode of spatial presentation, greatly influence the orientation of foraging parasitoids. By moving out of areas or wind plumes carrying high odour diversity, parasitoids have a better chance of perceiving the odour of their host's food plant against a less complex background. Short flights or other movements can change the angle at which two odour sources are presented to the parasitoids. When the odour sources are separated widely enough, so that distinct odour bouquets can be perceived by the parasitoids, they might be able to respond accordingly. The effect of spacing two odour sources on odour recognition has been shown in moths and beetles, both in the laboratory and in the field (Baker *et al.* 1998, Andersson *et al.* 2011). Increased spacing between pheromones and anti-attractants led to increased catches in the field. Beetles were affected by a separation of odours of some decimetres, moths were affected by separation distances of just a few centimetres (Andersson *et al.* 2011). The egg parasitoid *O. galerucivorus* was negatively affected by enhanced odour complexity (represented by the non-host plant odour of thyme) in the location of yarrow host plant odours in a four-chamber olfactometer (Randlkofer *et al.* 2007). Also, on a moving sphere (locomotion compensator), where both odours were blown together from two combined tubes at a distance of 3 centimetres in front of the parasitoids, the wasps did not walk towards the yarrow odour. However, when tubes delivering the host plant and non-host plant odours were separated by as little as 1 centimetre, the parasitoids were able to separate the odours and walked towards the host plant odour (T. Meiners, unpublished results). Although it is not known whether this laboratory response is as accurately expressed in a structurally complex field situation, it demonstrates the potential increase in discrimination ability that the parasitoids can gain by movement.

3.3.2 Learning, sensory filters and neural constraints affecting strategies for dealing with complexity

The advantage of gaining flexibility in responses by learning is most probably greatest for parasitoids that are not strictly host-specific. Adjusting their behaviour to certain chemical compounds would help these parasitoids take advantage of changes in host and habitat composition. Many parasitoids vary in their behaviour depending on internal factors and environmental conditions. When female parasitoids respond to host stimuli more strongly after they have parasitized a host, this may lead to faster host finding and/or more intensive searching. Associative learning allows parasitoids to focus on the most reliable cues (Vet *et al.* 1990, De Jong & Kaiser 1991, Papaj & Lewis 1993). Parasitoids can learn a wide range of volatiles, including ecologically relevant (Vet & Groenewold 1990, Kester & Barbosa 1991) as well as novel or even artificial odours (Lewis & Takasu 1990, Olson *et al.* 2003). Despite the broad range of studies on parasitoid learning, there are very few studies that examined the learning ability of single compounds within blends (Fukushima *et al.* 2002). For example, the parasitoid *Leptopilina heterotoma* (Thomson) (Hymenoptera: Eucoilidae) learned to respond to single C6 compounds in a complex background of yeast odour (Vet & Groenewold 1990, Vet *et al.* 1998). Meiners *et al.* (2003) showed that *M. croceipes* can learn to respond to individual compounds following experience with an odour mixture, although the olfactory background can affect the recognition of single compounds. Learning can be a helpful mechanism when dealing with complex odour bouquets and certain compounds from a host plant might be detected, even when they are presented in another background. However, for certain compounds of a mixture, learning can be blocked by other mixture components. Thus, learning can explain successful host location of parasitoids in environments with enhanced odour complexity or variability only to a certain degree. Parasitoids must use additional behavioural or physiological mechanisms to locate their hosts. Plants emit numerous volatiles with various functional groups and ranging in structure from short, straight carbon chains to aromatic hydrocarbons and complex multi-ring sesquiterpenes (Visser 1979, Paré & Tumlinson 1997, Knudsen *et al.* 2006). Thus, the natural volatile compounds that parasitoids are exposed to are potentially numerous and diverse. However, the volatile complexity that can be measured by volatile analysis and the complexity that is experienced by a parasitoid is not necessarily the same. Antennae and brains act as filters of complexity and enable the insects to reduce the complexity on the sensory or on the neural level (Matthews & Matthews 2010).

  Insects' sensory receptors for non-pheromonal odorants can respond to many compounds and exhibit complex interactions in response to compounds in blends (Ache 1989,

Atema *et al.* 1989). Mixture suppression (certain compounds in a blend are less well recognized due to the presence of other compounds) is a common finding in recordings of sensory cell responses to pure compounds and to blends (Ache 1989, Akers & Getz 1993) and may explain why interactions in some blends are stronger than in others. Consistent with the data obtained from mixture recognition studies in honey bees (Pham-Delégue *et al.* 1993, Wadhams *et al.* 1994, Laloi *et al.* 2000), Meiners *et al.* (2003) confirmed, for the parasitoid *M. croceipes* in wind tunnel studies, a hierarchy within the components of a blend, with some components eliciting blend recognition better than others. This means that some compounds are more representative of the conditioning blend than others and it might be due to this sensory filtering that parasitoids cope with certain odorous environments better than others during host searching.

Responses to a diverse group of compounds in *M. croceipes* suggest that their receptors are broadly tuned. The ability to detect a broad array of chemicals and to show rapid adaptation, suggested by the speed with which some parasitoids can be trained, may be related to their ecology, where they must locate specific host species within highly variable and dynamic foraging environments. Park *et al.* (2001) showed, in electroantennogram studies, that the antenna of *M. croceipes* is more sensitive to less-volatile compounds (e.g. methyl jasmonate, β-caryophyllene) than to more-volatile ones (e.g. β-ocimene) and thus the low volatility of some compounds is counterbalanced by this higher sensitivity.

As postulated for herbivores (Bruce *et al.* 2005), parasitoids may also recognize host plants on the basis of specific cues, on the ratio of components, or on the whole blend (McCormick *et al.* 2012). Although for specialist herbivores the specificity of a blend is often determined by the ratio of key volatile compounds (Visser 1986, Bruce *et al.* 2005), they are able to accommodate to variation in the compounds (Najar-Rodriguez *et al.* 2010, Cha *et al.* 2011). Such plasticity would also be crucial for the host location process of parasitoids. Very often it might be favourable for a parasitoid to generalize very similar odours and not to invest too much time or sensory-physiological effort in differentiating between several odours that might be promising indicators of hosts in natural environments. For example, when host plant varieties show similar, but not identical, changes in emitted volatiles, *L. heterotoma*, for example, can adjust its degree of discrimination between similar odours (e.g. apple varieties) according to the profitability of the information in terms of the host encounter rate that is connected with the odour (Vet *et al.* 1998). Coupling rewarding and unrewarding experiences with different odours allows the parasitoids to differentiate between similar odour bouquets; a process that might occur frequently in nature. This helps parasitoids to cope with variable host plant odours (see Section 3.2.2) as well as to locate a certain host plant odour blend accompanied by slightly changed or more complex background odours from the varying vegetation.

3.3.3 Influences of life history traits on foraging strategy

The strategy that a female parasitoid uses to cope with a chemically complex or varying environment may depend on her particular life history traits, since traits such as body size, degree of specialization and reproductive lifestyle will influence properties such as movement or learning abilities.

Body size can have an important effect on the ability of parasitoids to move actively and to respond to chemical complexity in a self-determined way. In particular, small egg parasitoids are restricted to passive habitat searching due to their low spreading/flying abilities.

They might concentrate on host cues and ignore environmental complexity altogether (Fatouros *et al.* 2008). For long-range dispersal, small egg parasitoids often use phoresy (the transportation of one organism by another, more mobile, one) via the adult flying host to disperse and to reach the suitable host stage (see Huigens & Fatouros, Chapter 5, this volume). This strategy could also be a consequence of neural restriction due to small brain size. However, at the moment there are no studies indicating that small brain size might determine the neural capacities of parasitoids that affect learning or discriminatory ability, although sometimes traits such as behavioural flexibility are associated with larger size of the relevant brain parts in insects such as *Drosophila* (Dukas 2008).

The degree of host specialization of parasitoids and the dietary specialization of their herbivorous hosts should be reflected in a parasitoid's infochemical use during the host location process. While parasitoids specialized on monophagous herbivores should utilize innate responses to chemical stimuli from the host plant, parasitoids of polyphagous herbivores should employ learning processes during host foraging to cope with a high variability in plant odours (Vet *et al.* 1995). These predictions are supported by studies on learning in differently specialized larval and egg parasitoids. The generalist parasitoid *M. croceipes* is challenged during host location with a multitude of blends with different components, since its larval noctuid hosts of *Helicoverpa* and *Heliothis* are polyphagous (Fitt 1989). *M. croceipes* has been successfully used for the study of various learning paradigms (see Section 3.3.2). *Oomyzus gallerucae* (Fonscolombe) (Hymenoptera: Eulophidae), an egg parasitoid of the monophagous elm leaf beetle, is also monophagous and responds innately during host location to terpenoids emitted from elm, induced by beetle oviposition (Büchel *et al.* 2011). The oligophagous parasitoid *Closterocerus ruforum* (Krausse) (Hymenoptera: Eulophidae) parasitizes the eggs of various diprionid sawflies on *Pinus sylvestris* L. (Pinaceae) (Mumm *et al.* 2005). It displays an unusual learning strategy by which common information is filtered out of similar but differing odour blends (on pine twigs tested 1–4 days after odour was induced), but a behavioural response to this learned information becomes evident only in a specific odour context (on twigs tested 3 days after odour induction) (Schröder *et al.* 2008). This indicates that the variability in the learning process is adjusted to the variability in the chemical environment.

Parasitoids have evolved different lifestyles concerning their reproductive strategy: known as idiobiosis and koinobiosis. Idiobionts are parasitoids that arrest the development of the host, so that the larvae are faced with an immobilized and static host such as insect eggs or pupae. On the other hand, hosts of koinobiont parasitoids are not static since the hosts stay alive, feed and grow further after parasitization (Mills 2009). Comparing the foraging strategies of parasitoids with these different reproductive lifestyles suggests differences in the utilization of chemical or visual cues. Koinobiont parasitoids with mobile hosts are expected to rely more on visual cues, such as the host's movement, in close-range host location (see Section 3.3.1). On the other hand, idiobiont parasitoids should also be adapted to cues giving indirect hints of the host's presence, for example footprints on the plant surface of the adult host stage acting as kairomones for egg parasitoids. Pupal parasitoids might also use cues from last instar-feeding host larvae, which indicate the presence of pupae in the near future (Gohole *et al.* 2003b). Differences during foraging of parasitoids with those two lifestyles might also become manifest in the evaluation of cues (Steidle & van Loon 2002). As outlined above, for koinobiont parasitoids it might be advantageous to evaluate the host plant quality from a distance by olfactory cues during habitat location. Koinobionts are more dependent on host plant quality than idiobionts, since their host will feed further on the host plant and take up nutrients that are then allocated to the developing parasitoid. These examples indicate that the foraging strategy of a parasitoid species should not be discussed without taking its reproductive lifestyle into account.

3.4 Conclusions

To cope with chemical complexity during host plant and host location, parasitoids display different strategies such as ignoring, preferring or avoiding chemically complex environments. They have developed different behavioural, sensory and neurophysiological adaptations to successfully locate their hosts in heterogeneous microhabitats, habitats, or even in the landscape as a whole. We propose that the success of parasitoids depends not on one but on multiple mechanisms which they can flexibly utilize as a strategy to deal with complex chemical environments (see Fig. 3.2). In addition, foraging strategies used for successful host location are shaped by different life history traits such as body size, degree of host and/or host plant specialization, and reproductive lifestyle. We suggest that all the mechanisms involved need to be taken into account when elucidating the foraging strategy of a specific parasitoid species.

So far, parasitoids have shown different responses to enhanced plant diversity in agricultural compared with natural environments. In agricultural situations, the effects of plant diversity on parasitoid abundance and parasitization rates have been variable, while in natural systems it has been found that plant diversification may have a disruptive effect on parasitoid host location. Studies on diversity effects on insect abundance are still mainly conducted in agroecosystems and there are few studies concerning plant diversity effects on parasitoids' behaviour and orientation from natural tritrophic systems (Hare 2002, Gols & Harvey 2009, Unsicker et al. 2009, Bezemer et al. 2010, Hare 2011). The insights achieved in agroecosystems are not easily transferable to natural settings, since naturally evolved ecosystems may function differently and coevolutionary processes develop under natural selection pressures, whereas crop plants are created by artificial selection (Gols & Harvey 2009). It is known that laboratory populations of predatory arthropods can change their responses to plant cues (Dicke et al. 2000). Agricultural populations of parasitoids might also change their responses to plant volatiles depending on their degree of isolation and genetic exchange with populations from (semi)natural environments. More studies in natural ecosystems are needed in order to understand how parasitoids cope with plant chemical diversity while foraging, and to be able to assess the transferability of findings from natural to agricultural environments and vice versa. Plant epicuticular waxes can play an important role in the short-range host location of parasitoids by affecting the recognition of host contact cues. More studies are also needed here, to elucidate the role of plant chemical diversity and variability on parasitoid host location.

Future directions
Host-seeking parasitoids are able to recognize and respond to chemical cues and to distinguish them from background odours, despite remarkable similarities in the bouquets of volatiles that are emitted from different herbivore-damaged plants (McCormick et al. 2012). Parasitoids have to cope with sensory conflicts arising from the simultaneous input of potentially misleading information. In complex chemical environments, these sensory conflicts might be more prevalent because more information is available. An insect's

decision to respond to a certain stimulus, while ignoring others, depends on the interplay between sensory and neuronal processes, on innate and learned preferences, on its current physiological state, and on the actual meaning of the stimuli (Martin *et al.* 2011).

Field studies on relations between parasitoid responses and plant diversity co-occurring with various other environmental factors have a correlative character. It is necessary to study the mechanisms behind these responses under controlled and simplified conditions in the laboratory (Schmitz 2007). Behavioural responses of insects to volatiles may differ between studies conducted in the laboratory and those in the field (Gohole *et al.* 2003a, Randlkofer *et al.* 2007; but see Dicke *et al.* 2003), since odour bouquets might be experienced differently by parasitoids in the laboratory (Knudsen *et al.* 2008). Combined experimental and observational field and laboratory work with both specialist and generalist parasitoids is needed for a better understanding of the strategies that parasitoids use to cope with the chemical complexity and variability of the environment within trophic webs.

To relate parasitoid responses to environmental factors, field measurements are needed that identify the chemical complexity of the environment and the variability in the host plant. This requires the development of new volatile sampling techniques that take into account the effects of different spatial scales, abiotic conditions and biotic factors on odour emission. In the laboratory, new assays mimicking the odour situation in the field are needed in order to obtain a more realistic picture of parasitoid responses to odour diversity and to understand their host location strategies in complex surroundings. Finally, we need to know whether plants themselves respond to enhanced plant diversity in their surroundings by changing their production of compounds and whether parasitoids change their host location strategy according to the perceived changes in complexity.

References

Ache, B.W. (1989) Central and peripheral bases for mixture suppression in olfaction. In: Laing, D.G. (ed) *Perception of Complex Smells and Tastes*. Academic Press, Marickville, NSW, pp. 101–14.

Akers, R.P. and Getz, W.M. (1993) Response of olfactory receptor neurons in honeybees to odorants and their binary mixtures. *Journal of Comparative Physiology A* **173**: 169–85.

Akino, T., Nakamura, K. and Wakamura, S. (2004) Diet-induced chemical phytomimesis by twig-like caterpillars of *Biston robustum* Butler (Lepidoptera: Geometridae). *Chemoecology* **14**: 165–74.

Andersson, M.N., Binyameen, M., Sadek, M.M. and Schlyter, F. (2011) Attraction modulated by spacing of pheromone components and anti-attractants in a bark beetle and a moth. *Journal of Chemical Ecology* **37**: 899–911.

Andow, D.A. (1991) Vegetational diversity and arthropod population response. *Annual Review of Entomology* **36**: 561–86.

Atema, J., Borroni, P., Johnson, B., Voigt, R. and Handrich, L. (1989) Adaption and mixture interaction in chemoreceptor cells: mechanisms for diversity and contrast enhancement. In: Laing, D.G. (ed) *Perception of Complex Smells and Tastes*. Academic Press, Marickville, NSW, pp. 83–100.

Attaran, E., Rostás, M. and Zeier, J. (2008) *Pseudomonas syringae* elicits emission of the terpenoid (*E,E*)-4,8,12-trimethyl-1,3,7,11-tridecatetraene in *Arabidopsis* leaves via jasmonate signaling and expression of the terpene synthase TPS4. *Molecular Plant–Microbe Interactions* **21**: 1482–97.

Attygalle, A.B., Aneshansley, D.J., Meinwald, J. and Eisner, T. (2000) Defense by foot adhesion in a chrysomelid beetle (*Hemisphaerota cyanea*): characterization of the adhesive oil. *Zoology – Analysis of Complex Systems* **103**: 1–6.

Badenes-Perez, F.R., Reichelt, M., Gershenzon, J. and Heckel, D.G. (2011) Phylloplane location of glucosinolates in *Barbarea* spp. (Brassicaceae) and misleading assessment of host suitability by a specialist herbivore. *New Phytologist* **189**: 549–56.

Baker, T.C., Fadamiro, H.Y. and Cosse, A.A. (1998) Moth uses fine tuning for odour resolution. *Nature* **393**: 530.

Bezemer, T.M., Harvey, J.A., Kamp, A.F.D., Wagenaar, R., Gols, R., Kostenko, O., Fortuna, T., Engelkes, T., Vet, L.E.M., van der Putten, W.H. and Soler, R. (2010) Behaviour of male and female parasitoids in the field: influence of patch size, host density, and habitat complexity. *Ecological Entomology* **35**: 341–51.

Blassioli Moraes, M.C., Laumann, R., Sujii, E.R., Pires, C. and Borges, M. (2005) Induced volatiles in soybean and pigeon pea plants artificially infested with the neotropical brown stink bug, *Euschistus heros*, and their effect on the egg parasitoid, *Telenomus podisi*. *Entomologia Experimentalis et Applicata* **115**: 227–37.

Broz, A.K., Broeckling, C.D., De-la-Peña, C., Lewis, M.R., Greene, E., Callaway, R.M., Sumner, L.W. and Vivanco, J.M. (2010) Plant neighbor identity influences plant biochemistry and physiology related to defense. *BMC Plant Biology* **10**: 115.

Bruce, T.J.A., Wadhams, L.J. and Woodcock, C.M. (2005) Insect host location: a volatile situation. *Trends in Plant Science* **10**: 269–74.

Büchel, K., Malskies, S., Mayer, M., Fenning, T., Gershenzon, J., Hilker, M. and Meiners, T. (2011) How plants give early herbivore alert: volatile terpenoids emitted from elm attract egg parasitoids to plants laden with eggs of the elm leaf beetle. *Basic and Applied Ecology* **12**: 403–12.

Bukovinszky, T., Gols, R., Hemerik, L., van Lenteren, J.C. and Vet, L.E.M. (2007) Time allocation of a parasitoid foraging in heterogeneous vegetation: implications for host–parasitoid interactions. *Journal of Animal Ecology* **76**: 845–53.

Bukovinszky, T., Poelman, E.H., Kamp, A., Hemerik, L., Prekatsakis, G. and Dicke, M. (2012) Plants under multiple herbivory: consequences for parasitoid searching behaviour and foraging efficiency. *Animal Behaviour* **83**: 501–9.

Cardoza, Y.J., Alborn, H.T. and Tumlinson, J.H. (2002) In vivo volatile emissions from peanut plants induced by simultaneous fungal infection and insect damage. *Journal of Chemical Ecology* **28**: 161–74.

Cardoza, Y.J., Teal, P.E.A. and Tumlinson, J.H. (2003) Effect of peanut plant fungal infection on oviposition preference by *Spodoptera exigua* and on host-searching behavior by *Cotesia marginiventris*. *Environmental Entomology* **32**: 970–6.

Cha, D.H., Linn, C.E., Teal, P.E.A., Zhang, A., Roelofs, W.L. and Loeb, G.M. (2011) Eavesdropping on plant volatiles by a specialist moth: significance of ratio and concentration. *PLoS ONE* **6**(2): e17033.

Chang, G.C. and Eigenbrode, S.D. (2004) Delineating the effects of a plant trait on interactions among associated insects. *Oecologia* **139**: 123–30.

Chen, Y.G., Schmelz, E.A., Wäckers, F. and Ruberson, J. (2008) Cotton plant, *Gossypium hirsutum* L., defense in response to nitrogen fertilization. *Journal of Chemical Ecology* **34**: 1553–64.

Colazza, S., Aquila, G., De Pasquale, C., Peri, E. and Millar, J.G. (2007) The egg parasitoid *Trissolcus basalis* uses *n*-nonadecane, a cuticular hydrocarbon from its stink bug host *Nezara viridula*, to discriminate between female and male hosts. *Journal of Chemical Ecology* **33**: 1405–20.

Colazza, S., Lo Bue, M., Lo Giudice, D. and Peri, E. (2009) The response of *Trissolcus basalis* to footprint contact kairomones from *Nezara viridula* females is mediated by leaf epicuticular waxes. *Naturwissenschaften* **96**: 975–81.

Collatz, J. and Steidle, J.L.M. (2008) Hunting for moving hosts: *Cephalonomia tarsalis*, a parasitoid of free-living grain beetles. *Basic and Applied Ecology* **9**: 452–7.

De Jong, R. and Kaiser, L. (1991) Odor learning by *Leptopilina boulardi*, a specialist parasitoid (Hymenoptera: Eucoilidae). *Journal of Insect Behavior* **4**: 743–50.

de Moraes, C.M., Lewis, W.J., Paré, P.W., Alborn, H.T. and Tumlinson, J.H. (1998) Herbivore infested plants selectively attract parasitoids. *Nature* **393**: 570–3.

Degen, T., Dillmann, C., Marion-Poll, F. and Turlings, T.C.J. (2004) High genetic variability of herbivore-induced volatile emission within a broad range of maize inbred lines. *Plant Physiology* **135**: 1928–38.

Dicke, M., de Boer, J.G., Höfte, M. and Rocha-Granados, M.C. (2003) Mixed blends of herbivore-induced plant volatiles and foraging success of carnivorous arthropods. *Oikos* **101**: 38–48.

Dicke, M., Schütte, C. and Dijkman, H. (2000) Change in behavioural response to herbivore-induced plant volatiles in a predatory mite population. *Journal of Chemical Ecology* **26**: 1497–514.

Dicke, M., van Loon, J.J.A. and Soler, R. (2009) Chemical complexity of volatiles from plants induced by multiple attack. *Nature Chemical Biology* **5**: 317–24.

Dippel, C. and Hilker, M. (1998) Effects of physical and chemical signals on host foraging behavior of *Drino inconspicua* (Diptera: Tachinidae), a generalist parasitoid. *Environmental Entomology* **27**: 682–7.

Drost, Y.C. and Cardé, R.T. (1992) Use of learned visual cues during habitat location by *Brachymeria intermedia*. *Entomologia Experimentalis et Applicata* **64**: 217–24.

Drost, Y.C., Lewis, W.J., Zanen, P.O. and Keller, M.A. (1986) Beneficial arthropod behavior mediated by airborne semiochemicals. I. Flight behavior and influence of preflight handling of *Microplitis croceipes* (Cresson). *Journal of Chemical Ecology* **12**: 1247–62.

Dudareva, N., Pichersky, E. and Gershenzon, J. (2004) Biochemistry of plant volatiles. *Plant Physiology* **135**: 1893–902.

Dukas, R. (2008) Evolutionary biology of insect learning. *Annual Review of Entomology* **53**: 145–60.

Eigenbrode, S.D., White, C., Rhode, M. and Simon, C.J. (1998) Epicuticular wax phenotype of the *wel* mutation and its effect on pea aphid populations in the greenhouse and in the field. *Pisum Genetics* **29**: 13–17.

Espelie, K.E. and Brown, J.J. (1990) Cuticular hydrocarbons of species which interact on four trophic levels: apple, *Malus pumila* Mill.; codling moth, *Cydia pomonella* L.; a hymenopteran parasitoid, *Ascogaster quadridentata* Wesmael; and a hyperparasite, *Perilampus fulvicornis* Ashmead. *Comparative Biochemistry and Physiology Part B: Comparative Biochemistry* **95**: 131–6.

Fatouros, N.E., Dicke, M., Mumm, R., Meiners, T. and Hilker, M. (2008) Chemoecology of host foraging behaviour: infochemical-exploiting strategies used by foraging egg parasitoids. *Behavioral Ecology* **19**: 677–89.

Federle, W., Riehle, M., Curtis, A.S.G. and Full, R.J. (2002) An integrative study of insect adhesion: mechanics and wet adhesion of pretarsal pads in ants. *Integrative and Comparative Biology* **42**: 1100–6.

Finch, S. and Collier, R.H. (2000) Host-plant selection by insects: a theory based on 'appropriate/inappropriate landings' by pest insects of cruciferous plants. *Entomologia Experimentalis et Applicata* **96**: 91–102.

Fitt, G.P. (1989) The ecology of *Heliothis* in relation to agroecosystems. *Annual Review of Entomology* **34**: 17–52.

Fraser, S.E.M., Dytham, C. and Mayhew, P.J. (2007) Determinants of parasitoid abundance and diversity in woodland habitats. *Journal of Applied Ecology* **44**: 352–61.

Fukushima, J., Kainoh, Y., Honda, H. and Takabayashi, J. (2002) Learning of herbivore-induced and nonspecific plant volatiles by a parasitoid, *Cotesia kariyai*. *Journal of Chemical Ecology* **28**: 579–86.

Geiselhardt, S.F., Geiselhardt, S. and Peschke, K. (2011) Congruence of epicuticular hydrocarbons and tarsal secretions as a principle in beetles. *Chemoecology* **21**: 181–6.

Glinwood, R., Ninkovic, V. and Pettersson, J. (2011) Chemical interaction between undamaged plants: effects on herbivores and natural enemies. *Phytochemistry* **72**: 1683–89.

Godfray, H.C.J. (1994) *Parasitoids: Behavioral and Evolutionary Ecology*. Princeton University Press, NJ.

Gohole, L.S., Overholt, W.A., Khan, Z.R., Pickett, J.A. and Vet, L.E.M. (2003a) Effects of molasses grass, *Melinis minutiflora* volatiles on the foraging behavior of the cereal stemborer parasitoid, *Cotesia sesamiae. Journal of Chemical Ecology* **29**: 731–45.

Gohole, L.S., Overholt, W.A., Khan, Z.R. and Vet, L.E.M. (2003b) Role of volatiles emitted by host and non-host plants in the foraging behaviour of *Dentichasmias busseolae*, a pupal parasitoid of the spotted stemborer *Chilo partellus. Entomologia Experimentalis et Applicata* **107**: 1–9.

Gohole, L.S., Overholt, W.A., Khan, Z.R. and Vet, L.E.M. (2005) Close-range host searching behavior of the stemborer parasitoids *Cotesia sesamiae* and *Dentichasmias busseolae*: influence of a non-host plant *Melinis minutiflora. Journal of Insect Behavior* **18**: 149–69.

Gols, R. and Harvey, J.A. (2009) Plant-mediated effects in the Brassicaceae on the performance and behaviour of parasitoids. *Phytochemistry Reviews* **8**: 187–206.

Gols, R., van Dam, N.M., Raaijmakers, C.E., Dicke, M. and Harvey, J.A. (2009) Are population differences in plant quality reflected in the preference and performance of two endoparasitoid wasps? *Oikos* **118**: 733–43.

Gonzalez, J.M., Cusumano, A., Williams, H.J., Colazza, S. and Vinson, S.B. (2011) Behavioral and chemical investigations of contact kairomones released by the mud dauber wasp *Trypoxylon politum*, a host of the parasitoid *Melittobia digitata. Journal of Chemical Ecology* **37**: 629–39.

Halitschke, R., Kessler, A., Kahl, J., Lorenz, A. and Baldwin, I.T. (2000) Ecophysiological comparison of direct and indirect defenses in *Nicotiana attenuata. Oecologia* **124**: 408–17.

Hambäck, P.A., Björkman, M. and Hopkins, R.J. (2010) Patch size effects are more important than genetic diversity for plant–herbivore interactions in *Brassica* crops. *Ecological Entomology* **35**: 299–306.

Hare, J.D. (2002) Plant genetic variation in tritrophic interactions. In: Tscharntke, T. and Hawkins, B.A. (eds) *Multitrophic Level Interactions*. Cambridge University Press, Cambridge, pp. 8–43.

Hare, J.D. (2011) Ecological role of volatiles produced by plants in response to damage by herbivorous insects. *Annual Review of Entomology* **56**: 161–80.

Harvey, J.A. (2005) Factors affecting the evolution of development strategies in parasitoid wasps: the importance of functional constraints and incorporating complexity. *Entomologia Experimentalis et Applicata* **117**: 1–13.

Heil, M. and Karban, R. (2010) Explaining evolution of plant communication by airborne signals. *Trends in Ecology and Evolution* **25**: 137–44.

Hilker, M. and McNeil, J. (2008) Chemical and behavioral ecology in insect parasitoids: how to behave optimally in a complex odorous environment. In: Wajnberg, E., Bernstein, C. and van Alphen, J.J.M. (eds) *Behavioral Ecology of Insect Parasitoids: From Theoretical Approach to Field Application*. Blackwell Publishing Ltd., Oxford, pp. 92–112.

Hilker, M., Kobs, C., Varama, M. and Schrank, K. (2002) Insect egg deposition induces *Pinus sylvestris* to attract egg parasitoids. *Journal of Experimental Biology* **205**: 455–61.

Himanen, S.J., Blande, J.D. and Holopainen, J.K. (2010) Plant-emitted semi-volatiles shape the info-chemical environment and herbivore resistance of heterospecific neighbors. *Plant Signaling and Behavior* **5**: 1234–6.

Hoballah, M.E., Tamò, C. and Turlings, T.C.J. (2002) Differential attractiveness of induced odors emitted by eight maize varieties for the parasitoid *Cotesia marginiventris*: is quality or quantity important? *Journal of Chemical Ecology* **28**: 951–68.

Holopainen, J.K. and Gershenzon, J. (2010) Multiple stress factors and the emission of plant VOCs. *Trends in Plant Science* **15**: 176–84.

Jetter, R., Kunst, L. and Samuels, A.L. (2006) Composition of plant cuticular waxes. In: Riederer, M. and Müller, C. (eds) *Biology of the Plant Cuticle*. Blackwell Publishing Ltd., Oxford, pp. 145–81.

Johnson, M.T.J. (2008) Bottom-up effects of plant genotype on aphids, ants, and predators. *Ecology* **89**: 145–54.

Karg, G., Suckling, D.M. and Bradley, S.J. (1994) Absorption and release of pheromone of *Epiphyas postvittana* (Lepidoptera, Tortricidae) by apple leaves. *Journal of Chemical Ecology* **20**: 1825–41.

Kester, K.M. and Barbosa, P. (1991) Effects of post-emergence experience on searching and landing responses to plants in the insect parasitoid, *Cotesia congregata* (Say). *Journal of Insect Behavior* **5**: 301–20.

Keszei, A., Hassan, Y. and Foley, W.J. (2010) A biochemical interpretation of terpene chemotypes in *Melaleuca alternifolia*. *Journal of Chemical Ecology* **36**: 652–61.

Klomp, H. (1981) Parasitic wasps as sleuth-hounds: response of an ichneumon wasp to the trail of its host. *Netherlands Journal of Zoology* **31**: 762–72.

Knudsen, G.K., Bengtsson, M., Kobro, S., Jaastad, G., Hofsvang, T. and Witzgall, P. (2008) Discrepancy in laboratory and field attraction of apple fruit moth *Argyresthia conjugella* to host plant volatiles. *Physiological Entomology* **33**: 1–6.

Knudsen, J.T., Eriksson, R., Gershenzon, J. and Stahl, B. (2006) Diversity and distribution of floral scent. *Botanical Review* **72**: 1–120.

Kosaki, A. and Yamaoka, R. (1996) Chemical composition of footprints and cuticular lipids of three species of lady beetles. *Japanese Journal of Applied Entomology and Zoology* **40**: 47–53.

Laloi, D., Bailez, O., Blight, M.M., Roger, B., Pham-Delégue, M.H. and Wadhams, L.J. (2000) Recognition of complex odours by restrained and free-flying honey bees, *Apis mellifera*. *Journal of Chemical Ecology* **26**: 2307–19.

Langenheim, J.H. (1994) Higher-plant terpenoids: a phytocentric overview of their ecological roles. *Journal of Chemical Ecology* **20**: 1223–80.

Langer, V. (1996) Insect–crop interactions in a diversified cropping system: parasitism by *Aleochara bilineata* and *Trybliographa rapae* of the cabbage root fly, *Delia radicum*, on cabbage in the presence of white clover. *Entomologia Experimentalis et Applicata* **80**: 365–74.

Lei, H., Wang, Y., Liang, F., Su, W., Feng, Y., Guo, X. and Wang, N. (2010) Composition and variability of essential oils of *Platycladus orientalis* growing in China. *Biochemical Systematics and Ecology* **38**: 1000–6.

Lewis, W.J. and Takasu, K. (1990) Use of learned odours by a parasitic wasp in accordance with host and food needs. *Nature* **348**: 635–6.

Loivamäki, M., Mumm, R., Dicke, M. and Schnitzler, J.-P. (2008) Isoprene interferes with the attraction of bodyguards by herbaceous plants. *Proceedings of the National Academy of Sciences USA* **105**: 17430–5.

Loreto, F. and Schnitzler, J.P. (2010) Abiotic stresses and induced BVOCs. *Trends in Plant Science* **15**: 154–66.

Lou, Y.G. and Baldwin, I.T. (2004) Nitrogen supply influences herbivore-induced direct and indirect defenses and transcriptional responses to *Nicotiana attenuata*. *Plant Physiology* **135**: 496–506.

Lucas-Barbosa, D., van Loon, J.J.A. and Dicke, M. (2011) The effects of herbivore-induced plant volatiles on interactions between plants and flower-visiting insects. *Phytochemistry* **72**: 1647–54.

MacArthur, R.H. and Pianka, E.R. (1966) On optimal use of a patchy environment. *American Naturalist* **100**: 603–9.

Macel, M. and Klinkhamer, P.G.L. (2010) Chemotype of *Senecio jacobaea* affects damage by pathogens and insect herbivores in the field. *Evolutionary Ecology* **24**: 237–50.

Mann, R.S., Ali, J.G., Hermann, S.L., Tiwari, S., Pelz-Stelinski, K.S., Alborn, H.T. and Stelinski, L.L. (2012) Induced release of a plant-defense volatile 'deceptively' attracts insect vectors to plants infected with a bacterial pathogen. *PLoS Pathogens* **8**(3): e1002610.

Martin, J.P., Beyerlein, A., Dacks, A.M., Reisenman, C.E., Riffell, J.A., Lei, H. and Hildebrand, J.G. (2011) The neurobiology of insect olfaction: sensory processing in a comparative context. *Progress in Neurobiology* **95**: 427–47.

Matthews, R.W. and Matthews, J.R. (2010) *Insect Behavior*, 2nd edn. Springer, Dordrecht.

Mayer, C.J., Vilcinskas, A. and Gross, J. (2008) Pathogen-induced release of plant allomone manipulates vector insect behavior. *Journal of Chemical Ecology* **34**: 1518–22.

McCormick, A.C., Unsicker, S.B. and Gershenzon, J. (2012) The specificity of herbivore-induced plant volatiles in attracting herbivore enemies. *Trends in Plant Science* **17**: 303–10.

Meiners, T., Wäckers, F. and Lewis, W.J. (2003) Associative learning of complex odours in parasitoid host location. *Chemical Senses* **28**: 231–6.

Mills, N. (2009) Parasitoids. In: Resh, V.H. and Cardé, R.T. (eds) *Encyclopedia of Insects*, 2nd edn. Academic Press, Amsterdam, pp. 748–51.

Mills, N.J. and Wajnberg, E. (2008) Optimal foraging behaviour and efficient biological control methods. In: Wajnberg, E. Bernstein, C. and van Alphen, J.J.M. (eds) *Behavioral Ecology of Insect Parasitoids: From Theoretical Approaches to Field Application*. Blackwell Publishing Ltd., Oxford, pp. 3–30.

Mraja, A., Unsicker, S.B., Reichelt, M., Gershenzon, J. and Roscher, C. (2011) Plant community diversity influences allocation to direct chemical defence in *Plantago lanceolata*. *PLoS ONE* **6**(12): e28055.

Müller, C. and Riederer, M. (2005) Plant surface properties in chemical ecology. *Journal of Chemical Ecology* **31**: 2621–51.

Mumm, R. and Dicke, M. (2010) Variation in natural plant products and the attraction of bodyguards involved in indirect plant defense. *Canadian Journal of Zoology* **88**: 628–67.

Mumm, R., Tiemann, T., Varama, M. and Hilker, M. (2005) Choosy egg parasitoids: specificity of oviposition-induced pine volatiles exploited by an egg parasitoid of pine sawfly. *Entomologia Experimentalis et Applicata* **115**: 217–25.

Najar-Rodriguez, A.J., Galizia, C.G., Stierle, J. and Dorn, S. (2010) Behavioral and neurophysiological responses of an insect to changing ratios of constituents in host plant-derived volatile mixtures. *Journal of Experimental Biology* **213**: 3388–97.

Nakashima, Y., Birkett, M.A., Pye, B.J., Pickett, J.A. and Powell, W. (2004) The role of semiochemicals in the avoidance of the seven-spot ladybird, *Coccinella septempunctata*, by the aphid parasitoid, *Aphidius ervi*. *Journal of Chemical Ecology* **30**: 1103–16.

Nakashima, Y., Birkett, M.A., Pye, B.J. and Powell, W. (2006) Chemically mediated intraguild predator avoidance by aphid parasitoids: interspecific variability in sensitivity to semiochemical trails of ladybird predators. *Journal of Chemical Ecology* **32**: 1989–98.

Nilsson, U., Rännbäck, L.-M., Anderson, P., Eriksson, A. and Rämert, B. (2011) Comparison of nectar use and preference in the parasitoid *Trybliographa rapae* (Hymenoptera: Figitidae) and its host, the cabbage root fly, *Delia radicum* (Diptera: Anthomyiidae). *Biocontrol Science and Technology* **21**: 1117–32.

Ninkovic, V., Al Abassi, S., Ahmed, E., Glinwood, R. and Pettersson, J. (2011) Effect of within-species plant genotype mixing on habitat preference of a polyphagous insect predator. *Oecologia* **166**: 391–400.

Noe, S.M., Copolovici, L., Niinemets, Ü. and Vaino, E. (2008) Foliar limonene uptake scales positively with leaf lipid content: 'non-emitting' species absorb and release monoterpenes. *Plant Biology* **10**: 129–37.

Noldus, L., Potting, R.P.J. and Barendregt, H.E. (1991) Moth sex-pheromone adsorption to leaf surface: bridge in time for chemical spies. *Physiological Entomology* **16**: 329–44.

Olson, D.M., Cortesero, A.M., Rains, G.C., Potter, T. and Lewis, W.J. (2009) Nitrogen and water affect direct and indirect plant systemic induced defense in cotton. *Biological Control* **49**: 239–44.

Olson, D.M., Rains, G.C., Meiners, T., Takasu, K., Tertuliano, M., Tumlinson, J.H., Wäckers, F.L. and Lewis, W.J. (2003) Parasitic wasps learn and report diverse chemicals with unique conditionable behaviors. *Chemical Senses* **28**: 545–9.

Opitz, S.E.W. and Müller, C. (2009) Plant chemistry and insect sequestration. *Chemoecology* **19**: 117–54.

Papaj, D.R. and Lewis, A. (1993) *Insect Learning: Ecological and Evolutionary Perspectives*. Chapman & Hall, New York.

Paré, P.W. and Tumlinson, J.H. (1997) De novo biosynthesis of volatiles induced by insect herbivory in cotton plants. *Plant Physiology* **114**: 1161–7.

Paré, P.W. and Tumlinson, J.H. (1999) Plant volatiles as a defense against insect herbivores. *Plant Physiology* **121**: 325–31.

Pareja, M., Qvarfordt, E., Webster, B., Mayon, P., Pickett, J., Birkett, M. and Glinwood, R. (2012) Herbivory by a phloem-feeding insect inhibits floral volatile production. *PLoS ONE* **7**(2): e31971.

Park, K.C., Zhu, J., Harris, J., Ochieng, S.A. and Baker, T.C. (2001) Electroantennogram responses of a parasitic wasp, *Microplitis croceipes*, to host-related volatile and anthropogenic compounds. *Physiological Entomology* **26**: 69–77.

Peng, J., van Loon, J.J.A., Zheng, S. and Dicke, M. (2011) Herbivore-induced volatiles of cabbage (*Brassica oleracea*) prime defence responses in neighbouring intact plants. *Plant Biology* **13**: 276–84.

Peñuelas, J. and Llusià, J. (2001) The complexity of factors driving volatile organic compound emission by plants. *Biologia Plantarum* **44**: 481–7.

Perfecto, I. and Vet, L.E.M. (2003) Effect of a nonhost plant on the location behavior of two parasitoids: the tritrophic system of *Cotesia* spp. (Hymenoptera: Braconidae), *Pieris rapae* (Lepidoptera: Pieridae), and *Brassica oleraceae*. *Environmental Entomology* **32**: 163–74.

Petermann, J.S., Müller, C.B., Roscher, C., Weigelt, A., Weisser, W.W. and Schmid, B. (2010) Plant species loss affects life-history traits of aphids and their parasitoids. *PLoS ONE* **5**(8): e12053.

Pham-Delégue, M.H., Bailez, O., Blight, M.M., Masson, C. and Picard-Nizou, A.L. (1993) Behavioural discrimination of oilseed rape volatiles by the honeybee *Apis mellifera* L. *Chemical Senses* **18**: 483–94.

Piskorski, R., Trematerra, P. and Dorn, S. (2010) Cuticular hydrocarbon profiles of codling moth larvae, *Cydia pomonella* (Lepidoptera: Tortricidae), reflect those of their host plant species. *Biological Journal of the Linnean Society* **101**: 376–84.

Portugal, A.H.A. and Trigo, J.R. (2005) Similarity of cuticular lipids between a caterpillar and its host plant: a way to make prey undetectable for predatory ants? *Journal of Chemical Ecology* **31**: 2551–61.

Quicke, D.L.J. (1997) *Parasitic Wasps*. Chapman & Hall, London.

Radhika, V., Kost, C., Bartram, S., Heil, M. and Boland, W. (2008) Testing the optimal defence hypothesis for two indirect defences: extrafloral nectar and volatile organic compounds. *Planta* **228**: 449–57.

Randlkofer, B., Obermaier, E., Casas, J. and Meiners, T. (2010a) Connectivity counts: disentangling effects of vegetation structure elements on the searching movement of a parasitoid. *Ecological Entomology* **35**: 446–55.

Randlkofer, B., Obermaier, E., Hilker, M. and Meiners, T. (2010b) Vegetation complexity: the influence of plant species diversity and plant structures on plant chemical complexity and arthropods. *Basic and Applied Ecology* **11**: 383–95.

Randlkofer, B., Obermaier, E. and Meiners, T. (2007) Mother's choice of the oviposition site: balancing risk of egg parasitism and need of food supply for the progeny with an infochemical shelter? *Chemoecology* **17**: 177–86.

Rasmann, S. and Turlings, T.C.J. (2007) Simultaneous feeding by aboveground and belowground herbivores attenuates plant-mediated attraction of their respective natural enemies. *Ecology Letters* **10**: 926–36.

Rodriguez-Saona, C., Crafts-Brandner, S.J. and Canas, L.A. (2003) Volatile emissions triggered by multiple herbivore damage: beet armyworm and whitefly feeding on cotton plants. *Journal of Chemical Ecology* **29**: 2539–50.

Rogers, M.E. and Potter, D.A. (2002) Kairomones from scarabaeid grubs and their frass as cues in below-ground host location by the parasitoids *Tiphia vernalis* and *Tiphia pygidialis*. *Entomologia Experimentalis et Applicata* **102**: 307–314.

Romeis, J., Shanower, T.G. and Zebitz, C.P.W. (1997) Volatile plant infochemicals mediate plant preference of *Trichogramma chilonis*. *Journal of Chemical Ecology* **23**: 2455–65.

Rostás, M. and Eggert, K. (2008) Ontogenetic and spatio-temporal patterns of induced volatiles in *Glycine max* in the light of the optimal defence hypothesis. *Chemoecology* **18**: 29–38.

Rostás, M. and Wölfling, M. (2009) Caterpillar footprints as host location kairomones for *Cotesia marginiventris*: persistence and chemical nature. *Journal of Chemical Ecology* **35**: 20–7.

Rostás, M., Ruf, D., Zabka, V. and Hildebrandt, U. (2008) Plant surface wax affects parasitoid's response to host footprints. *Naturwissenschaften* **95**: 997–1002.

Rostás, M., Ton, J., Mauch-Mani, B. and Turlings, T.C.J. (2006) Fungal infection reduces herbivore-induced plant volatiles of maize but does not affect naïve parasitoids. *Journal of Chemical Ecology* **32**: 1897–909.

Rutledge, C.E., Eigenbrode, S.D. and Ding, H.J. (2008) A plant surface mutation mediates predator interference among ladybird larvae. *Ecological Entomology* **33**: 464–72.

Saleh, N., Scott, A.G., Bryning, G.P. and Chittka, L. (2007) Distinguishing signals and cues: bumble-bees use general footprints to generate adaptive behaviour at flowers and nest. *Arthropod–Plant Interactions* **1**: 119–27.

Salerno, G., Frati, F., Conti, E., De Pasquale, C., Peri, E. and Colazza, S. (2009) A finely tuned strategy adopted by an egg parasitoid to exploit chemical traces from host adults. *Journal of Experimental Biology* **212**: 1825–31.

Sarfraz, M., Dosdall, L.M. and Keddie, B.A. (2009) Host plant nutritional quality affects the performance of the parasitoid *Diadegma insulare*. *Biological Control* **51**: 34–41.

Scherling, C., Roscher, C., Giavalisco, P., Schulze, E.-D. and Weckwerth, W. (2010) Metabolomics unravel contrasting effects of biodiversity on the performance of individual plant species. *PLoS ONE* **5**(9): e12569.

Schmelz, E.A., Alborn, H.T., Engelberth, J. and Tumlinson, J.H. (2003) Nitrogen deficiency increases volicitin-induced volatile emission, jasmonic acid accumulation, and ethylene sensitivity in maize. *Plant Physiology* **133**: 295–306.

Schmitz, O.J. (2007) From mesocosms to the field: the role and value of cage experiments in understanding top-down effects in ecosystems. In: Weisser, W.W. and Siemann, E. (eds) *Insects and Ecosystem Function*. Springer, Berlin, pp. 277–302.

Schröder, R. and Hilker, M. (2008) The relevance of background odor in resource location by insects: a behavioral approach. *Bioscience* **58**: 308–16.

Schröder, R., Wurm, L., Varama, M., Meiners, T. and Hilker, M. (2008) Unusual mechanisms involved in learning of oviposition-induced host plant odours in an egg parasitoid? *Animal Behaviour* **75**: 1423–30.

Shiojiri, K., Takabayashi, J., Yano, S. and Takafuji, A. (2001) Infochemically mediated tritrophic interaction webs on cabbage plants. *Population Ecology* **43**: 23–9.

Smith, J., Tucker, D., Alter, D., Watson, K. and Jones, G. (2010) Intraspecific variation in essential oil composition of *Eremophila longifolia* F. Muell. (Myoporaceae): evidence for three chemotypes. *Phytochemistry* **71**: 1521–7.

Soler, R., Bezemer, T.M., van der Putten, W.H., Vet, L.E.M. and Harvey, J.A. (2005) Root herbivore effects on above-ground herbivore, parasitoid and hyperparasitoid performance via changes in plant quality. *Journal of Animal Ecology* **74**: 1121–30.

Soler, R., Harvey, J.A., Kamp, A.F.D., Vet, L.E.M., van der Putten, W.H., van Dam, N.M., Stuefer, J.F., Gols, R., Hordijk, C.A. and Bezemer, T.M. (2007) Root herbivores influence the behaviour of an aboveground parasitoid through changes in plant-volatile signals. *Oikos* **116**: 367–76.

Steidle, J.L.M. and van Loon, J.J.A. (2002) Chemoecology of parasitoid and predator oviposition behaviour. In: Hilker, M. and Meiners, T. (eds) *Chemoecology of Insect Eggs and Egg Deposition.* Blackwell Publishing Ltd., Berlin, pp. 291–319.

Sun, Y.-C., Feng, L., Gao, F. and Ge, F. (2011) Effects of elevated CO_2 and plant genotype on interactions among cotton, aphids and parasitoids. *Insect Science* **18**: 451–61.

Tamiru, A., Bruce, T.J.A., Woodcock, C.M., Caulfield, J.C., Midega, C.A.O., Ogol, C., Mayon, P., Birkett, M.A., Pickett, J.A. and Khan, Z.R. (2011) Maize landraces recruit egg and larval parasitoids in response to egg deposition by a herbivore. *Ecology Letters* **14**: 1075–83.

Theis, N. and Lerdau, M. (2003) The evolution of function in plant secondary metabolites. *International Journal of Plant Sciences* **164**: 93–102.

Thoss, V., O'Reilly-Wapstra, J. and Iason, G.R. (2007) Assessment and implications of intraspecific and phenological variability in monoterpenes of Scots pine (*Pinus sylvestris*) foliage. *Journal of Chemical Ecology* **33**: 477–91.

Ton, J., D'Alessandro, M., Jourdie, V., Jakab, G., Karlen, D., Held, M., Mauch-Mani, B. and Turlings, T.C.J. (2007) Priming by airborne signals boosts direct and indirect resistance in maize. *Plant Journal* **49**: 16–26.

Turlings, T.C.J., Tumlinson, J. and Lewis, W.J. (1990) Exploitation of herbivore-induced plant odors by host-seeking parasitic wasps. *Science* **250**: 1251–3.

Unsicker, S.B., Kunert, G. and Gershenzon, J. (2009) Protective perfumes: the role of vegetative volatiles in plant defense against herbivores. *Current Opinion in Plant Biology* **12**: 479–85.

Vanbergen, A.J., Hails, R.S., Watt, A.D. and Jones, T.H. (2006) Consequences for host–parasitoid interactions of grazing-dependent habitat heterogeneity. *Journal of Animal Ecology* **75**: 789–801.

Vet, L.E.M. and Dicke, M. (1992) Ecology of infochemical use by natural enemies in a tritrophic context. *Annual Review of Entomology* **37**: 141–72.

Vet, L.E.M. and Groenewold, A.W. (1990) Semiochemicals and learning in parasitoids. *Journal of Chemical Ecology* **16**: 3119–35.

Vet, L.E.M., De Jong, R., van Giessen, W.A. and Visser, J.H. (1990) A learning- related variation in electronanntenogram responses of a parasitic wasp. *Physiological Entomology* **15**: 243–7.

Vet, L.E.M., De Jong, A.G., Franchi, E. and Papaj, D.R. (1998) The effect of complete versus incomplete information on odour discrimination in a parasitic wasp. *Animal Behaviour* **55**: 1271–9.

Vet, L.E.M., Lewis, W. and Cardé, R. (1995) Parasitoid foraging and learning. In: Cardé, R. and Bell, W. (eds) *Chemical Ecology of Insects.* Second Edition, Chapman & Hall, New York, pp. 65–101.

Vinson, S.B. (1976) Host selection by insect parasitoids. *Annual Review of Entomology* **21**: 109–133.

Visser, J.H. (1979) Electroantennogram responses of the Colorado beetle (*Leptinotarsa decemlineata*) to plant volatiles. *Entomologia Experimentalis et Applicata* **25**: 86–97.

Visser, J.H. (1986) Host odor perception in phytophagous insects. *Annual Review of Entomology* **31**: 121–44.

Voigt, D. and Gorb, S.N. (2012) Attachment ability of sawfly larvae to smooth surfaces. *Arthropod Structure and Development* **41**: 145–53.

Votsch, W., Nicholson, G., Muller, R., Stierhof, Y.D., Gorb, S. and Schwarz, U. (2002) Chemical composition of the attachment pad secretion of the locust *Locusta migratoria*. *Insect Biochemistry and Molecular Biology* **32**: 1605–13.

Wadhams, L.J., Blight, M.M., Kerguelen, V., Le Métayer, M., Marion, P.F., Masson, C., Pham-Delégue, M.H. and Woodcock, C.M. (1994) Discrimination of oilseed rape volatiles by honey bee: novel combined gas chromatographic-electrophysiological behavioral assay. *Journal of Chemical Ecology* **20**: 3221–31.

Wajnberg, E. (2006) Time allocation strategies in insect parasitoids: from ultimate predictions to proximate behavioral mechanisms. *Behavioral Ecology and Sociobiology* **60**: 589–611.

Wilms, J. and Eltz, T. (2008) Foraging scent marks of bumblebees: footprint cues rather than pheromone signals. *Naturwissenschaften* **95**: 149–53.

Winter, T.R. and Rostás, M. (2010) Nitrogen deficiency affects bottom-up cascade without disrupting indirect plant defense. *Journal of Chemical Ecology* **36**: 642–51.

Yuan, J.S., Himanen, S.J., Holopainen, J.K., Chen, F. and Stewart, C.N., Jr (2009) Smelling global climate change: mitigation of function for plant volatile organic compounds. *Trends in Ecology and Evolution* **24**: 323–31.

Zhang, Q.-H. and Schlyter, F. (2004) Olfactory recognition and behavioural avoidance of angiosperm nonhost volatiles by conifer-inhabiting bark beetles. *Agricultural and Forest Entomology* **6**: 1–19.

4

Chemical ecology of insect parasitoids in a multitrophic above- and below-ground context

Roxina Soler, T. Martijn Bezemer and Jeffrey A. Harvey

Department of Terrestrial Ecology, Netherlands Institute of Ecology (NIOO-KNAW), Wageningen, The Netherlands

Abstract

The chemical ecology of plant–insect interactions has received increasing attention over the past 30 years. While the focus in chemical ecology has traditionally been on above-ground plant–insect interactions, more recently researchers have begun to explore processes involving both plant shoots and roots and the plant-associated biota that inhabit the above- and below-ground domains. The discovery that soil-dwelling organisms can impact on the fitness of above-ground insect herbivores opened up a new challenge for the study of multitrophic plant–insect interactions. In this chapter, we present and discuss what is currently known about this exciting field of research. We first describe the general effects and mediating mechanisms that root feeders and soil-borne symbionts can exert on above-ground insect herbivores. We then discuss how the growth, development and host selection behaviour of parasitoids can be influenced by these soil-dwelling functional groups. In most empirical studies, the effects on parasitoids appear to be mediated by changes in primary and secondary above-ground plant metabolites and in herbivore-induced plant volatiles (HIPVs), all induced by soil organisms. Finally, we discuss how starting from a reductionist perspective and then expanding the spatial and temporal scales up to the level of communities and ecosystems can contribute to a better understanding of the assemblage and functioning of above–below-ground interactions.

Chemical Ecology of Insect Parasitoids, First Edition. Eric Wajnberg and Stefano Colazza.
© 2013 John Wiley & Sons, Ltd. Published 2013 by John Wiley & Sons, Ltd.

4.1 Introduction

Foliage-feeding insects are the most species-rich animal group on earth, and the study of their intricate interactions with their host plants has been of particular interest to ecologists for at least the past five decades (e.g. Ehrlich & Raven 1964, Breedlove & Ehrlich 1968). It is well known that plants have evolved a diverse array of defence strategies that enable them to resist herbivore attack and to recruit the herbivore's natural enemies, for example parasitoids of the herbivores (Karban & Baldwin 1997). One very effective way for plants to defend themselves against herbivores is through the production and/or constitutive expression of allelochemicals or phytotoxins. Phytotoxins function by reducing the feeding activity of the attacking herbivore or by being directly toxic and thus increasing mortality and/ or reducing rate of growth and size of the attacker (Schoonhoven et al. 2005). In response, many insects have evolved ways of dealing with plant phytotoxins, such as by evading or detoxifying them. In some cases, herbivores can even sequester phytotoxins as a putative defence against their own natural enemies. Increased plant resistance in reaction to herbivory, termed as 'herbivore-induced direct plant defence', is not only induced locally in the plant tissues that are subject to attack but may also be systemically induced to protect the (still) undamaged leaves. As a consequence, the physiological changes induced in the plant by certain attackers can establish strong interactions with diverse organisms in contact with different parts of the plant and/or later in time during the season (Kaplan & Denno 2007).

Just as insect herbivores exploit plants, many other arthropods have evolved to attack and consume the herbivores. One of the most important groups of herbivore consumers are parasitoid wasps (Godfray 1994). Parasitoids are insects whose larvae develop in ('endoparasitoids') or on ('ectoparasitoids') the bodies of invertebrates, usually other insects, while the adults are free-living (Godfray 1994). Endoparasitoids, or parasitoids whose larvae develop inside the host and thus live in a complex biochemical milieu, are highly susceptible to very small changes in the quality of the host's internal environment and thus the development of many species is closely coordinated with the development of their hosts (Harvey 2005). For many years the study of trophic interactions was confined to two levels, for instance plant–herbivore or herbivore–parasitoid interactions. However, a seminal paper by Price et al. (1980) argued that, in order to better understand the rules governing the structure and function of communities, it is necessary to incorporate and combine all links in food chains up to, and even beyond, the third trophic level. This idea stimulated new approaches to research on multitrophic interactions. Soon thereafter, it was discovered that, in reaction to herbivory (Turlings et al. 1990, Vet & Dicke 1992, De Moraes et al. 1998, Dicke 1999) and egg deposition (Fatouros et al. 2008, Hilker & Meiners 2010), plants emit volatile secondary metabolites, herbivore-induced plant volatiles (HIPVs) that are exploited as host location cues by female parasitoids. Therefore, by attracting natural enemies that kill the herbivores, HIPVs can function as a form of 'induced indirect plant resistance' (Dicke 1999).

The actual function of HIPVs in the field remains largely unknown and is the subject of considerable debate (Kessler & Heil 2011). HIPVs are beneficial to parasitoids because they can indicate the presence of a host on the food plant (Vet et al. 1991). One of the problems in reconciling the potential benefits of HIPVs on plant fitness is that plants that are attractive to parasitoids may also produce higher concentrations of phytotoxins, particularly if the volatile breakdown products of these toxins are the main attractants (Gols *et al.* 2011). It is known that phytotoxins consumed by the herbivores may negatively affect the developing offspring of their parasitoids, by negatively affecting herbivore performance (Barbosa *et al.* 1991, Harvey *et al.* 2003, 2011, Harvey 2005, Ode 2006, Gols *et al.* 2008, 2009, Gols & Harvey 2009). How these potential conflicts between direct and indirect plant defences are resolved in natural communities has been little studied. This clearly exemplifies the complexity that plant–herbivore–parasitoid interactions encapsulate.

Plants are also exposed to a second trophic level of consumers below-ground, where micro-invertebrates (such as nematodes) and larger arthropods (such as insects) function as the soil counterparts of above-ground insect herbivores. Pioneering studies from the early 1990s (Gange & Brown 1989, Moran & Whitham 1990, Masters *et al.* 1993) revealed that root-feeding insects and nematodes can have a significant impact on interactions between plants and above-ground insect herbivores. Plant responses to root feeders can even cascade up the above-ground trophic chain (van der Putten *et al.* 2001, Bezemer & van Dam 2005, Soler *et al.* 2005, Rasmann & Turlings 2007, Erb *et al.* 2009) and this has opened up new challenges for the study of multitrophic plant–insect interactions. As well as root consumers, plant roots also interact with beneficial soil-borne symbionts such as arbuscular mycorrhizal fungi and rhizobacteria, and with soil ecosystem 'engineers' such as earthworms, and these interactions can also influence above-ground multitrophic interactions (Masters *et al.* 2001, Poveda *et al.* 2003, Hempel *et al.* 2009, Pineda *et al.* 2010, Wurst 2010). In this chapter, we describe and discuss the influence that the most studied functional groups of root attackers and beneficial soil-borne symbionts can exert on the chemical ecology of above-ground plant–herbivore–parasitoid interactions (Fig. 4.1). Finally,

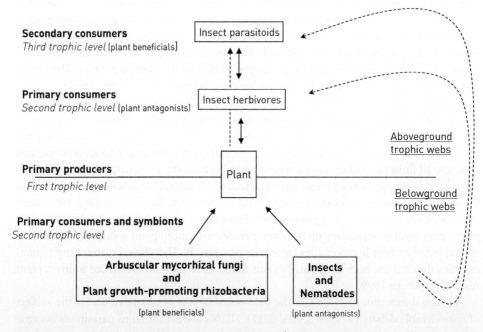

Figure 4.1 Simple scheme linking above-ground/below-ground interactions of root and shoot antagonists and beneficials. *Solid arrows* represent interactions mediated via nutrients and phytotoxins and *dashed arrows* represent interactions linked via herbivore-induced plant volatiles.

we present some aspects that we think could contribute to furthering our understanding of how plants accommodate their responses to multiple above- and below-ground organisms.

4.2 Influence of root feeders on above-ground insect herbivores

Soil-dwelling herbivores constitute a diverse range of subterranean fauna including about 10 orders of the class Insecta and the phylum Nematoda. It is widely acknowledged that the most important root attackers in terms of abundance and impact on plant fitness are plant parasitic nematodes (Stanton 1988). However, their impact on above-ground insects has been less well studied than the effects of root-feeding insects. Below we summarize the most commonly observed effects that below-ground root-feeding insects and nematodes can exert on above-ground insect herbivores, and the potential mechanisms that have been proposed to explain these linkages (Fig. 4.2).

Root-feeding insects
Early studies investigating whether root-feeding insects influence the performance of their above-ground counterparts all reported positive effects (Gange & Brown 1989, Moran & Whitham 1990, Masters & Brown 1992). As a consequence, it was initially believed that damage inflicted by below-ground insects enhances the growth and development of above-ground insect herbivores. Masters *et al.* (1993) proposed the first hypothesis to explain the mechanistic basis for this facilitation. Described as the 'stress response hypothesis' (Fig. 4.2, arrow number 1), this early model proposes that the capacity of roots to acquire sufficient

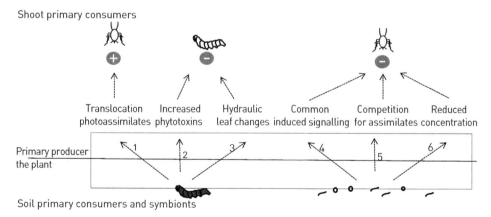

Figure 4.2 Plant-mediated effects of root-feeding insects and nematodes on above-ground insect herbivores. The *white insects* represent above-ground aphids and caterpillars, the *black caterpillar* represents root-chewing insects, and the *circles* and *curved lines* represent ecto- and migratory endoparasitic nematodes and root-knot or cyst-forming nematodes, respectively. The symbols '–' and '+' represent reduced and enhanced insect performance, respectively, compared with those on control root-undamaged plants. *Numbers* link the local effects on roots with the systemic effects on shoots.

water and nutrients from the soil is constrained due to removal of root tissue by root-feeding insects, creating an effect similar to water stress. Water stress is predicted by the 'plant stress hypothesis' to induce susceptibility to above-ground insect herbivores due to a temporarily increased proportion of nitrogen and carbon that is mobilized from the site of attack to sites of new growth and storage (White 1984). The supporting evidence for the 'stress response hypothesis' was derived from plant–aphid systems. Subsequently, the effects on leaf chewers began to be explored and, in contrast with earlier studies, negative effects of root herbivory were also found on above-ground insects (Tindall & Stout 2001, Bezemer et al. 2003, van Dam et al. 2003, 2005, Soler et al. 2005, Staley et al. 2007). This negative impact was explained by the 'defence induction hypothesis' (Fig. 4.2, arrow number 2), which proposed that root-feeding insects negatively impact the growth and development of leaf-chewing insects via induced increases in foliar phytotoxins (Bezemer et al. 2003). More recently, mechanistic physiological knowledge of herbivore-inducible long-distance signals in plants started to be incorporated into the above–below-ground models (Fig. 4.2, arrow number 3). For example, increased levels of abscisic acid (ABA) and reduced water contents were recorded on leaves of root-infested plants (maize, Zea mays), leading to the hypothesis that ABA signalling and/or hydraulic changes can mediate increased resistance to leaf chewers (Erb et al. 2011). Both ABA and low leaf water content are known to negatively impact the performance of leaf chewers (Thaler & Bostock 2004, Bodenhausen & Reymond 2007). It is difficult to separate these effects because ABA is a central chemical signal involved in the control of stomatal opening and closing in response to water stress. The mechanisms of above–below-ground defence signalling have only recently started to be explored (Erb et al. 2011).

Nematodes

Plant-feeding and root-knot or cyst-forming nematodes have been reported to negatively affect the performance of above-ground phloem feeders, which contrasts with the pattern observed for root-feeding insects (Bezemer et al. 2005, Wurst & van der Putten 2007, Kaplan et al. 2009, 2011, Hol et al. 2010, Vandegehuchte et al. 2010, Kabouw et al. 2011). The effects of nematodes on the performance of leaf chewers are less well studied and more variable. For instance, effects can be positive (Alston et al. 1991, Kaplan et al. 2009), neutral (Wurst & van der Putten 2007), or even negative (van Dam et al. 2005). Several mechanisms have been proposed to link the consistent negative impact of nematodes on aphid fitness. The first proposed explanation (Fig. 4.2, arrow number 4) was that nematodes and phloem-feeding insects trigger a common defence signalling pathway (Kaplan et al. 2009). This hypothesis is based on studies with solanaceous plants, where the defence gene Mi-1 medi-ates resistance to both root-knot nematodes and aphids (Li et al. 2006, Bhattarai et al. 2007). Thus, above-ground phloem-feeding and below-ground root-feeding nematodes nega-tively interact via changes induced in a common defence pathway. Subsequent studies showed that this gene is involved in the activation of distinct signalling pathways (Mantelin et al. 2007), challenging the basis of this hypothesis. Thus far, no empirical evidence has linked changes in the performance of aphids on nematode-attacked plants with defence signalling changes. Recently, the 'sink competition hypothesis' was proposed, which argues that above-ground phloem feeders and root-feeding nematodes compete for assimilates in the phloem, and this has been tested to explain negative interactions between foliar aphids and nematodes (Kaplan et al. 2011). Root-knot nematodes and aphids feed from vascular tissues and attract photoassimilates to the feeding site, representing a sink for the plant.

When aphids arrive on the plant shoots after nematodes have been feeding on the roots, the strength of the sink that aphids may initiate in the shoots (Guerrieri & Digilio 2008) could be limited by the sink previously created by nematodes in the roots (Fig. 4.2, arrow number 5). In particular, cyst- or gall-forming species are able to feed from the phloem, which makes them potential competitors of aphids. However, empirical evidence for this mechanism is currently lacking. The concentration of amino acids in the phloem of plants infested by root-feeding nematodes has been reported to be lower than on plants without nematodes (Fig. 4.2, arrow number 6), and this change has been correlated with reduced aphid fitness (Bezemer *et al.* 2005). Again, how widespread this response to nematode attack is remains unknown.

4.3 Influence of soil-borne symbionts on above-ground insect herbivores

A wide variety of soil-borne microbes promote plant growth and increase resistance to soil-dwelling organisms resulting in beneficial effects for the plant (van der Heijden *et al.* 2008). Two well-studied groups of beneficial soil-borne microbes are mycorrhizas and rhizosphere bacteria (Pieterse *et al.* 2001, Pozo & Azcon-Aguilar 2007). Arbuscular mycorrhizal fungi (AMF) are obligate biotrophs that cannot survive in the absence of plants. More than 80% of terrestrial plants establish a mutualistic association with AMF (Pozo & Azcon-Aguilar 2007). Several species of *Glomus*, particularly *Glomus intradices*, have been shown to significantly benefit plant fitness, providing plants with enhanced mineral nutrition and increased resistance to soil-borne pathogens (Whipps 2004) and parasitic nematodes (de la Peña *et al.* 2006). Rhizosphere bacteria are commonly abundant on the surface of roots, and obtain nutrients from root exudates (Lynch & Whipps 1990), although some strains are endophytes (Ramamoorthy *et al.* 2001). A number of strains of rhizosphere bacteria also enhance plant growth, and are called plant growth-promoting rhizobacteria (PGPR). The most well-studied taxon is *Pseudomonas* spp. As well as promoting plant growth, PGPR also confer resistance to a wide range of soil-dwelling organisms such as nematodes (Sikora *et al.* 1988), root-feeding insects (Zehnder *et al.* 1997), and soil-borne pathogens (Pieterse *et al.* 2001). By promoting plant growth and improving shoot nutritional quality, soil-borne symbionts such as AMF and PGPR can enhance the fitness of insect herbivores that feed above-ground (Fig. 4.3) (Koricheva *et al.* 2009, Pineda *et al.* 2010).

Besides promoting growth, soil-borne symbionts can also influence plant shoots by inducing systemic resistance to foliar pathogens (Ramamoorthy *et al.* 2001, Pozo & Azcon-Aguilar 2007, van Wees *et al.* 2008). A pioneer and well-documented example of induced systemic resistance (ISR) corresponds to *Pseudomonas fluorescens*, and involves components of the jasmonic acid (JA) and ethylene (ET) signalling pathways (Pieterse *et al.* 2001). ISR is commonly induced by distinct soil-borne beneficials such as PGPR, AMF and plant growth-promoting and endophytic fungi, and is associated with priming for enhanced JA/ET defences (Segarra *et al.* 2007, van Wees *et al.* 2008). ISR resistance against foliar pathogenic bacteria has been broadly documented, and more recently it has been shown that it can also increase resistance against insect herbivores (van Oosten *et al.* 2008). Meta-analyses have shown that soil-borne beneficial microbes can counterbalance the negative effects that the major plant threats exert on growth and fitness of individual plants (Morris *et al.* 2007). How the combined effects of enhancing shoot quality and increasing resistance

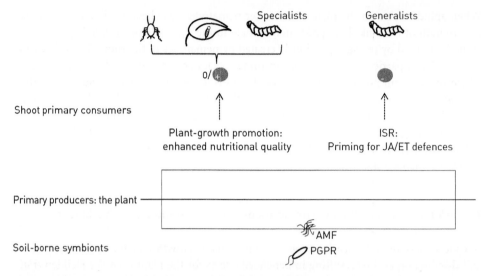

Figure 4.3 Plant-mediated effects of soil-borne beneficial microbes on above-ground insect herbivores. The *aphid* represents a guild of phloem feeders, the *caterpillar on the leaf* represents a guild of leaf-miners and the *caterpillar alone* represents a guild of leaf chewers, specialists and generalists. The drawings below represent plant growth-promoting rhizobacteria (PGPR) and arbuscular mycorrhizal fungi (AMF). The symbols '−' and '+' represent reduced and enhanced insect performance, respectively, compared with those on control plants without the symbiont association.

translate into a benefit or a detriment for the fitness of above-ground insects, however, remains unclear. The overall effects of AMF on the fitness of insect herbivores depend on the feeding mode and dietary breadth of the insect herbivore feeding above-ground (Koricheva *et al.* 2009) (Fig. 4.3). AMF have been shown to consistently increase resistance against generalist leaf-chewing insects, while beneficial or neutral effects have been reported on specialist leaf chewers, phloem feeders and mesophyll feeders (Gehring & Bennett 2009, Hartley & Gange 2009, Koricheva *et al.* 2009, Pineda *et al.* 2010). Similarly, a recent study showed that PGPR can also benefit the growth and development of above-ground insect herbivores, and the authors speculated that PGPR may specifically benefit certain guilds of above-ground herbivorous insects (Pineda *et al.* 2012). Taken together, it seems that soil-borne beneficial microbes, independent of their identity, could have similar effects on plant–insect interactions via plant growth promotion and ISR. However, within the AMF group, the identity of the mycorrhizal species can also influence the effect(s) above-ground (Koricheva *et al.* 2009). How similar the effects of AMF and PGPR – and soil-borne symbionts in general – are on plant–herbivore interactions above-ground remains to be seen.

4.4 Plant-mediated effects of root feeders and soil-borne symbionts on growth and development of parasitoids

Parasitoid development is limited to finite resources contained in its single herbivore insect host individual. The quality of the food plant of the host can have profound effects on the development of parasitoids, mediated through the host (Barbosa *et al.* 1986, 1991, Gunasena *et al.* 1990, Harvey 2005, Ode 2006, Gols *et al.* 2008, Morse 2009). The quality of the host, which can be indirectly evaluated through its fitness, can greatly affect the rate of parasitoid growth, adult body mass and survival, and therefore this can be a major determinant of parasitoid fitness (Mackauer & Sequeira 1993). Independent of the nature of the mediating mechanisms, below-ground plant antagonists and beneficials can clearly enhance or hamper the fitness of above-ground insect herbivores, which are the hosts of the parasitoids. Soil-dwelling organisms have been shown to influence a wide variety of fitness correlates of above-ground insect herbivores (Fig. 4.4). For example, larvae of the leaf-chewing *Pieris brassicae* (Lepidoptera: Pieridae) develop faster, although their final body size is not compromised, on plants previously attacked by the root-chewing insect *D. radicum* (Diptera: Anthomyiidae) (Fig. 4.4a; Soler *et al.* 2005). Similarly, the leaf-chewing *Spodoptera exigua* (Lepidoptera: Noctuidae) grows more slowly on plants previously infested by the root chewer *Agriotes lineatus* (Coleoptera: Helateridae). This effect is also reflected in the food consumption rate of the leaf chewer (Fig. 4.4b; Bezemer *et al.* 2003). An example that shows how complex these above-ground/below-ground interactions can become is provided by the interactions between the foliar aphid *Rhopalosiphum padi* (Hemiptera: Aphididae) and root-feeding nematodes. In a long-term microcosm experiment with 12 plant species, nematodes significantly reduced aphid reproduction. However, individual adult aphids attained a larger size later in the experiment. The reason for this could be that, due to reduced population sizes in microcosms with nematodes, there was less crowding, which may have resulted in larger individual aphids (Fig. 4.4c; Bezemer *et al.* 2005). The effects of nematodes on aphids also depended on the host plant the aphid was feeding on. Therefore, it remains difficult to accurately translate effects from single fitness correlates into overall herbivore fitness in above–below-ground frameworks. Another example of how complex the determination of general effects in these framework can be is provided by the interactions between PGPR *Pseudomonas fluorescens* and the foliar aphids *Myzus persicae* and *Brevicoryne brassicae* (Hemiptera: Aphididae) (Fig. 4.4d; Pineda *et al.* 2012). While the size of the specialist crucifer feeder *B. brassicae* (right) is not influenced by the presence of PGPR, the size of the generalist *M. persicae* (left) is. Here it seems that dietary specialization and/or species-specific effects are determining the outcome of the interaction. All these examples illustrate how root feeders and soil-borne symbionts can impact diverse aspects of the performance of above-ground insect herbivores, although whether these effects turn out to be beneficial or detrimental for overall insect fitness cannot always be accurately determined. In this regard, combining multiple fitness correlates may help us to more precisely interpret above–below-ground plant-mediated effects. In any case, it is clear that below-ground organisms can influence herbivore fitness and quality, and this can be expected to have consequences for parasitoid fitness.

The effect of plant and host quality on parasitoid fitness is often strongly correlated with developmental strategies exhibited by the parasitoid. A major life history dichotomy has been described for insect parasitoids, based on the way in which host resources are utilized by the larval parasitoid stages. So-called 'idiobiont' parasitoids attack non-growing host stages, such as eggs or pupae, or hosts that are paralysed prior to oviposition, whereas 'koinobiont' parasitoids attack hosts that continue to feed, grow and remain active during the parasitism phase (Askew & Shaw 1986). For idiobiont parasitoids, the host represents an ostensibly static resource to which no new biomass will be added during the interaction (Mackauer & Sequeira 1993, Godfray 1994, Harvey 2005). In contrast, for koinobionts the host represents a potentially dynamic resource which at the start of parasitism may contain

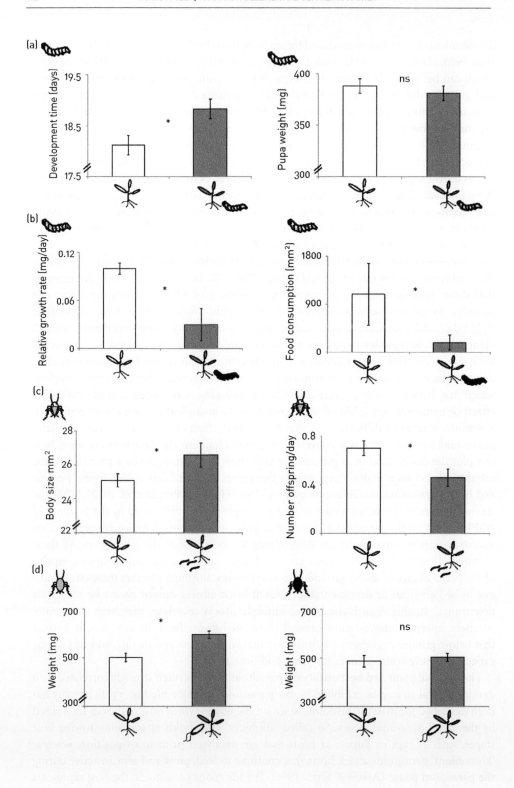

only a fraction of the resources necessary for successful development of the parasitoid progeny (Mackauer & Sequeira 1993, Harvey 2005). Furthermore, many idiobionts are ectoparasitoids, thus developing on the external host surface, whereas most koinobionts are endoparasitoids and develop within host tissues. Endoparasitoids are highly susceptible to very small changes in the quality of the host (Lawrence 1990) and their development is often strongly coordinated with the development of their hosts (Harvey 2005). The fitness of ectoparasitic idiobionts appears to be more affected by host size, which is imposed by the plant, whereas that of endoparasitic koinobionts is affected by both host size and quality (reviewed by Harvey 2005; see also Harvey *et al.* 2007, 2011). The larvae of many koino-biont parasitoids start their destructive feeding phase, leading to exponential growth, only when the host enters its final instar, in order to ensure sufficient resources to complete their development (Harvey *et al.* 1994). At the end of the host's larval development, allelochemi-cals present in the haemolymph are rapidly consumed by the koinobiont parasitoid (Harvey 2005). Therefore, koinobiont parasitoids are often highly susceptible to very small changes in the quality of their hosts (Harvey 2005) and potentially even more affected than their hosts by the quality of the host plant (Harvey *et al.* 2003). Considering that soil-dwelling organisms can influence host size, growth rate and internal host quality, koinobiont para-sitoids may therefore be expected to be more susceptible to changes in host quality/fitness correlates induced by root feeders than idiobionts are. Effects on the hosts of koinobiont parasitoids induced by below-ground organisms can be predicted to be passed on to the developing parasitoid larvae, and may even be magnified in the parasitoid. However, the question of whether the impact that soil-dwelling organisms exert on the fitness of above-ground herbivores is simply transferred to the subsequently developing parasitoid, or whether the strength of the effects is magnified or diluted, is still at an early stage of investigation.

Root feeders
So far, general patterns still cannot be identified because only three studies have explored the effects of root-feeding herbivores on the performance of above-ground herbivores and their primary parasitoids, always using koinobiont parasitoids (Bezemer *et al.* 2005, Soler *et al.* 2005, Kabouw *et al.* 2011). In microcosm experiments, Bezemer *et al.* (2005) found

Figure 4.4 Examples of diverse aspects of insect performance influenced by below-ground organisms. Illustration exemplifying the effects that distinct soil-dwelling organisms, for example root-feeding insects from the orders Diptera (a) and Coleop-tera (b), root-feeding nematodes (c), and soil-borne symbionts such as PGPR (d), can have on herbivore fitness correlates. The drawings on the *Y*-axes represent the feeding guild of the foliar insect: aphids or caterpillars. Means ± SE are reported. The draw-ings on the *X*-axes represent the functional groups of soil organisms. Different insect colours (above-ground: *white* and *white with black dots* for caterpillars/*white*, *black* and *grey* for aphids; below-ground: *grey* and *black* for root-feeding insects) represent different species. *Grey bars* indicate that the foliar insects are feeding on plants previ-ously colonized by a soil-dwelling organism, and *white bars* represent control unin-fested/non-colonized plants. *Asterisks* and *ns* indicate significant and non-significant differences ($P < 0.05$), respectively. Based on data from Bezemer *et al.* (2003, 2005), Soler *et al.* (2005) and Pineda *et al.* (2012).

that the solitary endoparasitoid *Aphidius colemani* performed better in the presence of nematodes. This positive effect on parasitoid performance could be explained by the fact that colonies of the host, the aphid *Rhopalosiphum padi*, were smaller in microcosms with nematodes, but that individual *R. padi* aphids were larger in the presence of nematodes. These larger aphids were better hosts for the parasitoids. Another study, with the host plant *Brassica oleracea*, the aphid *Brevicoryne brassicae*, and the parasitoid *Diaeretiella rapae*, reported no significant effects of nematodes on parasitoid development or adult size (Kabouw *et al.* 2011). In this study, nematodes also tended to have a negative effect on aphid population growth, but unfortunately individual aphid size was not reported. Soler *et al.* (2005) reported that the development time of caterpillars of *Pieris brassicae* was slower on black mustard plants infested with cabbage root fly maggots than on uninfested plants, while adult size was unaffected by the presence of the root fly. The developing parasitoid larvae similarly developed more slowly, and also attained a smaller size, on plants infested with root flies. It is notable that, although the two studies in which effects of root herbivory on parasitoids were found show opposite outcomes, in both studies parasitoid performance can be related to root herbivore-induced changes in the performance of the above-ground herbivore host. However, it is, of course, impossible to draw general conclusions on the basis of these few studies, two of which used root-feeding nematodes and phloem-feeding herbivores while the other used root- and leaf-chewing insects.

Soil-borne symbionts
Soil-borne beneficial microbes have been shown to enhance the performance, attraction, and parasitism rates of parasitoids and predators of above-ground insect herbivores (Guerrieri *et al.* 2004, Saravanakumar *et al.* 2008, Hempel *et al.* 2009, Hoffmann *et al.* 2011). As a result of the JA/ET priming effect, soil-borne symbionts may influence host location and attraction of insect parasitoids via induced changes in HIPVs, but this remains to be empirically tested (Pineda *et al.* 2010). Interestingly, Schausberger *et al.* (2012) recently reported that the composition of HIPVs emitted by bean plants (*Phaseolus vulgaris*) was significantly affected by AMF, and that predatory mites, which are specialist natural enemies of spider mites, preferred the HIPVs of mycorrizal plants to those of non-mycorrhizal plants. There is also evidence that AMF can negatively impact the performance of insect parasitoids (Gange *et al.* 2003). As with root feeders, there is little empirical evidence about these interactions, thus making it difficult to draw conclusions about the general patterns of these effects.

4.5 Effects of root-feeding insects on HIPVs and host location of parasitoids

Just over a decade ago, the study of the potential effects that soil-dwelling organisms could exert on parasitoid host searching behaviour started to attract attention (e.g. Masters *et al.* 2001, Gange *et al.* 2003, Wurst & Jones 2003, Guerrieri *et al.* 2004). Laboratory-based studies showed that host attraction and behaviour in parasitoids could be influenced by soil-dwelling organisms, which initiated a search for the possible mechanisms mediating these effects (Rasmann & Turlings 2007, Soler *et al.* 2007a). Because in above-ground systems parasitoid host searching is primarily guided by volatile cues that are produced by the host-infested plant (Turlings *et al.* 1990, Vet & Dicke 1992), HIPVs were the primary mediating cues to be tested. There is empirical evidence indicating that volatile blends emitted by plants exposed both to foliar-feeding insects and to root-feeding insects or nematodes quantitatively and qualitatively differ from blends emitted by plants exposed to each herbivore in isolation (Rasmann & Turlings 2007, Soler *et al.* 2007a, Olson *et al.* 2008, Pierre *et al.* 2011). In this regard, the most studied soil functional group has been root-feeding insects. For example, increased levels of herbivore-induced plant volatiles (VOCs) were recorded when cotton plants (*Gossypium* spp.) were exposed to shoot and root feeding by the leaf-chewing insect *Helicoverpa zea* and the root-knot nematode *Meloidogyne incognita*, compared with plants that were only exposed to the leaf chewer (Olson *et al.* 2008). Similarly, in a non-related species, Pierre *et al.* (2011) showed that VOCs emitted by *Brassica napus* plants simultaneously infested by a root- and a leaf-feeding insect, *Delia radicum* and *Pieris brassicae*, respectively, differed from volatiles emitted when the plants were infested by a single herbivore only. In this case, for example, root herbivory specifically induced 4-methyltridecane and salicylaldehyde, and leaf herbivory specifically induced methylsalicylate, while the volatile blends of plants co-infested by the root- and leaf-feeding herbivores were characterized by the induction of the green leaf volatile, hexylacetate.

It seems clear that root herbivores can influence HIPVs. However, the consequences for parasitoid attraction remain largely unexplored. Changes in VOCs induced by root-feeding insects have been correlated with a decreased attraction of koinobiont parasitoids to plants infested by the leaf-chewing host (Rasmann & Turlings 2007, Soler *et al.* 2007a). In our own studies (Soler *et al.* 2007a), we have shown that volatile blends of *Brassica nigra* plants exposed to *Pieris brassicae*, a leaf-chewing host of the koinobiont parasitoid *Cotesia glomerata*, were characterized by high levels of volatile compounds that are reported to act as insect attractants, such as β-farnesene and dimethylnonatriene (Fukushima *et al.* 2002, Ansebo *et al.* 2005). In contrast, plants exposed to *Delia radicum*, a root-feeding insect, were characterized by high amounts of sulphides such as dimethyl disulphide and dimethyl trisulphide, which often act as repellents and/or toxins to insects (Dugravot *et al.* 2004). Plants co-infested by both herbivore species showed a relatively high level of these repellents and a low level of the attractants compared with conspecific plants with only the above-ground herbivore. *Cotesia glomerata* females were significantly less attracted to plants with hosts that were also infested with root herbivores, and this reduced preference was correlated with the distinct volatile blend that characterized this plant–host complex. Similarly, in maize plants (*Zea mays*), the emission of the principal attractant, (*E*) β-caryophyllene, was lower when leaf- and root-chewing herbivores infested the plant compared with the single infestations (Rasmann & Turlings 2007). Female *Cotesia marginiventris* koinobiont parasitoids also preferred plants that were solely infested by their host over plants with both hosts and root herbivores. Here, the females showed the ability to learn the differences between co-infested plants and plants infested only by hosts and they increased their response to doubly infested plants after experiencing this combination (Rasmann & Turlings 2007). Taken together, these results suggest that root-damaged plants convey chemical information that above-ground parasitoids can use to make oviposition decisions (but see Olson *et al.*, 2008). It remains unclear, however, how common it is that female parasitoids can exploit changes in plant volatile emission induced by root herbivory and efficiently integrate it into their foraging decisions.

Optimal foraging theory predicts that carnivores prefer host/prey species that are most rewarding for them in terms of their fitness (Krebs & Davies 1984, Wajnberg *et al.* 2008). Similarly, within a host species, parasitoid females are expected to select the most profitable individuals in order to maximize their own fitness (Godfray 1994, Wajnberg *et al.* 2008). Considering that foraging efficiency in parasitoids is directly linked with their reproductive success, we hypothesize that female koinobiont parasitoids of leaf chewers, whose performance is reduced when feeding on plants previously attacked by root-feeding insects, will preferentially select hosts on root-uninfested plants. Support for this 'below-ground root-feeding insect avoidance hypothesis' is provided by results of our own work, where we found that female *Cotesia glomerata* wasps parasitized significantly more *Pieris brassicae* hosts on *Brassica nigra* plants without the presence of the root herbivore *Delia radicum* (Soler *et al.* 2007a). An earlier study had shown that *C. glomerata* larvae developed faster and attained a larger size on root-uninfested plants (Soler *et al.* 2005). Body size is an important component of parasitoid fitness because it is correlated with longevity, fecundity and mate-finding capacity (Mackauer & Sequeira 1993). A suboptimal performance of parasitoid offspring on root-infested plants was correlated with a higher amount of sinigrin, an important aliphatic glucosinolate, recorded in the shoots of root-infested plants (Soler *et al.* 2005). These results show a clear correlation between preference and performance of the parasitoid: females avoid plants co-infested by hosts and antagonistic root-feeding insects by exploiting chemical information. Most studies linking above–below-ground multitrophic interactions have independently addressed the effects on parasitoid performance, attraction or changes in plant volatiles, but very little attention has been paid to combining these parameters together. As a consequence, support for the hypothesis is limited at present.

It is important to emphasize that parasitoids provided with a positive experience, such as the presence of a host, may develop an attraction for hosts feeding on root-infested plants (Rasmann & Turlings 2007). It is clear from above-ground studies that the innate responses of foraging parasitoids to plant odours can change with experience, leading to local or temporary specialization and a concomitant enhancement of foraging success (Turlings *et al.* 1993, Vet *et al.* 1995). This, however, is rarely considered in the study of above–below-ground multitrophic interactions. Parasitoids have the ability to learn to distinguish between volatile blends emitted by plants infested only by their hosts and plants infested with both their hosts and root-feeding insects (Rasmann & Turlings 2007). However, the relatively few studies exploring whether and how root herbivores influence parasitoid preferences have explored innate responses or else have not distinguished between innate and learned responses (Masters *et al.* 2001, Soler *et al.* 2007a). The role of parasitoid learning in dealing with natural variations in plant and host quality and plant volatiles induced by root herbivory or other soil-dwelling organisms, remains largely unexplored.

4.6 Expanding an above–below-ground bitrophic reductionist perspective

As already described, there is now compelling evidence that the effects of plant quality on above-ground insect performance can be influenced by the presence of soil-dwelling herbivores or symbionts. The field is now ripe for studies that expand from this 'static and linear' bitrophic perspective. We conclude this chapter by pointing out a few areas for future research that we believe will provide important and novel insights into above–below-ground interactions.

Incorporating parasitoid preferences
Until now, most attention in above–below-ground research has been paid to the effects on insect performance mediated by changes in host-plant quality. However, we expect

another major influence of below-ground herbivores or symbionts on above-ground multitrophic interactions to be due to effects on the oviposition preferences of parasitoids. Such changes in preferences can have far-reaching consequences for insect performance or fitness, emphasizing that the effects of below-ground organisms on above-ground organisms can be diverse and are not easy to predict. For example, plants that are of suboptimal nutritional quality for above-ground leaf-chewing insects due to the presence of root-chewing insects can still be good host plants if they enhance herbivore fitness by providing enemy-free space because parasitoids avoid these plants (Thompson & Pellmyr 1991, Ohsaki & Sato 1994, Singer *et al.* 2004, Rasmann & Turlings 2007, Soler *et al.* 2007a). Future models of above–below-ground interactions should therefore incorporate the consequences/impacts of changes in preferences by above-ground parasitoids.

Intrinsic and extrinsic ecological factors influencing parasitoid preferences
Besides the need for more studies linking the effects of root feeders on parasitoid performance, preferences and HIPVs, some important tenets need to be considered if broader conclusions are to be drawn. Clearly, the 'below-ground root-feeding insect avoidance hypothesis' can only be valid when the costs of avoiding root-infested plants outweigh the benefits. How this is played out in nature is difficult to predict. First of all, it may depend on how much of a fitness cost the parasitoid suffers in selecting hosts on plants with and without root herbivores. The negative effects of plant/host choice on indirect fitness correlates, such as body size or development time, may be less important than if the 'wrong' choice results in the host's early death, which affects fitness directly (Harvey *et al.* 2004, Harvey 2005). Intraspecific changes in the physiological state of the parasitoid, for instance in 'intrinsic' factors such as the female's age, physiological condition and egg load may also strongly affect plant/host choice. If parasitoids are time-limited (i.e. the number of hosts is limiting rather than the number of eggs), parasitoids should not discriminate against hosts of inferior quality such as those feeding on plants exposed to root herbivory. Extrinsic 'ecological' factors may also be very important. These include habitat structural and chemical complexity, as this affects foraging efficiency, host patch location, and patch residence time in parasitoids. Soil-dwelling organisms and above-ground parasitoids can also interact even when the soil organisms and the parasitoid host do not share the same plant, through changes induced in the surrounding environment where the organisms coexist (White & Andow 2006, Soler *et al.* 2007b). For example, root herbivores can influence host–parasitoid interactions above-ground via their effects on changes in the structure of the plants in their environment. In maize plants (*Z. mays*), the percentage of parasitism of the European corn borer *Ostrinia nubilalis* by its specialist parasitoid *Macrocentrus grandii* was significantly reduced in the presence of the corn rootworm *Diabrotica virgifera* sharing its habitat (White & Andow 2006). Plant height and density were reduced in habitats where the rootworm was present, resulting in more open habitats that are less preferred by female parasitoids of this species. All of the above factors need to be taken into account in order to fully understand how parasitoids respond to the presence of hosts relative to the presence and effects of root-chewing insects and soil-dwelling organisms in general.

Integrating natural selection and temporal scales
It is known that the identity of the attacking herbivore may also influence both the qualitative and quantitative responses of the plant to herbivory (Rostás & Eggert 2008, Poelman *et al.* 2010). Furthermore, there may also be considerable intraspecific genetic variation in

both constitutive and inducible chemical defences in plants (Marak *et al.* 2000, Gols *et al.* 2008, Newton *et al.* 2009). Presumably, differences in the levels of allelochemicals are the result of local selection pressures acting independently on plants within well-defined populations (Newton *et al.* 2009). This selection at the micro-evolutionary scale is often played out at local levels that, when expanded upon, form a larger geographical mosaic (Thompson 2005). This idea constitutes our understanding of local 'hot-spots' in which intense selection among plants, herbivores and parasitoids leads to strong reciprocal selection (Thompson 2005). However, so far, nothing is known about the intensity, or even frequency, of above–below-ground interactions as they are played out under natural conditions. If plants very infrequently share some kinds of interactions then these will be so diffuse that they will have little or no effect on selecting for phenotypic traits among the various players. Attention should therefore focus on finding out the variation in, and frequency of, interactions between different types of root and shoot consumers, and how this varies within and between habitats that differ in various attributes, such as structural heterogeneity and plant species richness. It is also important to note that many of the most important studies in multitrophic interactions have been based on research with short-lived annual plants and multivoltine insects. These studies do not implicitly argue that there is strong reciprocal selection between the plants and their consumers, but this is often implied. How this variation plays out in terms of above–below-ground interactions over variable spatial and temporal scales has yet to be determined. For instance, different generations of some insects may obligatorily develop on different plant species over the course of a single growing season, and these plant species also vary in their seasonal phenology. How and when root- and shoot-infesting herbivores and their parasitoids leave the natal patch and find suitable new plant species and patches is not well understood. Furthermore, each insect generation may differ in the intensity of their interactions with different plant species, depending upon plant species preferences and their ability to locate new plants and habitat patches.

Above–below-ground interactions in natural communities

There is ample scope for considerably expanding the field of above–below-ground interactions in population and community ecology. Little is known about the importance of above–below-ground interactions in naturally occurring plant communities and how these processes are played out over various scales of space or time (Bardgett & Wardle 2010). In natural communities, plants are always exposed to complex communities of above- and below-ground organisms. At the same time, plants compete above- and below-ground with co-occurring plants, and may interact, above- and below-ground, with the multitrophic community associated to the neighbouring plants via so-called associational effects (Bezemer *et al.* 2010). Disentangling the relative importance of direct above–below-ground effects and indirect effects via the surrounding plant community is a major challenge in this field of research. However, if the results of laboratory-based studies are to be extrapolated to natural communities, then it is clear that assembly rules governing the structure and function of both above- and below-ground communities should incorporate the biotic activities of organisms associating with plants both in the rhizosphere and on the shoots.

How intimate interactions between soil-borne symbionts on the one hand, antagonists on the other, and above-ground insect communities are played out in nature remains largely unexplored. As suggested by Partida-Martínez & Heil (2011), the positive or negative effects of processes in the rhizosphere may counteract and thus nullify effects occurring above the ground. This does not mean that these indirect interactions are not important, but that the outcome of these complex indirect interactions depends critically on the number, strength and direction of positive and negative effects mediated by many players. A comprehensive integrative approach of dynamic multispecies multitrophic above–belowground interactions, linking ecological processes with physiological mechanisms underlying plant responses to below- and above-ground attackers and beneficials, will contribute to our further understanding of these complex interactions. This may ultimately push soil-dwelling organisms into mainstream ecological theory.

Acknowledgement

We thank Ana Pineda for sharing her data for Figure 4.4, and for her valuable comments on Section 4.3.

References

Alston, D.G., Bradley, J.R., Schmitt, D.P. and Coble, H.D. (1991) Response of *Helicoverpa zea* (Lepidoptera: Noctuidae) populations to canopy development in soybean as influenced by *Heterodera glycines* (Nematoda: Heteroderidae) and annual weed population densities. *Journal of Economic Entomology* **84**: 267–76.

Ansebo, L., Ignell, R., Lofqvist, J. and Hansson, B.S. (2005) Responses to sex pheromone and plant odours by olfactory receptor neurons housed in *sensilla auricillica* of the codling moth, *Cydia pomonella* (Lepidoptera: Tortricidae). *Journal of Insect Physiology* **51**: 1066–74.

Askew, R. and Shaw, M. (1986) Parasitoid communities: their size, structure and development. In: Waage, J. and Greathead, D. (eds) *Insect Parasitoids*. Academic Press, London, pp. 225–64.

Barbosa, P., Gross, P. and Kemper, J. (1991) Influence of plant allelochemicals on the tobacco hornworm and its parasitoid, *Cotesia congregata*. *Ecology* **72**: 1567–75.

Barbosa, P., Saunders, J.A., Kemper, J., Trumbule, R., Olechno, J. and Martinat, P. (1986) Plant allelochemicals and insect parasitoids: effects of nicotine on *Cotesia congregata* (Say) (Hymenoptera: Braconidae) and *Hyposoter annulipes* (Cresson) (Hymenoptera: Ichneumonidae). *Journal of Chemical Ecology* **12**: 1319–28.

Bardgett, R. and Wardle, D. (2010) *Aboveground–Belowground Linkages: Biotic Interactions, Ecosystem Processes, and Global Change*. Oxford University Press, New York.

Bezemer, T.M. and van Dam, N.M. (2005) Linking aboveground and belowground interactions via induced plant defenses. *Trends in Ecology and Evolution* **20**: 617–24.

Bezemer, T.M., De Deyn, G.B., Bossinga, T.M., van Dam, N.M., Harvey, J.A. and van der Putten, W.H. (2005) Soil community composition drives aboveground plant–herbivore–parasitoid interactions. *Ecology Letters* **8**: 652–61.

Bezemer, T.M., Fountain, M.T., Barea, J.M., Christensen, S., Dekker, S.C., Duyts, H., van Hal, R., Harvey, J.A., Hedlund, K., Maraun, M., Mikola, J., Mladenov, A.G., Robin, C., de Ruiter, P.C., Scheu, S., Setala, H., Smilauer, P. and van der Putten, W.H. (2010) Divergent composition but similar function of soil food webs of individual plants: plant species and community effects. *Ecology* **91**: 3027–36.

Bezemer, T.M., Wagenaar, R., van Dam, N.M. and Wäckers, F.L. (2003) Interactions between above- and belowground insect herbivores as mediated by the plant defense system. *Oikos* **101**: 555–562.

Bhattarai, K.K., Li, Q., Liu, Y., Dinesh-Kumar, S.P. and Kaloshian, I. (2007) The *Mi-1*-mediated pest resistance requires *Hsp90* and *Sgt1*. *Plant Physiology* **144**: 312–23.

Bodenhausen, N. and Reymond, P. (2007) Signaling pathways controlling induced resistance to insect herbivores in *Arabidopsis*. *Molecular Plant–Microbe Interactions* **20**: 1406–20.

Breedlove, D.E. and Ehrlich, P.R. (1968) Plant–herbivore coevolution: lupines and lycaenids. *Science* **162**: 671–2.

de la Peña, E., Rodriguez-Echeverria, S., van der Putten, W.H., Freitas, H. and Moens, M. (2006) Mechanism of control of root-feeding nematodes by mycorrhizal fungi in the dune grass *Ammophila arenaria*. *New Phytologist* **169**: 829–40.

De Moraes, C.M., Lewis, W.J., Pare, P.W., Alborn, H.T. and Tumlinson, J.H. (1998) Herbivore-infested plants selectively attract parasitoids. *Nature* **393**: 570–3.

Dicke, M. (1999) Are herbivore-induced plant volatiles reliable indicators of herbivore identity to foraging carnivorous arthropods? *Entomologia Experimentalis et Applicata* **91**: 131–42.

Dugravot, S., Thibout, E., Abo-Ghalia, A. and Huignard, J. (2004) How a specialist and a non-specialist insect cope with dimethyl disulfide produced by *Allium porrum*. *Entomologia Experimentalis et Applicata* **113**: 173–9.

Ehrlich, P.R. and Raven, P.H. (1964) Butterflies and plants: a study in coevolution. *Evolution* **18**: 586–608.

Erb, M., Flors, V., Karlen, D., de Lange, E., Planchamp, C., D'Alessandro, M., Turlings, T.C.J. and Ton, J. (2009) Signal signature of aboveground-induced resistance upon belowground herbivory in maize. *Plant Journal* **59**: 292–302.

Erb, M., Kollner, T.G., Degenhardt, J., Zwahlen, C., Hibbard, B.E. and Turlings, T.C.J. (2011) The role of abscisic acid and water stress in root herbivore-induced leaf resistance. *New Phytologist* **189**: 308–20.

Fatouros, N.E., Dicke, M., Mumm, R., Meiners, T. and Hilker, M. (2008) Foraging behavior of egg parasitoids exploiting chemical information. *Behavioral Ecology* **19**: 677–89.

Fukushima, J., Kainoh, Y., Honda, H. and Takabayashi, J. (2002) Learning of herbivore-induced and nonspecific plant volatiles by a parasitoid, *Cotesia kariyai*. *Journal of Chemical Ecology* **28**: 579–86.

Gange, A.C. and Brown, V.K. (1989) Effects of root herbivory by an insect on a foliar-feeding species, mediated through changes in the host plant. *Oecologia* **81**: 38–42.

Gange, A.C., Brown, V.K. and Aplin, D.M. (2003) Multitrophic links between arbuscular mycorrhizal fungi and insect parasitoids. *Ecology Letters* **6**: 1051–5.

Gehring, C. and Bennett, A. (2009) Mycorrhizal fungal–plant–insect interactions: the importance of a community approach. *Environmental Entomology* **38**: 93–102.

Godfray, H.C.J. (1994) *Parasitoids: Behavioral and Evolutionary Ecology*. Princeton University Press, Princeton, NJ.

Gols, R. and Harvey, J.A. (2009) Plant-mediated effects in the Brassicaceae on the performance and behaviour of parasitoids. *Phytochemistry Reviews* **8**: 187–206.

Gols, R., Bullock, J.M., Dicke, M., Bukovinszky, T. and Harvey, J.A. (2011) Smelling the wood from the trees: non-linear parasitoid responses to volatile attractants produced by wild and cultivated cabbage. *Journal of Chemical Ecology* **37**: 795–807.

Gols, R., van Dam, N.M., Raaijmakers, C.E., Dicke, M. and Harvey, J.A. (2009) Are population differences in plant quality reflected in the preference and performance of two endoparasitoid wasps? *Oikos* **118**: 733–43.

Gols, R., Wagenaar, R., Bukovinszky, T., van Dam, N.M., Dicke, M., Bullock, J.M. and Harvey, J.A. (2008) Genetic variation in defense chemistry in wild cabbages affects herbivores and their endoparasitoids. *Ecology* **89**: 1616–26.

Guerrieri, E. and Digilio, M.C. (2008) Aphid–plant interactions: a review. *Journal of Plant Interactions* **3**: 223–32.

Guerrieri, E., Lingua, G., Digilio, M.C., Massa, N. and Berta, G. (2004) Do interactions between plant roots and the rhizosphere affect parasitoid behaviour? *Ecological Entomology* **29**: 753–6.

Gunasena, G.H., Vinson, S.B. and Williams, H.J. (1990) Effects of nicotine on growth, development, and survival of the tobacco budworm (Lepidoptera: Noctuidae) and the parasitoid *Campoletis sonorensis* (Hymenoptera: Ichneumonidae). *Journal of Economic Entomology* **83**: 1777–82.

Hartley, S.E. and Gange, A.C. (2009) Impacts of plant symbiotic fungi on insect herbivores: mutualism in a multitrophic context. *Annual Review of Entomology* **54**: 323–42.

Harvey, J.A. (2005) Factors affecting the evolution of development strategies in parasitoid wasps: the importance of functional constraints and incorporating complexity. *Entomologia Experimentalis et Applicata* **117**: 1–13.

Harvey, J.A., Bezemer, T.M., Elzinga, J.A. and Strand, M.R. (2004) Development of the solitary endoparasitoid *Microplitis demolitor*: host quality does not increase with host age and size. *Ecological Entomology* **29**: 35–43.

Harvey, J.A., Gols, R., Wagenaar, R. and Bezemer, T.M. (2007) Development of an insect herbivore and its pupal parasitoid reflect differences in direct plant defense. *Journal of Chemical Ecology* **33**: 1556–69.

Harvey, J.A., Harvey, I.F. and Thompson, D.J. (1994) Flexible larval growth allows use of a range of host sizes by a parasitoid wasp. *Ecology* **75**: 1420–8.

Harvey, J.A., van Dam, N.M. and Gols, R. (2003) Interactions over four trophic levels: foodplant quality affects development of a hyperparasitoid as mediated through a herbivore and its primary parasitoid. *Journal of Animal Ecology* **72**: 520–31.

Harvey, J.A., van Dam, N.M., Raaijmakers, C.E., Bullock, J.M. and Gols, R. (2011) Tri-trophic effects of inter- and intra-population variation in defence chemistry of wild cabbage (*Brassica oleracea*). *Oecologia* **166**: 421–31.

Hempel, S., Stein, C., Unsicker, S.B., Renker, C., Auge, H., Weisser, W.W. and Buscot, F. (2009) Specific bottom-up effects of arbuscular mycorrhizal fungi across a plant–herbivore–parasitoid system. *Oecologia* **160**: 267–77.

Hilker, M. and Meiners, T. (2010) How do plants 'notice' attack by herbivorous arthropods? *Biological Reviews* **85**: 267–80.

Hoffmann, D., Vierheilig, H. and Schausberger, P. (2011) Arbuscular mycorrhiza enhances preference of ovipositing predatory mites for direct prey-related cues. *Physiological Entomology* **36**: 90–5.

Hol, W.H.G., de Boer, W., Termorshuizen, A.J., Meyer, K.M., Schneider, J.H.M., van Dam, N.M., van Veen, J.A. and van der Putten, W.H. (2010) Reduction of rare soil microbes modifies plant–herbivore interactions. *Ecology Letters* **13**: 292–301.

Kabouw, P., Kos, M., Kleine, S., Vockenhuber, E.A., van Loon, J.J.A., van der Putten, W.H., van Dam, N.M. and Biere, A. (2011) Effects of soil organisms on aboveground multitrophic interactions are consistent between plant genotypes mediating the interaction. *Entomologia Experimentalis et Applicata* **139**: 197–206.

Kaplan, I. and Denno, R.F. (2007) Interspecific interactions in phytophagous insects revisited: a quantitative assessment of competition theory. *Ecology Letters* **10**: 977–94.

Kaplan, I., Sardanelli, S. and Denno, R.F. (2009) Field evidence for indirect interactions between foliar-feeding insect and root-feeding nematode communities on *Nicotiana tabacum*. *Ecological Entomology* **34**: 262–70.

Kaplan, I., Sardanelli, S., Rehill, B.J. and Denno, R.F. (2011) Toward a mechanistic understanding of competition in vascular-feeding herbivores: an empirical test of the sink competition hypothesis. *Oecologia* **166**: 627–36.

Karban, R. and Baldwin, I.T. (1997) *Induced Responses to Herbivory*. University of Chicago Press, Chicago.

Kessler, A. and Heil, M. (2011) The multiple faces of indirect defences and their agents of natural selection. *Functional Ecology* **25**: 348–57.

Koricheva, J., Gange, A.C. and Jones, T. (2009) Effects of mycorrhizal fungi on insect herbivores: a meta-analysis. *Ecology* **90**: 2088–97.

Krebs, J.R. and Davies, N.B. (1984) *Behavioural Ecology: An Evolutionary Approach*, 2nd edn. Blackwell Scientific Publications, Oxford.

Lawrence, P.O. (1990) The biochemical and physiological effects of insect hosts on the development and ecology of their insect parasites: an overview. *Archives of Insect Biochemistry and Physiology* **13**: 217–28.

Li, Q., Xie, Q.G., Smith-Becker, J., Navarre, D.A. and Kaloshian, I. (2006) *Mi-1*-mediated aphid resistance involves salicylic acid and mitogen-activated protein kinase signaling cascades. *Molecular Plant–Microbe Interactions* **19**: 655–64.

Lynch, J.M. and Whipps, J.M. (1990) Substrate flow in the rhizosphere. *Plant and Soil* **129**: 1–10.

Mackauer, M. and Sequeira, R. (1993) Patterns of development in insect parasites. In: Beckage, N.E., Thompson, S.N. and Federici, B.A. (eds) *Parasites and Pathogens of Insects*. Academic Press, Orlando, FL, pp. 1–23.

Mantelin, S., Bhattarai, K.K. and Kaloshian, I. (2007) Is ethylene a key player in the *Mi-1* mediated resistance to potato aphids and root knot nematodes in tomato? *Phytopathology* **97**: S70.

Marak, H.B., Biere, A. and van Damme, J.M.M. (2000) Direct and correlated responses to selection on iridoid glycosides in *Plantago lanceolata* L. *Journal of Evolutionary Biology* **13**: 985–96.

Masters, G.J. and Brown, V.K. (1992) Plant-mediated interactions between two spatially separated insects. *Functional Ecology* **6**: 175–9.

Masters, G.J., Brown, V.K. and Gange, A.C. (1993) Plant mediated interactions between aboveground and belowground insect herbivores. *Oikos* **66**: 148–51.

Masters, G.J., Jones, T.H. and Rogers, M. (2001) Host-plant mediated effects of root herbivory on insect seed predators and their parasitoids. *Oecologia* **127**: 246–50.

Moran, N.A. and Whitham, T.G. (1990) Interspecific competition between root-feeding and leaf-galling aphids mediated by host-plant resistance. *Ecology* **71**: 1050–8.

Morris, W.F., Hufbauer, R.A., Agrawal, A.A., Bever, J.D., Borowicz, V.A., Gilbert, G.S., Maron, J.L., Mitchell, C.E., Parker, I.M., Power, A.G., Torchin, M.E. and Vazquez, D.P. (2007) Direct and interactive effects of enemies and mutualists on plant performance: a meta-analysis. *Ecology* **88**: 1021–9.

Morse, D.H. (2009) Four-level interactions: herbivore use of ferns and subsequent parasitoid-hyperparasitoid performance. *Ecological Entomology* **34**: 246–53.

Newton, E.L., Bullock, J.M. and Hodgson, D.J. (2009) Glucosinolate polymorphism in wild cabbage (*Brassica oleracea*) influences the structure of herbivore communities. *Oecologia* **160**: 63–76.

Ode, P.J. (2006) Plant chemistry and natural enemy fitness: effects on herbivore and natural enemy interactions. *Annual Review of Entomology* **51**: 163–85.

Ohsaki, N. and Sato, Y. (1994) Food plant choice of *Pieris* butterflies as a trade-off between parasitoid avoidance and quality of plants. *Ecology* **75**: 59–68.

Olson, D.M., Davis, R.F., Wäckers, F.L., Rains, G.C. and Potter, T. (2008) Plant–herbivore–carnivore interactions in cotton, *Gossypium hirsutum*: linking belowground and aboveground. *Journal of Chemical Ecology* **34**: 1341–8.

Partida-Martínez, L. and Heil, M. (2011) The microbe-free plant: fact or artifact? *Frontiers in Plant–Microbe Interaction* **2**(100): 1–16.

Pierre, P.S., Jansen, J.J., Hordijk, C.A., van Dam, N.M., Cortesero, A.M. and Dugravot, S. (2011) Differences in volatile profiles of turnip plants subjected to single and dual herbivory above- and belowground. *Journal of Chemical Ecology* **37**: 368–77.

Pieterse, C.M.J., van Pelt, J.A., van Wees, S.C.M., Ton, J., Leon-Kloosterziel, K.M., Keurentjes, J.J.B., Verhagen, B.W.M., Knoester, M., van der Sluis, I., Bakker, P. and van Loon, L.C. (2001) Rhizobacteria-mediated induced systemic resistance: triggering, signalling and expression. *European Journal of Plant Pathology* **107**: 51–61.

Pineda, A., Zheng, S.J., van Loon, J.J.A., Pieterse, C.M.J. and Dicke, M. (2010) Helping plants to deal with insects: the role of beneficial soil-borne microbes. *Trends in Plant Science* **15**: 507–14.

Pineda, A., Zheng, S.J., van Loon, J.A.A. and Dicke, M. (2012) Rhizobacteria modify plant–aphid interactions: a case of induced systemic susceptibility. *Plant Biology* **14**: 83–90.

Poelman, E.H., van Loon, J.J.A., van Dam, N.M., Vet, L.E.M. and Dicke, M. (2010) Herbivore-induced plant responses in *Brassica oleracea* prevail over effects of constitutive resistance and result in enhanced herbivore attack. *Ecological Entomology* **35**: 240–7.

Poveda, K., Steffan-Dewenter, I., Scheu, S. and Tscharntke, T. (2003) Effects of below- and above-ground herbivores on plant growth, flower visitation and seed set. *Oecologia* **135**: 601–5.

Pozo, M.J. and Azcon-Aguilar, C. (2007) Unraveling mycorrhiza-induced resistance. *Current Opinion in Plant Biology* **10**: 393–8.

Price, P.W., Bouton, C.E., Gross, P., McPheron, B.A., Thompson, J.N. and Weis, A.E. (1980) Interactions among three trophic levels: influence of plants on interactions between insect herbivores and natural enemies. *Annual Review of Ecology, Evolution and Systematics* **11**: 41–65.

Ramamoorthy, V., Viswanathan, R., Raguchander, T., Prakasam, V. and Samiyappan, R. (2001) Induction of systemic resistance by plant growth-promoting rhizobacteria in crop plants against pests and diseases. *Crop Protection* **20**: 1–11.

Rasmann, S. and Turlings, T.C.J. (2007) Simultaneous feeding by aboveground and belowground herbivores attenuates plant-mediated attraction of their respective natural enemies. *Ecology Letters* **10**: 926–36.

Rostás, M. and Eggert, K. (2008) Ontogenetic and spatio-temporal patterns of induced volatiles in *Glycine max* in the light of the optimal defence hypothesis. *Chemoecology* **18**: 29–38.

Saravanakumar, D., Lavanya, N., Muthumeena, B., Raguchander, T., Suresh, S. and Samiyappan, R. (2008) *Pseudomonas fluorescens* enhances resistance and natural enemy population in rice plants against leaffolder pest. *Journal of Applied Entomology* **132**: 469–79.

Schausberger, P., Peneder, S., Jurschik, S. and Hoffmann, D. (2012) Mycorrhiza changes plant volatiles to attract spider mite enemies. *Functional Ecology* **26**: 441–9.

Schoonhoven, L.M., van Loon, J.J.A. and Dicke, M. (2005) *Insect-Plant Biology*, 2nd edn. Oxford University Press, Oxford.

Segarra, G., Casanova, E., Bellido, D., Odena, M.A., Oliveira, E. and Trillas, I. (2007) Proteome, salicylic acid, and jasmonic acid changes in cucumber plants inoculated with *Trichoderma asperellum* strain T34. *Proteomics* **7**: 3943–52.

Sikora, R.A., Reckhaus, P., Adamou, E. and Rijksuniv, G. (1988) Presence, distribution and importance of plant parasitic nematodes in irrigated agricultural crops in Niger. *Communications of the Faculty of Agricultural Sciences of the State University of Ghent*, Vol 53, Pts 2a, 2b, 3a, 3b: 1988 International Symposium on Crop Protection **40**: 821–34.

Singer, M.S., Rodrigues, D., Stireman, J.O. and Carriere, Y. (2004) Roles of food quality and enemy-free space in host use by a generalist insect herbivore. *Ecology* **85**: 2747–53.

Soler, R., Bezemer, T.M., van der Putten, W.H., Vet, L.E.M. and Harvey, J.A. (2005) Root herbivore effects on above-ground herbivore, parasitoid and hyperparasitoid performance via changes in plant quality. *Journal of Animal Ecology* **74**: 1121–30.

Soler, R., Harvey, J.A., Kamp, A.F.D., Vet, L.E.M., van der Putten, W.H., van Dam, N.M., Stuefer, J.F., Gols, R., Hordijk, C.A. and Bezemer, T.M. (2007a) Root herbivores influence the behaviour of an aboveground parasitoid through changes in plant-volatile signals. *Oikos* **116**: 367–76.

Soler, R., Harvey, J.A. and Bezemer, T.M. (2007b) Foraging efficiency of a parasitoid of a leaf herbivore is influenced by root herbivory on neighbouring plants. *Functional Ecology* **21**: 969–74.

Staley, J.T., Mortimer, S.R., Morecroft, M.D., Brown, V.K. and Masters, G.J. (2007) Summer drought alters plant-mediated competition between foliar- and root-feeding insects. *Global Change Biology* **13**: 866–77.

Stanton, N.L. (1988) The underground in grasslands. *Annual Review of Ecology, Evolution and Systematics* **19**: 573–89.

Thaler, J.S. and Bostock, R.M. (2004) Interactions between abscisic-acid-mediated responses and plant resistance to pathogens and insects. *Ecology* **85**: 48–58.

Thompson, J.N. (2005) Coevolution: the geographic mosaic of coevolutionary arms races. *Current Biology* **15**: 992–4.

Thompson, J.N. and Pellmyr, O. (1991) Evolution of oviposition behavior and host preference in Lepidoptera. *Annual Review of Entomology* **36**: 65–89.

Tindall, K.V. and Stout, M.J. (2001) Plant-mediated interactions between the rice water weevil and fall armyworm in rice. *Entomologia Experimentalis et Applicata* **101**: 9–17.

Turlings, T.C.J., Tumlinson, J.H. and Lewis, W.J. (1990) Exploitation of herbivore-induced plant odors by host-seeking parasitic wasps. *Science* **250**: 1251–3.

Turlings, T.C.J., Wäckers, F.L., Vet, L.E.M., Lewis, W.J. and Tumlinson, J.H. (1993) Learning of host-finding cues by hymenopterous parasitoids. In: Papaj, D.R. and Lewis, A.C. (eds) *Insect Learning: Ecological and Evolutionary Perspectives* Chapman and Hall, New York, pp. 51–78.

van Dam, N.M., Harvey, J.A., Wäckers, F.L., Bezemer, T.M., van der Putten, W.H. and Vet, L.E.M. (2003) Interactions between aboveground and belowground induced responses against phytophages. *Basic and Applied Ecology* **4**: 63–77.

van Dam, N.M., Raaijmakers, C.E. and van der Putten, W.H. (2005) Root herbivory reduces growth and survival of the shoot feeding specialist *Pieris rapae* on *Brassica nigra*. *Entomologia Experimentalis et Applicata* **115**: 161–70.

van der Heijden, M.G.A., Bardgett, R.D. and van Straalen, N.M. (2008) The unseen majority: soil microbes as drivers of plant diversity and productivity in terrestrial ecosystems. *Ecology Letters* **11**: 296–310.

van der Putten, W.H., Vet, L.E.M., Harvey, J.A. and Wäckers, F.L. (2001) Linking above- and belowground multitrophic interactions of plants, herbivores, pathogens, and their antagonists. *Trends in Ecology and Evolution* **16**: 547–54.

van Oosten, V.R., Bodenhausen, N., Reymond, P., van Pelt, J.A., van Loon, L.C., Dicke, M. and Pieterse, C.M.J. (2008) Differential effectiveness of microbially induced resistance against herbivorous insects in *Arabidopsis*. *Molecular Plant–Microbe Interactions* **21**: 919–30.

van Wees, S.C.M., van der Ent, S. and Pieterse, C.M.J. (2008) Plant immune responses triggered by beneficial microbes. *Current Opinion in Plant Biology* **11**: 443–8.

Vandegehuchte, M.L., de la Pena, E. and Bonte, D. (2010) Interactions between root and shoot herbivores of *Ammophila arenaria* in the laboratory do not translate into correlated abundances in the field. *Oikos* **119**: 1011–19.

Vet, L.E.M. and Dicke, M. (1992) Ecology of infochemical use by natural enemies in a tritrophic context. *Annual Review of Entomology* **37**: 141–72.

Vet, L.E.M., Lewis, W.J. and Cardé, R. (1995) Parasitoid foraging and learning. In: Cardé, R. and Bell, W.J. (eds) *Chemical Ecology of Insects*. Chapman and Hall, New York, pp. 65–101.

Vet, L.E.M., Wäckers, F.L. and Dicke, M. (1991) How to hunt for hiding hosts: the reliability–detectability problem in foraging parasitoids. *Netherlands Journal of Zoology* **41**: 202–13.

Wajnberg, E., Bernstein, C. and van Alphen, J. (2008) *Behavioural Ecology of Insect Parasitoids: From Theoretical Approaches to Field Applications*. Wiley-Blackwell, Oxford.

Whipps, J.M. (2004) Prospects and limitations for mycorrhizas in biocontrol of root pathogens. *Canadian Journal of Botany* **82**: 1198–227.

White, T.C.R. (1984) The abundance of invertebrate herbivores in relation to the availability of nitrogen in stressed food plants. *Oecologia* **63**: 90–105.

White, J.A. and Andow, D.A. (2006) Habitat modification contributes to associational resistance between herbivores. *Oecologia* **148**: 482–90.

Wurst, S. (2010) Effects of earthworms on above- and belowground herbivores. *Applied Soil Ecology* **45**: 123–30.

Wurst, S. and Jones, T.H. (2003) Indirect effects of earthworms (*Aporrectodea caliginosa*) on an above-ground tritrophic interaction. *Pedobiologia* **47**: 91–7.

Wurst, S. & van der Putten, W.H. (2007) Root herbivore identity matters in plant-mediated interactions between root and shoot herbivores. *Basic and Applied Ecology* **8**: 491–9.

Zehnder, G., Kloepper, J., Tuzun, S., Yao, C.B., Wei, G., Chambliss, O. and Shelby, R. (1997) Insect feeding on cucumber mediated by rhizobacteria-induced plant resistance. *Entomologia Experimentalis et Applicata* **83**: 81–5.

5

A hitch-hiker's guide to parasitism: the chemical ecology of phoretic insect parasitoids

Martinus E. Huigens[1] and Nina E. Fatouros[2]

[1] Dutch Butterfly Conservation, Wageningen, The Netherlands
[2] Laboratory of Entomology, Department of Plant Sciences, Wageningen University, The Netherlands

Abstract

Phoretic arthropods use other animals as vehicles to migrate to new environments. Among insect parasitoids, phoresy is almost exclusively restricted to minute wasp species that develop in or on the smallest and most inconspicuous life stage of their host: the egg. Females of about 35 egg parasitoid species are known to hitch-hike with adult hosts to reach their egg-laying sites. Recent studies suggest that phoretic parasitoids strongly rely on chemical espionage to locate their transporting host. These wasps have evolved intriguing ways to exploit cues that are part of their host's communication system, including sex, anti-sex and aggregation pheromones. Such a 'chemical-espionage-and-ride' strategy can be innate but it can also be learned. The extent and mechanisms by which hosts might avoid exploitation are poorly understood. Here we discuss why we expect phoresy to be much more widespread among egg parasitoids than is known so far. It is expected to be adaptive, especially in those species that have limited ability for directed flight, have a short time window available for parasitism, have a narrow host range, and parasitize abundant hosts that lay large eggs (or eggs in groups) with a large distance between them. We review some recently published examples of chemical espionage by phoretic egg parasitoids and discuss to what extent phoretic wasps represent a selective force against the use of chemical cues by their hosts. At the end of the chapter we identify unexplored aspects of the chemical ecology of phoretic insect parasitoids that warrant further investigation. In addition to the fundamental interest, research into the ways that phoretic parasitoids have evolved to locate their host may help improve the efficacy of using parasitoids as biological control agents against insect pests.

Chemical Ecology of Insect Parasitoids, First Edition. Eric Wajnberg and Stefano Colazza.
© 2013 John Wiley & Sons, Ltd. Published 2013 by John Wiley & Sons, Ltd.

5.1 Phoresy

The transport of certain organisms on the bodies of others for purposes other than direct parasitism of the transporting individual is termed 'phoresy' (Ferrière 1926, Howard 1927). This phenomenon is very common in the arthropod world (Ferrière 1926, Clausen 1976, Binns 1982, Fatouros & Huigens 2012). To end up on a transport vehicle, phoretic arthropod species have evolved various ways of exploiting the pheromones used by the transporting species. A fascinating example comprises larvae of the blister beetle *Meloe franciscanus*, which inhabit the deserts of the American Southwest (Saul-Gershenz & Millar 2006). The larvae feed on pollen, nectar and eggs of the solitary bee *Habropoda pallida* inside the bee's nest. To obtain access to the nest, a mass of hundreds to thousands of beetle larvae on a plant lure a male bee not only by visually mimicking a female bee but also by mimicking her sex pheromone blend. After attracting, 'jumping', and riding on a male bee, they transfer onto a female bee during mating, after which she transports them to her nest. There, the larvae complete their development into adults. Finally, the adult beetles leave the bee nest and lay their eggs at the base of plants to repeat the whole cycle (Saul-Gershenz & Millar 2006).

Such aggressive chemical mimicry does not seem to be used by tiny phoretic insect parasitoids of which the adult females are transported by a host insect. Instead, female parasitoids have a solitary lifestyle and seem to be attracted by a pheromone emitted by a transporting host vehicle; in other words, they make use of chemical espionage. In this chapter on the chemical ecology of such phoretic parasitoids, we particularly focus on the different types of chemical espionage that they use and the selective pressures that they may inflict on their hosts. But first we discuss the prevalence of phoresy among insect parasitoids and traits that are important for a phoretic relationship between parasitoids and hosts.

5.2 Prevalence of phoretic parasitoids

With respect to the study of insect parasitoids, the phoretic transport of adult females has been described almost exclusively for those species that attack and develop in or on insect eggs. Only one pupal parasitoid, *Pteromalus puparum*, has been observed to hitch a ride on late-instar caterpillars of cabbage white butterflies to eventually parasitize fresh pupae (Fig. 5.1) (Takagi 1986, Godfray 1994). Even though the parasitoid ultimately parasitizes the transporting host individual, it does not parasitize the developmental stage which it uses for transport. We consider this form of hitch-hiking as phoresy, and would modify the definitions of Ferrière (1926) and Howard (1927) accordingly:

> the transport of certain organisms on the bodies of others for purposes other than direct parasitism of the developmental stage of the transporting individual.

The generalist wasp *P. puparum* probably uses chemical cues from the caterpillars and the host-plant complex (Fig. 5.1). The fact that most phoretic parasitoids documented so far are egg parasitoids (Clausen 1976) does not mean that phoretic behaviour is non-adaptive for parasitoids that attack other life stages of their hosts. However, as there is simply much

Figure 5.1 The generalist pupal parasitoid *Pteromalus puparum* L. (Hymenoptera: Pteromalidae) is frequently phoretic on last-instar caterpillars of lepidopteran hosts to hitch a ride to pupation sites away from the plant. (a) A female wasp making antennal contact with the posterior end of an L5 instar caterpillar of the large cabbage white butterfly *Pieris brassicae*. The wasps can discriminate between chemical cues released by L4 and L5 caterpillars of *P. brassicae* and the small cabbage white butterfly *P. rapae* from a short distance. To locate their hosts from a longer distance, *P. puparum* females probably make use of volatile cues from the plants fed upon by caterpillars (N.E. Fatouros, unpublished data). (b) A wasp riding on the back of a caterpillar (photograph courtesy of N.E. Fatouros, http://www.bugsinthepicture.com).

less known about phoresy in other parasitoid wasps, we focus here on those that parasitize eggs. Since Howard's (1927) report of a few examples of phoretic parasitoids at the beginning of the 20th century, about 35 egg parasitoid species belonging to six hymenopteran families have been reported to use a variety of adult hosts as vehicles to reach their egg-laying sites (Table 5.1) (Fatouros & Huigens 2012). Transporting hosts include butterflies, moths, praying mantids, grasshoppers, crickets, dragonflies, beetles and bugs. Wasps may attach themselves to a host by firmly implanting their mandibles or using claws on their legs (Clausen 1976, Pinto 1994). Phoretic individuals are mostly transported by (gravid) female hosts. However, wasps are also occasionally found on male hosts and may transfer onto females during mating (Table 5.1).

Most phoretic species currently known belong to the families Scelionidae (nine genera) and Trichogrammatidae (four genera) (Table 5.1). One of the most likely reasons for this is that the life-history traits of members of these families have been more often investigated

Table 5.1 Overview of phoretic egg parasitoids (based on data from Clausen 1976). Family and genus names (and number of species with phoresy proven per genus in brackets) are provided for the parasitoids in the first column. Other columns include host information. N = number of known host species. For a more detailed overview including parasitoid and host species names, body part to which parasitoids attach themselves, attracting cues, and references, we refer to supplementary table 1 in Fatouros & Huigens (2012).

Phoretic wasps	Hosts	Gregarious hosts (*N*)	Solitary hosts (*N*)	Transporting sex
Scelionidae				
Synoditella (3 species)	Grasshoppers	3	–	Females
Sceliocerdo (1 species)	Grasshoppers	2	–	Females
Lepidoscelio (1 species)	Grasshopper	1	–	–
Mantibaria (2 species)	Praying mantis	1	–	Females and males
Telenomus (>8 species)	Moths	7	–	Females
	Stink bug	1	–	Females and males
	Butterfly	1	–	Females
Epinomus (1 species)	Coreid bug	1	–	–
Asolcus (1 species)	Shield bug	1	–	–
Calotelea (1 species)	Dragon fly	1	–	Females
Protelenomus (1 species)	Coreid bug	1	–	Females and males
Trichogrammatidae				
Oligosita (2 species)	Cricket	1	–	Females
	Grasshopper	1	–	
Pseudoxenufens (1 species) (=*Xenufens* sp.)	Butterflies	2	–	Females and males
Brachista (1 species)	Robber fly	1	–	Females
Trichogramma (>3 species)	Moths	–	3	Females
	Butterflies	1	4	Females and males
Encyrtidae				
Taftia (1 species)	Scale insect	1	–	Ant workers
Eupelmidae				
Anastatus (1 species)	Moth	1	–	Females
Eulophidae				
Emersonella (1 species)	Leaf beetle	1	–	Females
Chrysocharodes (1 species)	Leaf beetle	1	1	Females
Grassator (1 species)	Weevil	1	–	–
Torymidae				
Podagrion (1 species)	Praying mantis	1	1	Females

than those of other egg parasitoid families because they are frequently used in biological control programmes against insect pests (van Lenteren 2012). It is certainly not because these two families comprise more species. Other families with phoretic species such as the Torymidae (approx. 900 species), Encyrtidae (approx. 3300 species), Eupelmidae (approx. 850 species) and Eulophidae (approx. 3400 species) are equally, or sometimes even more, species-rich (see http://www.sel.barc.usda.gov/hym/chalcid.html for details). Given that the Scelionidae and Trichogrammatidae together comprise about 4000 described species (Austin *et al.* 2005, Querino *et al.* 2010) only about 1% of them are known to be phoretic.

Phoresy is expected to be much more common within these families and among egg parasitoids in general (Clausen 1976). It is a foraging strategy that is easily overlooked in traditional life-history studies and has not been thoroughly investigated for many egg parasitoid species. Demonstration of phoretic behaviour requires extensive field sampling of adult hosts at the peak moment of oviposition, with specific trapping equipment and proper control catches (Fatouros & Huigens 2012), combined with a set of behavioural assays in the laboratory. Phoretic egg parasitoids have certainly been overlooked in large field collections of adult hosts that were not aimed at finding these small insects (approximately 0.5–3.0 mm long). They are also very likely to get lost using conventional trapping methods. For example, phoretic wasps of the genus *Trichogramma* can only be collected on adult butterflies in the field with specific adaptations to a butterfly net, and not with regular nets (Fatouros & Huigens 2012). Moreover, the number of individuals hitching a ride on an adult host in the field may be low for quite a number of parasitoid species because the phoretic relationship is facultative, and not obligate such as between the European praying mantis (*Mantis religiosa*) and highly adapted phoretic *Mantibaria* wasps (Scelionidae) that remove their wings after ending up on a female mantis in order to be able to enter her frothy egg mass (Couturier 1941, Bin 1985). A facultative phoretic relationship, for example, applies to two closely related *Trichogramma* species, *T. brassicae* and *T. evanescens*, that are found at low prevalence on adult cabbage white butterflies in nature (Fatouros & Huigens 2012). Laboratory studies show that these species also exploit chemical cues of *Brassica* plants induced by egg deposition to find the eggs of these butterflies (Fatouros *et al.* 2005a, 2008a, 2009, 2012, Pashalidou *et al.* 2010). Phoresy could occasionally help females of both parasitoid species to reach a host habitat (long-range dispersal) and to have immediate access to eggs freshly laid by the transporting female host. Other plant- or host-related cues can be used to search for more host eggs within the habitat (short-range dispersal). All in all, the comparatively few instances of phoretic egg parasitoid species documented so far are likely to be a considerable underestimation. We are convinced that more extensive field surveys combined with detailed behavioural experiments will bring many more instances to light.

5.3 Important parasitoid and host traits

5.3.1 Parasitoid traits

Phoresy should be particularly adaptive in those species that meet one or more of the following four criteria (which apply to many parasitoid species):

1 limited dispersal capacity
2 a short time window for parasitism

3 a small egg load
4 a narrow host range.

Limited dispersal capacity
Trichogramma egg parasitoids mainly use walking and jumping for short-distance dispersal (Romeis *et al.* 2005). These wasps have also been shown to be capable of short flights depending on the density and type of the tested crop (see Suverkropp *et al.* 2009 and references therein). However, the wasps were incapable of directed upwind flight in the open field or in wind tunnel experiments (Keller *et al.* 1985, Noldus *et al.* 1991b). Dispersal by flight probably also depends on abiotic conditions such as wind and temperature (Romeis *et al.* 2005). Only few egg parasitoid species, for example *Oomyzus gallerucae*, have been shown to actively orientate towards an odour source in the field (Fatouros *et al.* 2008b).

Short time window for parasitism
Phoresy has frequently evolved in scelionid parasitoids (Table 5.1), which might reflect a physiological need to develop well in host eggs at an early stage of development. Indeed, their telescopic ovipositor is often less sclerotized and completely different from that of all other parasitic Hymenoptera (Field & Austin 1994) and thus does not seem to be well adapted to penetrate into hardened host eggs (Quicke 1997). The phoretic scelionid *Telenomus calvus*, for example, is restricted to ovipositing in the soft young eggs (less than 12 hours old) of the pentatomid bug *Podisus maculiventris* (Orr *et al.* 1986). Hitch-hiking with an egg-laying female host virtually ensures the availability of freshly laid eggs.

In contrast, *Trichogramma* wasps are able to parasitize older host eggs until the head capsule of the caterpillar has sclerotized (reviewed by Pak 1986). *T. brassicae* wasps, for example, produced the most offspring in 3-day-old eggs of the large cabbage white butterfly *Pieris brassicae* (Fatouros *et al.* 2005a). This may explain why this wasp is only occasionally found to be phoretic on butterflies (Fatouros & Huigens 2012). To locate older *P. brassicae* eggs, *T. brassicae* can also use volatile or contact cues of *Brassica* plants several days after egg deposition by *P. brassicae* (Fatouros *et al.* 2005a, 2012).

Small egg load
Another life-history trait related to phoretic behaviour seems to be the relatively small number of eggs produced. For example, *Telenomus calvus* females had the highest mean daily progeny production during the first 24 hours after emergence from the host egg (Orr *et al.* 1986). Most females deposit their entire complement of eggs (mean fecundity: 22.4 ± 3) in the first encountered host egg clutch. Other non-phoretic scelionids had a higher fecundity and their reproductive periods were markedly longer (see Orr *et al.* 1986 and references therein). However, this does not imply that species with a much larger egg load are never phoretic. Trichogrammatids are facultative gregarious parasitoids that produce a high number (sometimes more than 100) eggs per female (Jervis *et al.* 2008). The lifetime fecundity of *Trichogramma evanescens*, which is occasionally phoretic on cabbage white butterflies (Huigens *et al.* 2009, 2010, Fatouros & Huigens 2012), has been shown to be over 60 eggs per female (Doyon & Boivin 2005).

Narrow host range
An obligate phoretic relationship is likely to require specificity. Many of the egg parasitoid species showing a phoretic lifestyle appear to have relatively narrow dietary breadth. This,

for example, applies to the previously mentioned obligatory phoretic relationship between the praying mantis and *Mantibaria* egg parasitoids (Couturier 1941, Bin 1985). In general, scelionids are more specialized (Vinson 2010), and phoretic species in particular tend to have more narrow dietary breadth. Within the genus *Telenomus*, the phoretic *T. calvus*, for example, is specialized on eggs of *Podisus* bugs, whereas the non-phoretic *T. podisi* parasitizes the eggs of a wide range of pentatomid bugs (Orr *et al.* 1986, Bruni *et al.* 2000).

Trichogramma wasps are generally assumed to be very polyphagous (Wajnberg & Hassan 1994, Smith 1996). *T. evanescens*, for example, has been reported to parasitize the eggs of more than 65 different lepidopteran species (Hase 1925). Inexperienced females of this species are known to climb onto host butterflies, and preferentially mount hosts rather than non-host insects, but do not discriminate their sex or mating status. This species can learn to hitch-hike specifically with mated female host butterflies after one successful ride leading to an oviposition into fresh butterfly eggs (Huigens *et al.* 2009, 2010). Its ability to learn is probably related to *T. evanescens'* relatively wide dietary breadth. Polyphagy increases the chance to encounter hosts and decreases the adaptive value of being restricted to using very specific host-related cues (Vet & Dicke 1992, Vet *et al.* 1995). Instead, poly- phagous parasitoids can learn to associate specific cues with the presence of a rewarding host (Vet *et al.* 1995, Hoedjes *et al.* 2011).

5.3.2 Host traits

Whether phoresy is adaptive also depends to a large extent on the type of host that is used. For a female egg parasitoid, it should be most adaptive to ride on those hosts that meet one or both of the following criteria (which apply to many host species):

1 deposition of large single eggs or eggs in clusters
2 patchy egg distribution over a large area.

Deposition of large single eggs or eggs in clusters
For a facultative gregarious parasitoid species with a relatively large egg load, a trans- porting female host that lays an egg clutch (gregarious host) or a large single egg at a time (solitary host) immediately offers the female parasitoid an opportunity to lay many of her eggs. In the first case, the parasitoid should also have ample time to descend from the egg-laying host. Some obligatory phoretic wasps, such as *Mantibaria* wasps riding on the European mantis, even return to the same transport vehicle after parasitism (Coutu- rier 1941).

By far the majority of the 35 species (>90%) that are currently known to be phoretic, ride on gregarious hosts and only very few on solitary hosts that lay relatively small eggs (Table 5.1). The latter hosts only offer the female parasitoid an immediate opportunity to lay a small fraction of her eggs. It may, however, be rewarding to use such hosts when they are highly abundant. This is illustrated by the parasitoids *T. brassicae* and *T. evanescens*, which both have a lifetime fecundity of 45 or more eggs (Cerutti & Bigler 1995, Doyon & Boivin 2005). In the Netherlands, both species not only hitch-hike with the gregarious *P. brassicae* (which lays about 20–50 eggs per oviposition event) but also with the much more abundant, solitary small cabbage white *P. rapae* (Huigens *et al.* 2010, Fatouros & Huigens

2012). All eggs of a female parasitoid can in theory be laid into one or two *P. brassicae* egg clutches, whereas it only lays 2–3 eggs into a singly laid *P. rapae* egg (Fatouros & Huigens 2012, Kruidhof *et al.* 2012). Thus, from a parasitoid's perspective, not only a host's egg-laying behaviour is important but also its abundance.

Patchy egg distribution over a large area

Phoresy should particularly be adaptive for species with a limited dispersal capacity that live in an unreliable patchy environment with large distances between host egg patches (Vinson 1985). Many of the hosts that are known to carry phoretic wasps are highly mobile and their eggs are patchily distributed over a large area; examples include grasshoppers, dragonflies, butterflies and moths (Table 5.1).

5.4 Chemical espionage on host pheromones

Predators and parasitoids are known to exploit the intraspecific communication systems of other organisms to gain access to resources (see reviews by Vinson 1984, Dicke & Sabelis 1992, Stowe *et al.* 1995, Zuk & Kolluru 1998, Haynes & Yeargan 1999, Powell 1999, Wyatt 2003, Fatouros *et al.* 2008b, Colazza *et al.* 2010). A number of egg parasitoids have developed the ability to use host pheromones as foraging kairomones (in other words, chemical espionage) to overcome the problem of the low detectability of host eggs (Table 5.2) (Fatouros *et al.* 2008b, Colazza *et al.* 2010); they associate those adult host pheromones with areas where eggs are likely to be found. Phoretic egg parasitoids are also likely to eavesdrop on pheromones emitted by their transporting host. However, such an 'espionage-and-ride' strategy has so far only been described for four parasitoid species and three types of pheromones. Three different types of espionage have been observed so far:

1 espionage on male aggregation pheromone
2 espionage on sex pheromones
3 espionage on anti-sex pheromones.

5.4.1 Espionage on male aggregation pheromone

Aldrich *et al.* (1984) were the first to identify an attractant pheromone for a true bug, the spined soldier bug *Podisus maculiventris*. Males of this generalist predator of phytophagous insects call for mates by releasing a pheromone from their dorsal abdominal glands, which end in openings underneath the wings. Both female and male bugs are attracted to the odour blend of pheromone-calling males, which is a mixture of (*E*)-2-hexenal and α-terpineol. The synthetic pheromone was also shown to be attractive to egg parasitoids including the phoretic *Telenomus calvus* (Fig. 5.2a) (Aldrich *et al.* 1984, Bruni *et al.* 2000). The wasps become phoretic only on female bugs, to which they transfer during mating (Buschman & Whitcomb 1980, Orr *et al.* 1986). They probably recognize the female *P. maculiventris* from a female-specific blend produced by small dorsal glands (Aldrich *et al.* 1984, Aldrich 1995).

Table 5.2 Identified volatile kairomones that elicited a response in egg parasitoids in laboratory and/or field studies, and that sometimes lead to phoretic behaviour.

Type of infochemical	Chemical	Insect host	Egg parasitoid	Response	Phoresy	Reference
Attractant pheromone						
Male bug pheromone	(E)-2-hexenyl (Z)-3-hexenoate	Riptortus clavatus	Ooencyrtus nezarae [Encyrtidae]	Attraction in field	Not tested	Leal et al. 1995, Mizutani et al. 1997
Male bug pheromone	(E)-2-hexenal, α-terpineol, benzyl alcohol	Podisus maculiventis	Telenomus calvus [Scelionidae]	Observations, field catches, attraction in field	Phoretic	Buschman & Whitcomb 1980, Aldrich et al. 1984, Orr et al. 1986, Aldrich 1995, Bruni et al. 2000
Beetle aggregation pheromone	4-methyl-3-heptanol, multistriatin	Scolytus multistriatus	Entedon leucogramma [Eulophidae]	Attraction in field	Not tested	Kennedy 1979, 1984
Sex pheromone						
Virgin female moth pheromone	(Z)-13-hexadecen-11-inyl acetate	Thaumetopoea pityocampa	O. pityocampae [Encyrtidae]	Attraction in field	Suggested	Battisti 1989
Virgin female moth pheromone	(Z)-11-hexadecenyl acetate, (Z)-11-hexadecenol, (Z)-11-hexadecenal, dodecyl acetate	Sesamia nonagrioides	T. busseolae	Attraction	Not tested	Colazza et al. 1997

Virgin female moth pheromone	(9Z)-16-methyl-9-heptadecenyl isobutyrate	Orvasca taiwana	T. euproctidis	Observations, field catches, attraction in field	Phoretic	Arakaki 1990, Arakaki et al. 1996, 1997, Arakaki & Wakamura 2000
Virgin female moth pheromone	(10R)-10-14-dimethylpentadecyl isobutyrate	Arna pseudoconspersa	T. euproctidis	Observations, field catches, attraction in field	Phoretic	Arakaki et al. 1995, 1997
Virgin female moth pheromone	(Z6,Z9,11S,12S)-11,12-epoxyheneicosa-6,9-diene	Orgyia postica	T. euproctidis	Observations, attraction in field	Non-phoretic	Arakaki et al. 2011
Virgin female moth pheromone	(Z)-9-tetradecene-l-ol acetate, (Z)-9-dodecene-l-ol acetate	Spodoptera frugiperda	T. remus	Attraction	Not tested	Nordlund et al. 1983
Virgin female moth pheromone	(Z)-11-hexadecenal, (Z)-11-hexadecenyl acetate	Plutella xylostella	Trichogramma chilonis (Trichogrammatidae)	Attraction	Not tested	Reddy et al. 2002
Virgin female moth pheromone	(9Z,12E)-9,12-tetra-decadienyl acetate	Ephestia spp.	T. evanescens	Attraction	Not tested	Schöller & Prozell 2002
Virgin female moth pheromone	(E)-12-tetradecenyl acetate	Ostrinia furnacalis	T. ostriniae, T. chilonis	Arrestment, attraction	Not tested	Boo & Yang 2000, Bai et al. 2004

(cont'd)

Table 5.2 (cont'd)

Type of infochemical	Chemical	Insect host	Egg parasitoid	Response	Phoresy	Reference
Virgin female moth pheromone	(E)-11-tetradecenyl acetate	Ostrinia nubilalis	T. ostriniae, T. brassicae	Attraction	Not tested	Kaiser et al. 1989, Frenoy et al. 1991, 1992, Yong et al. 2007
Virgin female moth pheromone	(Z)-11-hexadecenyl acetate	Helicoverpa assulta	T. chilonis	Attraction	Not tested	Boo & Yang 2000
Virgin female moth pheromone	(Z)-7-tetradecenal	Prays oleae	T. oleae, T. cacoeciae, T. bourarachae	Attraction	Suggested	Milonas et al. 2009a, 2009b
Virgin female moth pheromone	(E)-11-hexadecenal	Palpita unionalis	T. oleae	Attraction	Suggested	Milonas et al. 2009a
Virgin female moth pheromone	(Z)-7-hexadecenal, (Z)-9-hexadecenal, (Z)-1 1-hexadecenal	Heliothis zea	T. pretiosum	Attraction	Not tested	Lewis et al. 1982, Noldus 1988, Noldus et al. 1990, 1991b
Virgin female moth pheromone	(Z)-11-hexadecenyl acetate	Mamestra brassicae	T. pretiosum, T. evanescens	Attraction	Not tested	Noldus et al. 1991a, 1991b
Virgin female moth pheromone	(Z)-11-tetradecen-(-ol acetate	Rhopobota naevana	T. sibericum	Arrestment	Not tested	McGregor & Henderson 1998

Virgin male bug pheromone	methyl 2,6,10-trimethyltridecanoate	*Euschistus heros*	*Telenomus podisi*	Attraction	Not tested	Borges *et al.* 1998, 1999, Silva *et al.* 2006
Virgin female sawfly pheromone	(2*S*,3*R*,7*R*)-3,7-dimethyl-2-tridecanol	*Diprion pini*	*Chrysonotomyia ruforum* (Eulophidae)	Arrestment	Not tested	Hilker *et al.* 2000
Virgin female sawfly pheromone	(2*S*,3*S*,7*S*)-3,7-dimethyl-2-pentadecyl acetate	*Neodiprion sertifer*	*C. ruforum*	Arrestment	Not tested	Hilker *et al.* 2000
Anti-sex pheromone						
Mated female butterfly pheromone	benzyl cyanide	*Pieris brassicae*	*T. brassicae, T. evanescens*	Arrestment, observations, field catches	Phoretic	Fatouros *et al.* 2005b, Huigens *et al.* 2009, 2011, Fatouros & Huigens 2012
Mated female butterfly pheromone	indole, methyl salicylate	*Pieris rapae*	*T. brassicae, T. evanescens*	Arrestment, observations, field catches	Phoretic	Huigens *et al.* 2010, Fatouros & Huigens 2012
Allomone						
Bug defensive allomone	(*E*)-2-decenal, (*E*)-4-oxo-2-hexenal	*Nezara viridula*	*Trissolcus basalis* (Scelionidae)	Attraction	Not tested	Mattiacci *et al.* 1993, Laumann *et al.* 2009
Bug defensive allomone	(*E*)-4-oxo-2-hexenal, (*E*)-2-hexenal, tridecane	*Euschistus heros*	*Telenomus podisi*	Attraction	Not tested	Laumann *et al.* 2009

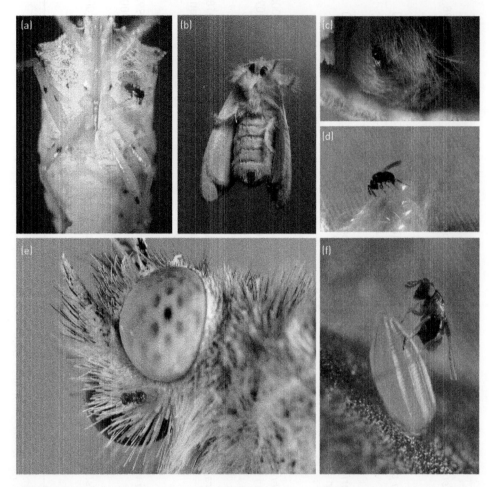

Figure 5.2 Phoretic egg parasitoids. (a) *Telenomus calvus* on a female spined soldier bug, *Podisus maculiventris* (photograph courtesy of J.R. Aldrich). (b) Female moth of *Orvasca taiwana*. (c) Female *Telenomus euproctidis* on anal tuft of an adult female moth. (d) Female *Telenomus euproctidis* parasitizing freshly laid eggs of *O. taiwana* (b–d: photographs courtesy of N. Arakaki). (e) Female *Trichogramma evanescens* on a female of the large cabbage white butterfly *Pieris brassicae*. (f) Female *T. evanescens* parasitizing an egg of the small cabbage white butterfly *Pieris rapae* (e–f: photographs courtesy of N.E. Fatouros, http://www.bugsinthepicture.com).

5.4.2 Espionage on sex pheromones

The different sex pheromones of three tussock moths from Japan, *Orvasca* (=*Euproctis*) *taiwana* (Fig. 5.2b) using (9*Z*)-16-methyl-9-heptadecenyl isobutyrate, *Arna* (=*Euproctis*) *pseudoconspersa* using (10*S*)-10,14-dimethylpentadecyl isobutyrate, and *Orgyia postica* using (6*Z*,9*Z*,11*S*,12*S*)-11,12-epoxyheneicosa-6,9-diene, were shown to lure the specialist egg parasitoid *Telenomus euproctidis* Wilcox (Arakaki 1990, Arakaki *et al.* 1995, 1996, 1997, 2011, Arakaki & Wakamura 2000). The wasps are phoretic on virgin females of *Orvasca taiwana* and *A. pseudoconspersa*. Up to one-third of the moths carried parasitoids that hide in the dense scale hairs of the long anal tuft (Fig. 5.2c). Once the moths have mated and moved to oviposition sites, the wasps dismount and start parasitizing the freshly laid eggs (Fig. 5.2d) (Arakaki 1990, Arakaki *et al.* 1995). In contrast, no parasitoids were found to be phoretic on *Orgyia postica* females. The female moths are wingless, inactive and immobile, and therefore not suitable for transportation. Moreover, the calling sites of *O. postica* are identical to their oviposition sites (Arakaki *et al.* 2011). An extensive body of literature has documented chemical espionage by different egg parasitoid families on female sex pheromones of moths (Powell 1999, Fatouros *et al.* 2008b, Colazza *et al.* 2010). These examples are an excellent starting point to demonstrate hitherto unknown cases of phoresy. However, it remains unclear how the wasps solve the spatial (in other words, the calling sites of host females are different from the oviposition sites) and the temporal (in other words, the nocturnal activity of the moths and the diurnal activity of the wasps) gaps that are apparent when using host pheromones as an 'infochemical detour' to locate host eggs (Fatouros *et al.* 2008b). Arakaki & Wakamura (2000) revealed that *Telenomus euproctidis* makes use of scale hairs from the anal tufts of female *Orvasca taiwana* moths. They indicated that trace amounts of the sex pheromone were retained on, and released from, the hairs up to 48 hours after calling and enabled the parasitoids to locate the host females. Moreover, the wasps could directly locate the egg masses by pheromone traces adsorbed by scale hairs covering the eggs after oviposition.

5.4.3 Espionage on anti-sex pheromones

To enforce female monogamy, various insects such as butterflies, bees, bugs, beetles and flies utilize anti-sex pheromones (anti-aphrodisiacs) transferred from males to females during mating (Happ 1969, Kukuk 1985, Zawistowski & Richmond 1986, Scott & Richmond 1987, Andersson *et al.* 2000, 2003, 2004, Zhang & Aldrich 2003, Kyriacou 2007, Schulz *et al.* 2008). The unattractiveness of a just-mated female may benefit both males and females: males are ensured of their sperm fertilizing the eggs and females are less subjected to harassment by other males and thus save energy and time for egg deposition (Forsberg & Wiklund 1989). In cabbage white butterflies (*Pieris* spp.), anti-aphrodisiac compounds are species-specific (Andersson *et al.* 2003). We have recently shown that there may also be costs of using these pheromones for both male and female butterflies inflicted by eavesdropping parasitoids. For a female parasitoid, an anti-aphrodisiac pheromone emitted by a mated, and thus egg-producing, female host is a more reliable indicator of host presence than a sex pheromone emitted by a virgin female host. Both the anti-aphrodisiac of the gregarious *P. brassicae* (benzyl cyanide) and of the solitary, much more abundant, *P. rapae* (methyl salicylate and indole) can lure *Trichogramma* wasps, whereupon the wasps hitch a ride on the mated female butterfly to reach the oviposition site and parasitize her freshly laid eggs (Fig. 5.2e–f) (Fatouros *et al.* 2005b, Huigens *et al.* 2009, 2010). In bioassays performed in the laboratory, inexperienced *T. brassicae* wasps preferably mounted mated (and thus egg-producing) female butterflies or virgin females treated with a synthetic anti-aphrodisiac. The wasps, moreover, preferred the odours of mated females or virgin females treated with an anti-aphrodisiac when offered against the odours of virgin females or males (Fatouros *et al.* 2005b, Huigens *et al.* 2010). However, the wasps do not respond to virgin female butterflies treated with the anti-aphrodisiac of the other *Pieris* species. This indicates that the wasps detect the anti-aphrodisiac compounds as part of a whole odour blend of mated female *Pieris* butterflies. Benzyl cyanide, methyl salicylate and indole are ubiquitous compounds that are also emitted by other organisms such as wild and cultivated cabbage plants (see Huigens *et al.* 2010 and references therein).

Females of the closely related extreme generalist *T. evanescens* exploit the anti-aphrodisiac pheromones of both *Pieris* species in a similar way, but only after learning. They do not inherently discriminate between male, virgin female and mated female butterflies. Mated female butterflies are, however, preferred over non-hosts (Huigens *et al.* 2009). However, after a single ride on a mated female butterfly leading to successful parasitism of a host egg, the wasps learn to associate the odour blend of a mated female butterfly with the reward of parasitizing the host egg. Such experienced wasps show a preference for the odours of mated female butterflies over males and virgin females and also prefer to mount them. Depending on the reward value constituted by the host eggs, *T. evanescens* wasps can form a protein-synthesis-dependent long-term memory for mated female butterfly odours (which lasts for more than 24 hours after a single rewarding hitch-hiking experience on *P. brassicae*) or a shorter-lasting memory (after a similar experience on *P. rapae*) (Huigens *et al.* 2009, 2010, Kruidhof *et al.* 2012). All in all, these results highlight that multiple parasitoid species can exploit an anti-aphrodisiac pheromone of a single host species, and that a single parasitoid species can exploit different anti-aphrodisiacs from different host species.

5.5 Coevolution between phoretic spies and hosts

In theory, when the costs inflicted on the host by natural enemies spying on them outweigh the benefits of using a pheromone, natural selection should favour mechanisms in the host to avoid the exploiters, which in turn could result in counter-adaptations in the chemical spies, and so on: in other words, a coevolutionary arms race (Ehrlich & Raven 1964, Dawkins & Krebs 1979). At the moment, it is still unknown whether phoretic parasitoids exert selective pressures that are strong enough to induce changes in their hosts' communication systems. To investigate this, one first needs to determine whether the parasitoids represent a significant mortality factor. This would probably apply to phoretic wasps of neotropical owl butterflies (*Caligo* spp.). Indeed, up to 250 individuals of the trichogrammatid *Xenufens* sp. can be found on a single male or female *C. eurilochus* butterfly, resulting in a mean parasitism rate of 78% of the butterfly's eggs (Malo 1961). This indicates that the transporting owl butterflies are tremendously attractive to the parasitoids, most probably due to their body odours. Indeed, male butterflies have a large scent organ which emits volatiles that can even be perceived by human olfaction and which are reminiscent of smoked ham (Wasserthal & Wasserthal 1977). The volatiles could also be used by the egg parasitoids to find their transporting host. Of the phoretic species that are known to eavesdrop on their hosts' pheromones, *Trichogramma* wasps can reach high parasitism rates on cabbage white butterfly eggs in nature. Mean parasitism rates of the eggs of *Pieris* butterflies laid on the wild crucifer *Brassica nigra* in the Netherlands vary around 30–40% over different years (Fatouros *et al.* 2012). *Trichogramma* wasps can also exploit *B. nigra* volatiles induced by *Pieris* egg deposition (Fatouros *et al.* 2012). Therefore, it is difficult to determine to what extent the exploitation of an anti-aphrodisiac of a mated *P. rapae* female and subsequent hitch-hiking contribute to their parasitism success.

Very little is known about the mechanisms used by hosts to avoid phoretic chemical spies. An ovipositing adult female host might protect her offspring from egg parasitoids in general by hiding them in plant tissue, laying eggs in stacks, by transferring defensive compounds to the egg, or by covering them with faeces, hairs, scales or secretions (Hinton 1981, Blum & Hilker 2002, Deas & Hunter 2011). Mechanisms that are specifically used to avoid hitch-hiking enemies are limited. Mated female *Pieris* butterflies can successfully kick off *T. brassicae* wasps when they try to climb onto their legs, which are covered with highly sensitive hairs. *P. brassicae* needs more kicks than *P. rapae* to remove the wasps (Fig. 5.3). The underlying mechanism for this difference remains unknown. To avoid being kicked off, wasps tend to climb onto the butterflies' wings instead. Most phoretic parasitoid species were indeed found on the wings or abdomen (Fatouros *et al.* 2005b, Fatouros & Huigens 2012).

Hosts might not always be attractive to the hitch-hikers. Although it is not a direct mechanism aimed at avoiding phoretic *Trichogramma* wasps, female *Pieris* butterflies become less attractive, and eventually unattractive, to the wasps over time after mating due to a reduction in anti-aphrodisiac titre (Huigens *et al.* 2011). The titre is depleted by the frequent display of a so-called 'mate-refusal posture' which indicates a female's unwillingness to mate (Obara 1964, Huigens *et al.* 2011). Hosts might also shift pheromone release to times when egg parasitoids are not active. However, a few species of moths, of which the females mostly release a sex pheromone at night, can also be exploited by diurnal, phoretic parasitoids that spy on the pheromone (Arakaki *et al.* 1995, 1996, 1997, Fatouros & Huigens 2012).

It seems that the best way to investigate the extent to which an espionage-and-ride strategy of phoretic wasps selects for changes in pheromone use by their hosts is to correlate geographical variation in pheromone composition with local parasitoid selection pressure: in other words, to egg parasitism rates, the prevalence of phoretic wasps on adult hosts, and behavioural responses of phoretic wasps to host pheromones in laboratory assays. One interesting study by Arakaki *et al.* (1997) on two regional strains of the rather specialized phoretic *Telenomus euproctidis* shows that female wasps strongly prefer the sex pheromone of the local tussock moth species occurring at their location, either *A. pseudoconspersa* or *Orvasca taiwana*. This suggests that the wasps have locally adapted to their host's pheromone and do not (yet) represent sufficient selective pressure for the host to change its pheromone. One example of a change in pheromone composition induced by non-phoretic chemical spies includes geographical variation in an aggregation pheromone blend of *Ips pini* bark beetles that is related to local predation and response by generalist predators; for instance, clerid beetles. Local predators show a stronger attraction to the pheromone blend of the allopatric *I. pini* population than to that of the local conspecific bark beetle population, suggesting that *I. pini* has evolved blends to minimize local predator pressure (Raffa & Dahlsten 1995, Aukema *et al.* 2000). In contrast, a highly specialized parasitoid shows a stronger attraction to the local than to the distant *I. pini* population (Raffa & Dahlsten 1995). Specialist parasitoids are expected to evolve specialized abilities to detect and follow changes in local host pheromones because their reproduction depends on finding the limited number of host species in which their offspring develop (Thompson 1994). Counter-adaptations to a change in host pheromone composition should be delayed in generalist parasitoids by the availability of alternative hosts. Obviously, we still need a better understanding of potential coevolutionary arms races between phoretic chemical spies and their hosts.

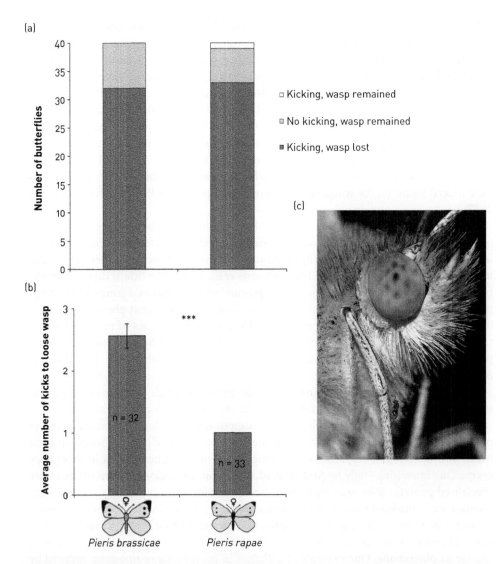

(a)

(b)

(c)

Figure 5.3 Kicking behaviour of mated females of the large cabbage white butterfly, *Pieris brassicae*, and the small cabbage white butterfly, *P. rapae*, when a female *Trichogramma brassicae* wasp climbs onto its leg. The experiment was carried out in a plastic container (Huigens *et al.* 2009) with one wasp and one butterfly under standard laboratory conditions of 22 ± 1°C and 60% RH (for more details see Fatouros *et al.* 2005b, Huigens *et al.* 2010). The butterflies were placed in the container after cooling for about 10 minutes in a refrigerator (4°C) to decrease their mobility. (a) Number of *P. brassicae* and *P. rapae* females that either kicked or did not kick when a wasp mounted its leg. In most cases the wasp was successfully kicked off. (b) Mean (±SE) number of kicks a single *P. brassicae* or *P. rapae* female needed to kick off a wasp. Means were compared using a non-parametric Mann–Whitney U test, ***$P < 0.001$. (c) Female wasp on the leg of a mated *P. brassicae* female (photograph courtesy of N.E. Fatouros, http://www.bugsinthepicture.com).

5.6 Biological control

In addition to the intrinsic interest in intricate phoretic relationships between egg parasitoids and their hosts, knowledge of whether an egg parasitoid strongly relies on phoresy to find host eggs can be important in order to determine the correct timing and type of parasitoid release in a biological control programme against a given pest species. Egg parasitoids are frequently used in biological control as they attack a host stage before actual crop damage occurs (Wajnberg & Hassan 1994). If an egg parasitoid relies heavily on phoresy to find the eggs of a target pest species, it would probably be essential to release such parasitoids before most adult pests start laying their eggs. Moreover, it might be recommended to release adult hosts that are infected with phoretic parasitoids as well as releasing the parasitoids alone (Clausen 1976). It can also be important to understand which chemical cues are used by phoretic egg parasitoids to find transporting hosts, in order to increase their efficacy as biological control agents. The use of dispensers with sex-, aggregation- or anti-aphrodisiac pheromones might, for example, disturb mating of a pest insect, and help to attract wasps from a distance to a crop field. On the other hand, this may also reduce efficient host finding by the parasitoids within the crop field.

To the best of our knowledge, it has never been conclusively tested whether the tiny egg parasitoid species that are commercially available (see van Lenteren 2012 for a list of species) strongly rely on phoresy to find the eggs of a certain target pest. Only for *T. brassicae* and *T. evanescens* is it known that they can occasionally hitch-hike with adults of host species against which they have never been released in a biological control programme, such as *Pieris* butterflies, the painted lady *Vanessa cardui*, and the moth *Xestia c-nigrum* (Fatouros & Huigens 2012). However, we do not expect that generalist *Trichogramma* wasps, which are the most frequently used egg parasitoids in biological control worldwide, heavily rely on transporting hosts to find their host eggs. They can parasitize 'older' host eggs and also use induced plant cues several days after host egg deposition. Other, more specialized, biological control agents that need to be in time to parasitize freshly laid host eggs, such as certain *Telenomus* species, are more likely to rely heavily on hitch-hiking with adult hosts.

5.7 Future perspectives

Despite almost a century of research on phoretic parasitoids whose adult females ride on transporting hosts, we are only just beginning to understand the chemical intricacies of their interactions with their hosts. Over the last three decades, a limited number of studies have shown that four tiny phoretic egg parasitoid species use complicated ways to exploit various chemical communication systems of transporting hosts in order to reach their egg-laying sites. Is chemical espionage on host pheromones indeed a common way of finding a transporting host vehicle among phoretic egg parasitoids, and can such espionage induce changes in host pheromones? Ideally, this should not only be investigated in behavioural assays in the laboratory but also in (semi-)field studies. We expect chemical espionage on host communication systems to be widespread among phoretic egg parasitoids, as it is among egg parasitoids in general (for a review, see Fatouros *et al.* 2008b). However, inexperienced *T. evanescens* wasps do not discriminate between climbing onto male, virgin and mated female *Pieris* butterflies, but prefer them over non-hosts (desert locusts) in a two-choice setup. This suggests that phoretic egg parasitoids might also use other more general chemical and/or visual cues to discriminate potential hosts from non-hosts from a short distance (Huigens *et al.* 2009).

Besides the cues that a female parasitoid uses to locate a transporting host, we still lack knowledge about the cues used to descend from an egg-laying host to parasitize freshly laid host eggs. Moreover, virtually no information exists about where phoretic parasitoids and hosts meet and how they locate those meeting sites. Flowers were shown to serve as meeting points for crab spiders, mites and even pathogens to transfer onto flower visitors (for example, bumble bees and butterflies) during their nectar feeding (Treat 1969, Morse 1986, Durrer & Schmid-Hempel 1994, Schwarz & Huck 1997). For most parasitoid species, sugars obtained from floral nectar are an essential energy source to sustain many physiological processes. Flower location in egg parasitoids is mediated by both olfactory and visual stimuli (Romeis *et al.* 2005). Thus, flowers may be used for horizontal transfer, because many hosts and parasitoids use floral nectar. On the practical side, the host is stationary during nectar feeding, making it easier for the parasitoid to mount.

After discovering that a given phoretic parasitoid uses chemical cues to find a host or a meeting site (either innately or through learning), the next step would be to investigate how it processes such cues. A huge body of literature on traditional model organisms indicates that the organization of insect brains, and mechanisms underlying chemoreception and learning in the brain, have striking similarities with those in higher animals (Strausfeld & Hildebrand 1999, Dubnau 2003, Strausfeld *et al.* 2009). Only a few studies have investigated the brain tissues (for example, antennal lobes, mushroom bodies, ventral unpaired median neurons) involved in olfactory learning in non-phoretic parasitoid wasps of caterpillars and fly pupae (Smid *et al.* 2003, Bleeker *et al.* 2006, Hoedjes *et al.* 2011). We believe that investigating the mechanisms underlying olfaction and learning in parasitoids is a highly exciting research field, with phoretic egg parasitoid wasps being of particular interest because of their minute size. How do 0.5-mm-long phoretic *T. evanescens* wasps, for example, learn to respond to the odours of mated female *Pieris* butterflies after a rewarding hitch-hiking experience on such females, with so little brain tissue? Modern advanced microscopic and molecular techniques offer excellent means to investigate such an intriguing research question. Moreover, more studies on the ecological relevance of learning for phoretic parasitoids are needed.

In this chapter, we focused on the chemical ecology of phoretic parasitoids of insect eggs, simply because almost no information exists about phoresy in parasitoids of other host stages. Phoresy can, however, also be adaptive for species that parasitize another inconspicuous and immobile host stage, the pupa (Fig. 5.1). Actually, it might be adaptive in any given non-egg parasitoid species of which the females are much smaller than a potential transporting host and/or are rather immobile. Clearly, more investigation on the prevalence of phoresy among parasitoids other than those that attack host eggs is needed, as well as research on their chemical ecology.

Acknowledgements

We thank Foteini Pashalidou for her help with the experiments on female cabbage white butterflies that are trying to remove *Trichogramma brassicae* wasps from their legs (Fig. 5.3). Marcel Dicke is acknowledged for his constructive reading of an earlier version of this chapter. We also thank the NWO/ALW for funding (VENI grants 863.05.020 and 863.09.002, and the open competition grant 820.01.012).

References

Aldrich, J.R. (1995) Chemical communication in the true bugs and parasitoid exploitation. In: Cardé, R.T. and Bell, W.J. (eds) *Chemical Ecology of Insects II*. Chapman and Hall, New York, pp. 318–63.

Aldrich, J.R., Kochansky, J.P. and Abrams, C.B. (1984) Attractant for beneficial insect and its parasitoids: pheromone of the predatory spined soldier bug, *Podisus maculiventris* (Hemiptera: Pentatomidae). *Environmental Entomology* **13**: 1031–6.

Andersson, J., Borg-Karlson, A.K. and Wiklund, C. (2000) Sexual cooperation and conflict in butterflies: a male-transferred anti-aphrodisiac reduces harassment of recently mated females. *Proceedings of the Royal Society: Biological Sciences* **267**: 1271–5.

Andersson, J., Borg-Karlson, A.K. and Wiklund, C. (2003) Antiaphrodisiacs in pierid butterflies: a theme with variation! *Journal of Chemical Ecology* **29**: 1489–99.

Andersson, J., Borg-Karlson, A.K. and Wiklund, C. (2004) Sexual conflict and anti-aphrodisiac titre in a polyandrous butterfly: male ejaculate tailoring and absence of female control. *Proceedings of the Royal Society: Biological Sciences* **271**: 1765–70.

Arakaki, N. (1990) Phoresy of *Telenomus* sp. (Scelionidae: Hymenoptera), an egg parasitoid of the tussock moth *Euproctis taiwana*. *Journal of Ethology* **8**: 1–3.

Arakaki, N. and Wakamura, S. (2000) Bridge in time and space for an egg parasitoid: kairomonal use of trace amount of sex pheromone adsorbed on egg mass scale hair of the tussock moth, *Euproctis taiwana* (Shiraki) (Lepidotera: Lymantriidae), by an egg parasitoid, *Telenomus euproctidis* Wilcox (Hymenoptera: Scelionidae), for host location. *Entomological Science* **3**: 25–31.

Arakaki, N., Wakamura, S. and Yasuda, T. (1995) Phoresy by an egg parasitoid, *Telenomus euproctidis* (Hymenoptera: Scelionidae), on the tea tussock moth, *Euproctis pseudoconspersa* (Lepidoptera: Lymantriidae). *Applied Entomology and Zoology* **30**: 602–3.

Arakaki, N., Wakamura, S. and Yasuda, T. (1996) Phoretic egg parasitoid, *Telenomus euproctidis* (Hymenoptera: Scelionidae), uses sex pheromone of tussock moth *Euproctis taiwana* (Lepidoptera: Lymantriidae) as a kairomone. *Journal of Chemical Ecology* **22**: 1079–85.

Arakaki, N., Wakamura, S. Yasuda, T. and Yamagishi, K. (1997) Two regional strains of a phoretic egg parasitoid, *Telenomus euproctidis* (Hymenoptera: Scelionidae), that use different sex pheromones of two allopatric tussock moth species as kairomones. *Journal of Chemical Ecology* **23**: 153–61.

Arakaki, N., Yamazawa, H. and Wakamura, S. (2011) The egg parasitoid *Telenomus euproctidis* (Hymenoptera: Scelionidae) uses sex pheromone released by immobile female tussock moth *Orgyia postica* (Lepidoptera: Lymantriidae) as kairomone. *Applied Entomology and Zoology* **46**: 195–200.

Aukema, B.H., Dahlsten, D.L. and Raffa, K.F. (2000) Exploiting behavioral disparities among predators and prey to selectively remove pests: maximizing the ratio of bark beetles to predators removed during semiochemically based trap-out. *Environmental Entomology* **29**: 651–60.

Austin, A.D., Johnson, N.F. and Dowton, M. (2005) Systematics, evolution, and biology of scelionid and platygastrid wasps. *Annual Review of Entomology* **50**: 553–82.

Bai, S.X., Wang, Z.Y., He, K.L. and Zhou, D.R. (2004) Olfactory responses of *Trichogramma ostriniae* Pang et Chen to kairomones from eggs and different stages of adult females of *Ostrinia furnacalis* (Guenee). *Acta Entomologica Sinica* **47**: 48–54.

Battisti, A. (1989) Field studies on the behaviour of two egg parasitoids of the pine processionary moth *Thaumetopoea pityocampa*. *Entomophaga* **34**: 29–38.

Bin, F. (1985) Phoresy in an egg parasitoid: *Mantibaria seefelderiana* (De Stef.-Per.) (Hym. Scelionidae). *Atti XIV Congresso Nazionale Italiano di Entomologia*: 901–2.

Binns, E.S. (1982) Phoresy as migration: some functional aspects of phoresy in mites. *Biological Reviews* **57**: 571–620.

Bleeker, M.A.K., Smid, H.M., Steidle, J.L.M., Kruidhof, H.M., van Loon, J.J.A. and Vet L.E.M. (2006) Differences in memory dynamics between two closely related parasitoid wasp species. *Animal Behaviour* **71**: 1343–50.

Blum, M.S. and Hilker, M. (2002) Chemical protection of insect eggs. In: Hilker, M. and Meiners, T. (eds) *Chemoecology of Insect Eggs and Egg Deposition*. Blackwell Publishing Ltd., Oxford, pp. 61–90.

Boo, K.S. and Yang, J.P. (2000) Kairomones used by *Trichogramma chilonis* to find *Helicoverpa assulta* eggs. *Journal of Chemical Ecology* **26**: 359–75.

Borges, M., Costa, M.L.M., Sujii, E.R., Cavalcanti, M. das G., Redígolo, G.F., Resck, I.S. and Vilela, E.F. (1999) Semiochemical and physical stimuli involved in host recognition by *Telenomus podisi* (Hymenoptera: Scelionidae) toward *Euschistus heros* (Heteroptera: Pentatomidae). *Physiological Entomology* **24**: 227–33.

Borges, M., Schmidt, F.G.V., Sujii, E.R., Medeiros, M.A., Mori, K., Zarbin, P.H.G. and Ferreira, J.T.B. (1998) Field responses of stink bugs to the natural and synthetic pheromone of the Neotropical brown stink bug, *Euschistus heros* (Heteroptera: Pentatomidae). *Physiological Entomology* **23**: 202–7.

Bruni, R., Sant'Ana, J., Aldrich, J.R. and Bin, F. (2000) Influence of host pheromone on egg parasitism by scelionid wasps: comparison of phoretic and nonphoretic parasitoids. *Journal of Insect Behavior* **12**: 165–73.

Buschman, L.L. and Whitcomb, W.H. (1980) Parasites of *Nezara viridula* (Hemiptera: Pentatomidae) and other Hemiptera in Florida. *Florida Entomologist* **63**: 154–62.

Cerutti, F. and Bigler, F. (1995) Quality assessment of *Trichogramma brassicae* in the laboratory. *Entomologia Experimentalis et Applicata* **75**: 19–26.

Clausen, C.P. (1976) Phoresy among entomophagous insects. *Annual Review of Entomology* **21**: 343–68.

Colazza, S., Peri, E., Salerno, G. and Conti, E. (2010) Host searching by egg parasitoids: exploitation of host chemical cues. In: Consôli, F.L., Parra, J.R.P. and Zucchi, R. (eds) *Egg Parasitoids in Agroecosystems with Emphasis on Trichogramma*. Springer, Dordrecht, pp. 97–147.

Colazza, S., Rosi, C.M. and Clemente, A. (1997) Response of egg parasitoid *Telenomus busseolae* to sex pheromone of *Sesamia nonagrioides*. *Journal of Chemical Ecology* **23**: 2437–44.

Couturier, A. (1941) Nouvelles observations sur *Rielia manticida* Kief., hyménoptère parasite de la mante religieuse. II. Comportement de l'insecte parfait. *Revue de Zoologie Agricole et Appliquée* **40**: 49–62.

Dawkins, R. and Krebs, J.R. (1979) Arms races between and within species. *Proceedings of the Royal Society of London, Series B: Biological Sciences* **205**: 489–511.

Deas, J.B. and Hunter, M.S. (2011) Mothers modify eggs into shields to protect offspring from parasitism. *Proceedings of the Royal Society B: Biological Sciences* **279**: 847–53.

Dicke, M. and Sabelis, M.W. (1992) Costs and benefits of chemical information conveyance: proximate and ultimate factors. In: Roitberg, B.D. and Isman, M.B. (eds) *Insect Chemical Ecology: An Evolutionary Approach*. Chapman & Hall, New York, pp. 122–55.

Doyon, J. and Boivin, G. (2005) The effect of development time on the fitness of female *Trichogramma evanescens*. *Journal of Insect Science* **5**: 1–5.

Dubnau, J. (2003) Neurogenetic dissection of conditioned behavior: evolution by analogy or homology? *Journal of Neurogenetics* **17**: 295–326.

Durrer, S. and Schmid-Hempel, P. (1994) Shared use of flowers leads to horizontal pathogen transmission. *Proceedings of the Royal Society B: Biological Sciences* **258**: 299–302.

Ehrlich, P.R. and Raven, P.H. (1964) Butterflies and plants: a study in coevolution. *Evolution* **18**: 586–608.

Fatouros, N.E. and Huigens, M.E. (2012) Phoresy in the field: natural occurrence of *Trichogramma* egg parasitoids on butterflies and moths. *BioControl* **57**: 493–502.

Fatouros, N.E., Broekgaarden, C., Bukovinszkine'Kiss, G., van Loon, J.J.A., Mumm, R., Huigens, M.E., Dicke, M. and Hilker, M. (2008a) Male-derived butterfly anti-aphrodisiac mediates induced indirect plant defense. *Proceedings of the National Academy of Sciences USA* **105**: 10033–8.

Fatouros, N.E., Bukovinszkine'Kiss, G., Kalkers, L.A., Soler Gamborena, R., Dicke, M. and Hilker, M. (2005a) Oviposition-induced plant cues: do they arrest *Trichogramma* wasps during host location? *Entomologia Experimentalis et Applicata* **115**: 207–15.

Fatouros, N.E., Dicke, M., Mumm, R., Meiners, T. and Hilker, M. (2008b) Foraging behavior of egg parasitoids exploiting chemical information. *Behavioral Ecology* **19**: 677–89.

Fatouros, N.E., Huigens, M.E., van Loon, J.J.A., Dicke, M. and Hilker, M. (2005b) Chemical communication: butterfly anti-aphrodisiac lures parasitic wasps. *Nature* **433**: 704.

Fatouros, N.E., Lucas-Barbosa, D., Weldegergis, B.T., Pashalidou, F.G., van Loon, J.J.A., Dicke, M., Harvey, J.A., Gols, R. and Huigens, M.E. (2012) Plant volatiles induced by herbivore egg deposition affect insects of different trophic levels. *PLoS ONE* **7**(8): e43607.

Fatouros, N.E., Pashalidou, F.G., Aponte Cordero, W.V., van Loon, J.J.A., Mumm, R., Dicke, M., Hilker, M. and Huigens, M.E. (2009) Anti-aphrodisiac compounds of male butterflies increase the risk of egg parasitoid attack by inducing plant synomone production. *Journal of Chemical Ecology* **35**: 1373–81.

Ferrière, C. (1926) La phorésie chez les insectes. *Mitteilungen der Schweizerischen Entomologischen Gesellschaft* **13**: 489–96.

Field, S.A. and Austin, A.D. (1994) Anatomy and mechanics of the telescopic ovipositor system of *Scelio latreille* (Hymenoptera: Scelionidae) and related genera. *International Journal of Insect Morphology and Embryology* **23**: 135–58.

Forsberg, J. and Wiklund, C. (1989) Mating in the afternoon: time-saving in courtship and remating by females of a polyandrous butterfly *Pieris napi* L. *Behavioral Ecology and Sociobiology* **25**: 349–56.

Frenoy, C., Durier, C., and Hawlitzky, N. (1992) Effect of kairomones from egg and female adult stages of *Ostrinia nubilalis* (Hübner) (Lepidoptera, Pyralidae) on *Trichogramma brassicae* Bezdenko (Hymenoptera, Trichogrammatidae) female kinesis. *Journal of Chemical Ecology* **18**: 761–73.

Frenoy, C., Farine, J.P., Hawlitzky, N. and Durier, C. (1991) Role of kairomones in the relations between *Ostrinia nubilalis* Hübner (Lep., Pyralidae) and *Trichogramma brassicae* Bezdenko (Hym, Trichogrammatidae). *Redia* **74**: 143–51.

Godfray, H.C.J. (1994) *Parasitoids: Behavioral and Evolutionary Ecology*. Princeton University Press, Princeton, NJ.

Happ, G.M. (1969) Multiple sex pheromones of the mealworm beetle, *Tenebrio molitor* L. *Nature* **222**: 180–1.

Hase, A. (1925) Beiträge zur Lebensgeschichte der Schlupfwespe *Trichogramma evanescens* Westwood. *Arbeiten aus der Biologischen Reichsanstalt für Land- und Forstwirtschaft Berlin-Dahlem* **14**: 171–224.

Haynes, K.F. and Yeargan, K.V. (1999) Exploitation of intraspecific communication systems: illicit signalers and receivers. *Annals of the Entomological Society of America* **92**: 960–70.

Hilker, M., Blaeske, V., Kobs, C. and Dippel, C. (2000) Kairomonal effects of sawfly sex pheromones on egg parasitoids. *Journal of Chemical Ecology* **26**: 2591–601.

Hinton, H.E. (1981) *Biology of Insect Eggs*. Pergamon Press, Oxford.

Hoedjes, K.M., Kruidhof, H.M., Huigens, M.E., Dicke, M., Vet, L.E.M. and Smid, H.M. (2011) Natural variation in learning rate and memory dynamics in parasitoid wasps: opportunities for converging ecology and neuroscience. *Proceedings of the Royal Society B: Biological Sciences* **278**: 889–97.

Howard, L.O. (1927) Concerning phoresy in insects. *Entomological News* **38**: 145–7.

Huigens, M., de Swart, E. and Mumm, R. (2011) Risk of egg parasitoid attraction depends on anti-aphrodisiac titre in the large cabbage white butterfly *Pieris brassicae*. *Journal of Chemical Ecology* **37**: 364–7.

Huigens, M.E., Pashalidou, F.G., Qian, M.H., Bukovinszky, T., Smid, H.M., van Loon, J.J.A., Dicke, M. and Fatouros, N.E. (2009) Hitch-hiking parasitic wasp learns to exploit butterfly antiaphro-disiac. *Proceedings of the National Academy of Sciences USA* **106**: 820–5.

Huigens, M.E., Woelke, J.B., Pashalidou, F.G., Bukovinszky, T., Smid, H.M. and Fatouros, N.E. (2010) Chemical espionage on species-specific butterfly anti-aphrodisiacs by hitchhiking *Trichogramma* wasps. *Behavioral Ecology* **21**: 470–8.

Jervis, M.A., Ellers, J. and Harvey, J.A. (2008) Resource acquisition, allocation, and utilization in parasitoid reproductive strategies. *Annual Review of Entomology* **53**: 361–85.

Kaiser, L., Pham-Delegue, M.H., Bakchine, E. and Masson, C. (1989) Olfactory responses of *Tricho-gramma maidis* Pint. et Voeg.: effects of chemical cues and behavioral plasticity. *Journal of Insect Behavior* **2**: 701–12.

Keller, M.A., Lewis, W.J. and Stinner, R.E. (1985) Biological and practical significance of movement by *Trichogramma* species: a review. *Southwestern Entomologist* **8**: 138–55.

Kennedy, B.H. (1979) The effect of multilure on parasites of the European elm bark beetle, *Scolytus multistriatus*. *Bulletin of the Entomological Society of America* **25**: 116–18.

Kennedy, B.H. (1984) Effect of multilure and its components on parasites of *Scolytus multistriatus* (Coleoptera: Scolytidae). *Journal of Chemical Ecology* **10**: 373–85.

Kruidhof, H.M., Pashalidou, F.G., Fatouros, N.E., Figueroa, I.A., Vet, L.E.M., Smid, H.M. and Huigens, M.E. (2012) Reward value determines memory consolidation in parasitic wasps. *PLoS ONE* **7**(8): e39615.

Kukuk, P. (1985) Evidence for an antiaphrodisiac in the sweat bee *Lasioglossum zephyrum* (Dialictus). *Science* **227**: 656–7.

Kyriacou, C.P. (2007) Behavioural genetics: sex, flies and acetate. *Nature* **446**: 502–4.

Laumann, R.A., Aquino, M.F.S., Moraes, M.C.B., Pareja, M. and Borges, M. (2009) Response of the egg parasitoids *Trissolcus basalis* and *Telenomus podisi* to compounds from defensive secretions of stink bugs. *Journal of Chemical Ecology* **35**: 8–19.

Leal, W.S., Higushi, H., Mizutani, N., Nakamori, H., Kadosawa, T. and Ono, M. (1995) Multifunc-tional communication in *Riptortus clavatus* (Heteroptera: Alydidae): conspecific nymphs and egg parasitoid *Ooencyrtus nezarae* use the same adult attractant pheromone as chemical cue. *Journal of Chemical Ecology* **211**: 973–85.

Lewis, W.J., Nordlund, D.A., Gueldner, R.C., Teal, P.E.A. and Tumlinson, J.H. (1982) Kairomones and their use for management of entomophagous insects. XIII. Kairomonal activity for *Trichogramma* spp. of abdominal tips, excretion, and a synthetic sex pheromone blend of *Heliothis zea* (Boddie) moths. *Journal of Chemical Ecology* **8**: 1323–31.

Malo, F. (1961) Phoresy in *Xenufens* (Hymenoptera: Trichogrammatidae), a parasite of *Caligo eurilo-chus* (Lepidoptera: Nymphalidae). *Journal of Economic Entomology* **54**: 465–6.

Mattiacci, L., Vinson, S.B. and Williams, H.J. (1993) A long-range attractant kairomone for egg para-sitoid *Trissolcus basalis*, isolated from defensive secretion of its host, *Nezara viridula*. *Journal of Chemical Ecology* **19**: 1167–81.

McGregor, R. and Henderson, D. (1998) The influence of oviposition experience on response to host pheromone in *Trichogramma sibericum* (Hymenoptera: Trichogrammatidae). *Journal of Insect Behavior* **11**: 621–32.

Milonas, P.G., Martinou, A.F., Kontodimas, D.C., Karamaouna, F. and Konstantopoulou, M.A. (2009b) Attraction of different *Trichogramma* species to *Prays oleae* sex pheromone. *Annals of the Entomological Society of America* **102**: 1145–50.

Milonas, P., Mazomenos, B.E. and Konstantopoulou, M.A. (2009a) Kairomonal effect of sex pheromone components of two lepidopteran olive pests on *Trichogramma* wasps. *Insect Science* **16**: 131–6.

Mizutani, N., Wada, T., Higuchi, H., Ono, M. and Leal, S.W. (1997) A component of a synthetic aggregation pheromone of *Riptortus clavatus* (Thunberg) (Heteroptera: Alydidae), that attracts an egg parasitoid, *Ooencyrtus nezarae* Ishii (Hymenoptera: Encyrtidae). *Applied Entomology and Zoology* **32**: 504–7.

Morse, D.H. (1986) Predatory risk to insects foraging at flowers. *Oikos* **46**: 223–8.

Noldus, L.P.J.J. (1988) Response of the egg parasitoid *Trichogramma pretiosum* to the sex pheromone of its host *Heliothis zea*. *Entomologia Experimentalis et Applicata* **48**: 293–300.

Noldus, L.P.J.J., Lewis, W.J. and Tumlinson, J.H. (1990) Beneficial arthropod behavior mediated by airborne semiochemicals. IX. Differential response of *Trichogramma pretiosum*, an egg parasitoid of *Heliothis zea*, to various olfactory cues. *Journal of Chemical Ecology* **16**: 3531–44.

Noldus, L.P.J.J., Potting, R.P.J. and Barendregt, H.E. (1991a) Moth sex pheromone adsorption to leaf surface: bridge in time for chemical spies. *Physiological Entomology* **16**: 329–344.

Noldus, L.P.J.J., van Lenteren, J.C. and Lewis, W.J. (1991b) How *Trichogramma* parasitoids use moth sex pheromones as kairomones: orientation behaviour in a wind tunnel. *Physiological Entomology* **16**: 313–27.

Nordlund, D.A., Lewis, W.J. and Gueldner, R.C. (1983) Kairomones and their use for management of entomophagous insects. XIV. Response of *Telenomus remus* to abdominal tips of *Spodoptera frugiperda*, (Z)-9-tetradecene-1-ol acetate and (Z)-9-dodecene-1-ol acetate. *Journal of Chemical Ecology* **9**: 695–701.

Obara, Y. (1964) Mating behaviour of the cabbage white, *Pieris rapae crucivora*. II. The 'mate-refusal posture' of the female. *Zoological Magazine* **73**: 175–8.

Orr, D.B., Russin, J.S. and Boethel, D.J. (1986) Reproductive biology and behavior of *Telenomus calvus* (Hymenoptera: Scelionidae), a phoretic egg parasitoid of *Podisus maculiventris* (Hemiptera: Pentatomidae). *Canadian Entomologist* **118**: 1063–72.

Pak, G.A. (1986) Behavioural variations among strains of *Trichogramma* spp.: a review of the literature on host-age selection. *Journal of Applied Entomology* **101**: 55–64.

Pashalidou, F.G., Huigens, M.E., Dicke, M. and Fatouros, N.E. (2010) The use of oviposition-induced plant cues by *Trichogramma* egg parasitoids. *Ecological Entomology* **35**: 748–53.

Pinto, J.D. (1994) A taxonomic study of *Brachista* (Hymenoptera: Trichogrammatidae) with a description of two new species phoretic on robberflies of the genus *Efferia* (Diptera: Asilidae). *Proceedings of the Entomological Society of Washington* **96**: 120–32.

Powell, W. (1999) Parasitoid hosts. In: Hardie, J. and Minks, A.K. (eds) *Pheromones of Non-Lepidopteran Insects Associated with Agricultural Plants*. CABI Publishing, Wallingford, UK, pp. 405–27.

Querino, R.B., Zucchi, R.A. and Pinto, J.D. (2010) Systematics of the Trichogrammatidae (Hymenoptera: Chalcidoidea) with a focus on the genera attacking Lepidoptera. In: Consôli, F.L., Parra, J.R.P. and Zucchi, R. (eds) *Egg Parasitoids in Agroecosystems with Emphasis on Trichogramma*. Springer, Dordrecht, pp. 191–218.

Quicke, D.L.J. (1997) *Parasitic Wasps*. Chapman & Hall, London.

Raffa, K.F. and Dahlsten, D.L. (1995) Differential responses among natural enemies and prey to bark beetle pheromones. *Oecologia* **102**: 17–23.

Reddy, G.V.P., Holopainen, J.K. and Guerrero, A. (2002) Olfactory responses of *Plutella xylostella* natural enemies to host pheromone, larval frass, and green leaf cabbage volatiles. *Journal of Chemical Ecology* **28**: 131–43.

Romeis, J., Babendreier, D., Wäckers, F.L. and Shanower, T.G. (2005) Habitat and plant specificity of *Trichogramma* egg parasitoids: underlying mechanisms and implications. *Basic and Applied Ecology* **6**: 215–36.

Saul-Gershenz, L.S. and Millar, J.G. (2006) Phoretic nest parasites use sexual deception to obtain transport to their host's nest. *Proceedings of the National Academy of Sciences USA* **103**: 14039–44.

Schöller, M. and Prozell, S. (2002) Response of *Trichogramma evanescens* to the main sex pheromone component of *Ephestia* spp. and *Plodia interpunctella*, (Z,E)-9,12-tetra-decadenyl acetate (ZETA). *Journal of Stored Products Research* **38**: 177–84.

Schulz, S., Estrada, C., Yildizhan, S., Boppré, M. and Gilbert, L. (2008) An antiaphrodisiac in *Heliconius melpomene* butterflies. *Journal of Chemical Ecology* **34**: 82–93.

Schwarz, H.H. and Huck, K. (1997) Phoretic mites use flowers to transfer between foraging bumblebees. *Insectes Sociaux* **44**: 303–10.

Scott, D. and Richmond, R.C. (1987) Evidence against an antiaphrodisiac role for *cis*-vaccenyl acetate in *Drosophila melanogaster*. *Journal of Insect Physiology* **33**: 363–9.

Silva, C.C., Moraes, M.C.B., Laumann, R.A. and Borges, M. (2006) Sensory response of the egg parasitoid *Telenomus podisi* to stimuli from the bug *Euschistus heros*. *Pesquisa Agropecuaria Brasileira* **41**: 1093–8.

Smid, H.M., Bleeker, M.A.K., van Loon, J.J.A. and Vet, L.E.M. (2003) Three-dimensional organization of the glomeruli in the antennal lobe of the parasitoid wasps *Cotesia glomerata* and *C. rubecula*. *Cell and Tissue Research* **312**: 237–48.

Smith, S.M. (1996) Biological control with *Trichogramma*: advances, successes, and potential of their use. *Annual Review of Entomology* **41**: 375–406.

Stowe, M.K., Turlings, T.C.J., Loughrin, J.H., Lewis, W.J. and Tumlinson, J.H. (1995) The chemistry of eavesdropping, alarm, and deceit. *Proceedings of the National Academy of Sciences USA* **92**: 23–8.

Strausfeld, N.J. and Hildebrand, J.G. (1999) Olfactory systems: common design, uncommon origins? *Current Opinion in Neurobiology* **9**: 634–9.

Strausfeld, N.J., Sinakevitch, I., Brown, S.M. and Farris, S.M. (2009) Ground plan of the insect mushroom body: functional and evolutionary implications. *Journal of Comparative Neurology* **513**: 265–91.

Suverkropp, B.P., Bigler, F. and van Lenteren, J.C. (2009) Dispersal behaviour of *Trichogramma brassicae* in maize fields. *Bulletin of Insectology* **62**: 113–20.

Takagi, M. (1986) The reproductive strategy of the gregarious parasitoid, *Pteromalus puparum* (Hymenoptera: Pteromalidae). 2. Host size discrimination and regulation of the number and sex ratio of progeny in a single host. *Oecologia* **70**: 321–5.

Thompson, J.N. (1994) *The Coevolutionary Process*. University of Chicago Press, Chicago and London.

Treat, A.E. (1969) Behavioral aspects of the association of mites with noctuid moths. In: *2nd International Congress of Acarology*, Budapest, pp. 275–86.

van Lenteren, J.C. (2012) The state of commercial augmentative biological control: plenty of natural enemies, but a frustrating lack of uptake. *BioControl* **57**: 1–20.

Vet, L.E.M. and Dicke, M. (1992) Ecology of infochemical use by natural enemies in a tritrophic context. *Annual Review of Entomology* **37**: 141–72.

Vet, L.E.M., Lewis, W.J. and Cardé, R.T. (1995) Parasitoid foraging and learning. In: Cardé, R.T. and Bell, W.J. (eds) *Chemical Ecology of Insects*. Chapman & Hall, New York, pp. 65–101.

Vinson, S.B. (1984) How parasitoids locate their hosts: a case of insect espionage. In: Lewis, T. (ed) *Insect Communication*. Academic Press, London, pp. 325–48.

Vinson, S.B. (1985) The behavior of parasitoids. In: Kerkut, G.A. and Gilbert, L.I. (eds) *Comprehensive Insect Physiology, Biochemistry and Pharmacology*. Pergamon Press, New York, pp. 417–69.

Vinson, S.B. (2010) Nutritional ecology of insect egg parasitoids. In: Consôli, F.L., Parra, J.R.P. and Zucchi, R. (eds) *Egg Parasitoids in Agroecosystems with Emphasis on Trichogramma*. Springer, Dordrecht, pp. 25–55.

Wajnberg, E. and Hassan, S.A. (1994) *Biological Control with Egg Parasitoids*. CAB International, Wallingford, UK.

Wasserthal, L.T. and Wasserthal, W. (1977) Ultrastructure of a scent scale organ with pressure discharge in male *Caligo eurilochus brasiliensis* (Fldr.) (Lepidoptera: Brassolidae). *Cell and Tissue Research* **177**: 87–103.

Wyatt, T.D. (2003) *Pheromones and Animal Behaviour: Communication by Taste and Smell.* Cambridge University Press, Cambridge.

Yong, T.H., Pitcher, S., Gardner, J. and Hoffmann, M.P. (2007) Odor specificity testing in the assessment of efficacy and non-target risk for *Trichogramma ostriniae* (Hymenoptera: Trichogrammatidae). *Biocontrol Science and Technology* **17**: 135–53.

Zawistowski, S. and Richmond, R.C. (1986) Inhibition of courtship and mating of *Drosophila melanogaster* by the male-produced lipid, *cis*-vaccenyl acetate. *Journal of Insect Physiology* **32**: 189–92.

Zhang, Q.H. and Aldrich, J.R. (2003) Male-produced anti-sex pheromone in a plant bug. *Naturwissenschaften* **90**: 505–8.

Zuk, M. and Kolluru, G.R. (1998) Exploitation of sexual signals by predators and parasitoids. *Quarterly Review of Biology* **73**: 415–38.

6

Novel insights into pheromone-mediated communication in parasitic hymenopterans

Joachim Ruther

Institute of Zoology, Chemical Ecology Group, University of Regensburg, Germany

Abstract

The contribution of parasitic wasps to ecosystem services is enormous because of the pivotal role these insects play as natural enemies. To optimize the pest control capacity of parasitic wasps, all aspects that influence their reproductive success need to be understood, including the mechanisms controlling mate finding and intraspecific communication. As in most insects, the chemical sense is of crucial importance for these abilities. However, despite the importance of parasitic wasps in natural and agricultural ecosystems, our knowledge about their pheromone communication has lagged far behind our knowledge about other insect taxa for many decades. In this chapter, I review the tremendous progress that has been made during the past decade in our understanding of pheromone-mediated interactions in parasitic wasps. The focus is on chemical signals mediating mate finding and recognition, but some other groups of pheromones are also addressed; for instance, aggregation pheromones and marking pheromones used in the context of oviposition to avoid superparasitism or to optimize clutch size and offspring sex ratio. Sex pheromones are involved at three levels of sexual communication. First, male- or female-derived volatile compounds mediate the attraction of sexual mates to the site of release. Secondly, less-volatile compounds, such as lipids covering the female's cuticle, function as contact sex pheromones – enabling mate recognition and eliciting stereotypic courtship behaviour. Thirdly, aphrodisiac pheromones released by males during courtship from oral or antennal glands elicit female receptiveness. As well as intraspecifically acting sex pheromones, parasitic wasps also use host-associated semiochemicals to find mates. Apart from reviewing the

Chemical Ecology of Insect Parasitoids, First Edition. Eric Wajnberg and Stefano Colazza.
© 2013 John Wiley & Sons, Ltd. Published 2013 by John Wiley & Sons, Ltd.

progress made in the study of pheromone chemistry, biosynthetic pathways and production sites, I also focus on the plasticity of pheromone communication due to innate mechanisms, associative learning and abiotic factors. Finally, I review some recent studies showing that parasitic wasps are excellent models not only for studying the mechanisms of chemical communication but also for answering evolutionary questions. Thus, the study of parasitic wasp pheromones has great potential to pave the way for major advances in our understanding of animal communication in general.

6.1 Introduction

Without parasitic wasps our world would look very different because of the pivotal contribution these carnivorous insects provide to ecosystem services as natural enemies by effectively controlling potential pests and maintaining species diversity (Godfray 1994, Quicke 1997, Macfadyen *et al.* 2011). Therefore parasitic wasps have great potential to be used as pest management tools in both agricultural and natural ecosystems (see Chapters 9–13 of this volume). To optimize pest control services by parasitic wasps, all aspects that influence their reproductive success need to be understood, including their ability to find hosts and sexual mates. The chemical sense is of crucial importance for both these activities. Like most insects, parasitic wasps use pheromones for intraspecific communication and allelochemicals (kairomones and synomones) for foraging (Vet & Dicke 1992, Godfray 1994, Quicke 1997, Kainoh 1999, Steidle & van Loon 2002). Despite the great species diversity and biological importance of parasitic wasps, our knowledge about their pheromone communication has lagged far behind that for other insect taxa for many decades, as indicated by the limited number of bioactive compounds compiled in earlier reviews (see Quicke 1997, Kainoh 1999, Ayasse *et al.* 2001, Keeling *et al.* 2004). However, now that well-studied model organisms such as the jewel wasp *Nasonia vitripennis* have also been discovered by chemoecologists, enormous advances have been made in the investigation of parasitic wasp pheromones, with implications for a better understanding of animal communication in general.

In this chapter, I review these recent advances with particular emphasis on novel pheromones and mechanisms and those not covered by previous reviews (Quicke 1997, Kainoh 1999, Ayasse *et al.* 2001, Keeling *et al.* 2004). For more details on older examples going beyond the summary given in Table 6.1, reference will be made to earlier works. I not only cover the chemistry of novel parasitoid sex pheromones (Fig. 6.1) but also address topics such as production sites, plasticity in pheromone communication, biosynthetic pathways and evolutionary aspects. I also review some examples in which non-pheromonal semiochemicals are involved in the mate finding of parasitic wasps. The use of pheromones in the context of oviposition, for instance to avoid competition and superparasitism or to optimize clutch size and offspring sex ratio, is also briefly addressed, although the chemicals involved largely remain unknown. Pheromones of parasitoids from insect orders other than the Hymenoptera are not covered in this chapter.

Table 6.1 Outline of parasitoid species with identified pheromones.

Family Species	Chemical name[1]	Function[2]	Volatility	Source	Remarks	References
Bethylidae						
Cephalonomia tarsalis	dodecanal (**9**)	sex (f)	volatile	unknown	Males arrested; substrate-borne	Collatz et al. 2009
Cephalonomia stephanoderis	skatole (**26**)	alarm?[3] (f)	volatile	mandibular gland	Agitated running, interpreted as alarm behaviour; also in other Bethylinae	Gomez et al. 2005, Goubault et al. 2008
Goniozus legneri	2-methyl-1,7-dioxaspiro [5.5] undecane (**27**)	appeasement (f/m)	volatile	mandibular gland	Released by losers in female–female contests; also in other Epyrinae	Goubault et al. 2006, 2008
Braconidae						
Ascogaster quadridentata	(Z,Z)-9,12-octadecadienal (**11**)	sex (f)	volatile	unknown	Long-distance attraction in bioassay and field trials	Delury et al. 1999
Ascogaster reticulatus	(Z)-9-hexadecenal (**10**)	sex (f)	volatile	tibial gland	Males follow trail; substrate-borne	Kainoh et al. 1991, Kainoh & Oishi 1993
Cardiochiles nigriceps	(Z,Z)-7,13-heptacosadiene (**19**) (Z,Z)-7,15-hentriacontadiene (**20**)	sex (f)	contact	Dufour's gland	Male antennating; no full courtship toward synthetic compounds	Syvertsen et al. 1995

Dendrocerus carpenteri	juvenile hormone III (**25**)	host marking (f)	contact	ovaries	Oviposition deterring effect, but much higher doses necessary than found in the wasps	Höller *et al.* 1994
Macrocentris grandii	(Z)-4-tridecenal (**12**) (3S,5R,6S)-3,5-dimethyl-6-(methylethyl)-3,4,5,6-tetrahydropyran-2-one (**13**)	sex (f)	volatile	(**13**) in mandibular gland	Anemotaxis; synergism; aldehyde is oxidation product of 9,13-alkadienes; (**13**) occurs in both sexes	Swedenborg 1992, Swedenborg *et al.* 1993, 1994
Spathius agrili	dodecanal (**9**) (4R,11E)-tetradecen-4-olide (**29**) (Z)-10-heptadecen-2-one (**30**)	aggregation (m)	volatile	unknown	Female and male anemotaxis; female response stronger; four other male-specific compounds without known function	Cossé *et al.* 2012
Chalcididae						
Brachymeria intermedia	3-hexanone (**28**)	aggregation (m/f)	volatile	unknown	Filter paper exposed to wasps attractive for conspecifics of either sex	Mohamed & Coppel 1987
Encyrtidae						
Ooencyrtus kuvanae	cuticular hydrocarbons 5-methylheptacosane (**22**) 5,17-dimethylheptacosane (**23**)	sex (f)	contact	cuticle	Close-range attraction of males; antennation; stereoselective response to (R)-5-MeC27 and (5R,17S)-5,17-DimeC27	Ablard *et al.* 2012

(cont'd)

Table 6.1 *(cont'd)*

Family Species	Chemical name[1]	Function[2]	Volatility	Source	Remarks	References
Eulophidae						
Diglyphus isaea	series of 11-alkanol fatty acid esters (**24**)	sex (f)	contact	cuticle[4]	Polar cuticular lipid fraction elicited courtship in males	Finidori-Logli *et al.* 1996
Melittobia digitata	α- (**1**) and β-trans-bergamotene (**2**)	sex (m)	volatile	abdomen[5]	Attraction of virgin females; interspecific cross-attraction	Gonzalez *et al.* 1985, Cônsoli *et al.* 2002
Eurytomidae						
Eurytoma amygdali[6]	(6Z,9Z)-tricosadiene (**14**) (6Z,9Z)-pentacosadiene (**15**)	sex (f)	volatile	cuticle	Anemotaxis; male wing raising	Krokos *et al.* 2001, Mazomenos *et al.* 2004
Figitidae						
Leptopilina heterotoma	(−)-iridomyrmecin (**7**) (+)-isoiridomyrmecin (**8**)	sex (f)	volatile	cephalic gland	Attraction of males; wing fanning; further iridoids involved	Stökl *et al.* 2012, I. Weiss, J. Hofferberth, J. Ruther and J. Stökl, unpublished data

Ichneumonidae

Species	Compound	Type	Volatile/Contact	Source	Behaviour	Reference
Itoplectis conquisitor	geranial (31) neral (32)	sex (f)	volatile	unknown	Attraction of males, copulation attempts; other unknown compounds involved	Robacker & Hendry 1977
Syndipnus rubiginosus	ethyl (Z)-9-hexadecenoate (16)	sex (f)	volatile	unknown	Anemotaxis; wing fanning and copulation attempts	Eller et al. 1984
Platygastridae						
Trissolcus brochymenae	(Z)-11-hexadecen-1-yl acetate (17) tetradecyl acetate (18)	sex (f)	contact	unknown	Antennation and mounting of males; long-range attraction not tested	Salerno et al. 2012
Pteromalidae						
Dibrachys cavus	cuticular hydrocarbons	sex (f)	contact	cuticle	Male courtship behaviour; arrestment	Ruther et al. 2011a
Lariophagus distinguendus	cuticular hydrocarbons 3-methylheptacosane (21) triacylglycerides	sex (f)	contact	cuticle	Male courtship; wing fanning; arrestment; young males also attractive; (21) is key component	Steiner et al. 2005, Ruther & Steiner 2008, Kühbandner et al. 2012b
Nasonia giraulti	(4R,5S)-5-hydroxy-4-decanolide (3) 4-methylquinazoline (5)	sex (m)	volatile	rectal vesicle	Attraction and arrestment of virgin females; (5) is synergist	Niehuis et al. 2013
Nasonia longicornis	(4R,5S)-5-hydroxy-4-decanolide (3) 4-methylquinazoline (5)	sex (m)	volatile	rectal vesicle	Attraction and arrestment of virgin females; (5) is synergist	Niehuis et al. 2013

(cont'd)

Table 6.1 (cont'd)

Family Species	Chemical name[1]	Function[2]	Volatility	Source	Remarks	References
Nasonia vitripennis	(4R,5S)- 5-hydroxy-4-decanolide (3) (4R,5R)-5-hydroxy-4-decanolide (4) 4-methylquinazoline (5)	sex (m)	volatile	rectal vesicle	Attraction and arrestment of virgin females; (3) is major attractant (4) and (5) are synergists; (5) mediates site fidelity of releasing males	Ruther et al. 2007, 2008, 2011b
	cuticular hydrocarbons	sex (f)	contact	cuticle	Male courtship; arrestment	Steiner et al. 2006
Roptrocerus xylophagorum	cuticular hydrocarbons	sex (f)	contact	cuticle	Male courtship behaviour; arrestment	Sullivan 2002
Spalangia endius	methyl 6-methylsalicylate (6)	sex (f)	volatile	unknown	Arrestment of males wing fanning	Nichols et al. 2010
Trichomalopsis sarcophagae	(4R,5S)-5-hydroxy-4-decanolide (3) 4-methylquinazoline (5)	sex (m)	volatile	rectal vesicle	Attraction and arrestment of virgin females; (5) is synergist	Niehuis et al. 2013 Syvertsen et al. 1995

[1]Numbers in parentheses refer to the structures shown in Figure 6.1.
[2]Letters in parentheses indicate releasing sex (f = female, m = male).
[3]Question mark indicates a putative function; more experiments are necessary to finally prove this function.
[4]Exact production site unknown but pheromone is extractable from different tagmata.
[5]In *M. australica* and *M. femorata*.
[6]*E. amygdali* and several other species of the Eurytomidae are herbivores but are nevertheless listed.

6.2 Pheromones and sexual behaviour

Pheromones are defined as

> molecules that are evolved signals, in defined ratios in the case of multiple component pheromones, which are emitted by an individual and received by a second individual of the same species, in which they cause a specific reaction, for example, a stereotyped behaviour or a developmental process (Wyatt 2010).

In the case of sex pheromones, this behaviour or developmental process is related to mating. Sex pheromones are involved at three levels of sexual communication in parasitic hymenopterans.

1 Male- or female-derived volatile compounds mediate the attraction of the sexual mate to the site of release. I use the term 'volatile' for compounds eliciting the attraction of the responder from a distance (i.e. without establishing direct contact with the pheromone source).
2 Once potential mating partners have encountered each other, a stereotypic courtship sequence typically begins, in which less-volatile pheromones are involved that are used for mate recognition.
3 Aphrodisiac pheromones released by males elicit female receptiveness.

6.2.1 Volatile sex attractants

Depending on the mating system, volatile sex pheromones can be released by either sex of a parasitic wasp, but in most species studied so far they have been female-derived (Quicke 1997). It is predicted by theory that the limiting sex (in parasitoids usually the female) assumes the less costly position during sexual communication (Wyatt 2003). If travel costs are high because of energetic expenditures in relation to biosynthetic costs, it is the female that calls. If travel costs are low, for instance under conditions favouring local mate competition (Werren 1980), and/or metabolic costs are high because the chemical signal is derived from limited biosynthetic precursors, it might be the male that calls.

Male pheromones
The first known male-derived volatile sex attractant in a parasitic wasp was identified in the eulophid wasp *Melittobia digitata* (Cônsoli *et al.* 2002). Males of this species are blind and flightless and attract females by releasing the sesquiterpenes α-trans-bergamotene (**1**; bold numbers refer to those used in both Figure 6.1 and Table 6.1) and β-trans-bergamotene (**2**) in a ratio of approximately 5 : 95. The β-isomer is much more attractive and no synergism has been found between the two bergamotenes, suggesting that β-trans-bergamotene is the natural product and the 'α' isomer is an artefact formed by spontaneous rearrangement. It was found that pheromone release in *M. digitata* males peaked 2 days after emergence. The production site of the pheromone has been suggested to be the male abdomen, as has been found in the congeneric species *M. australica* and *M. femorata* (Gonzalez *et al.* 1985). Behavioural bioassays employing *M. digitata* and some congeneric species revealed a certain degree of cross-attraction, suggesting that the trans-bergamotenes also occur in the other species.

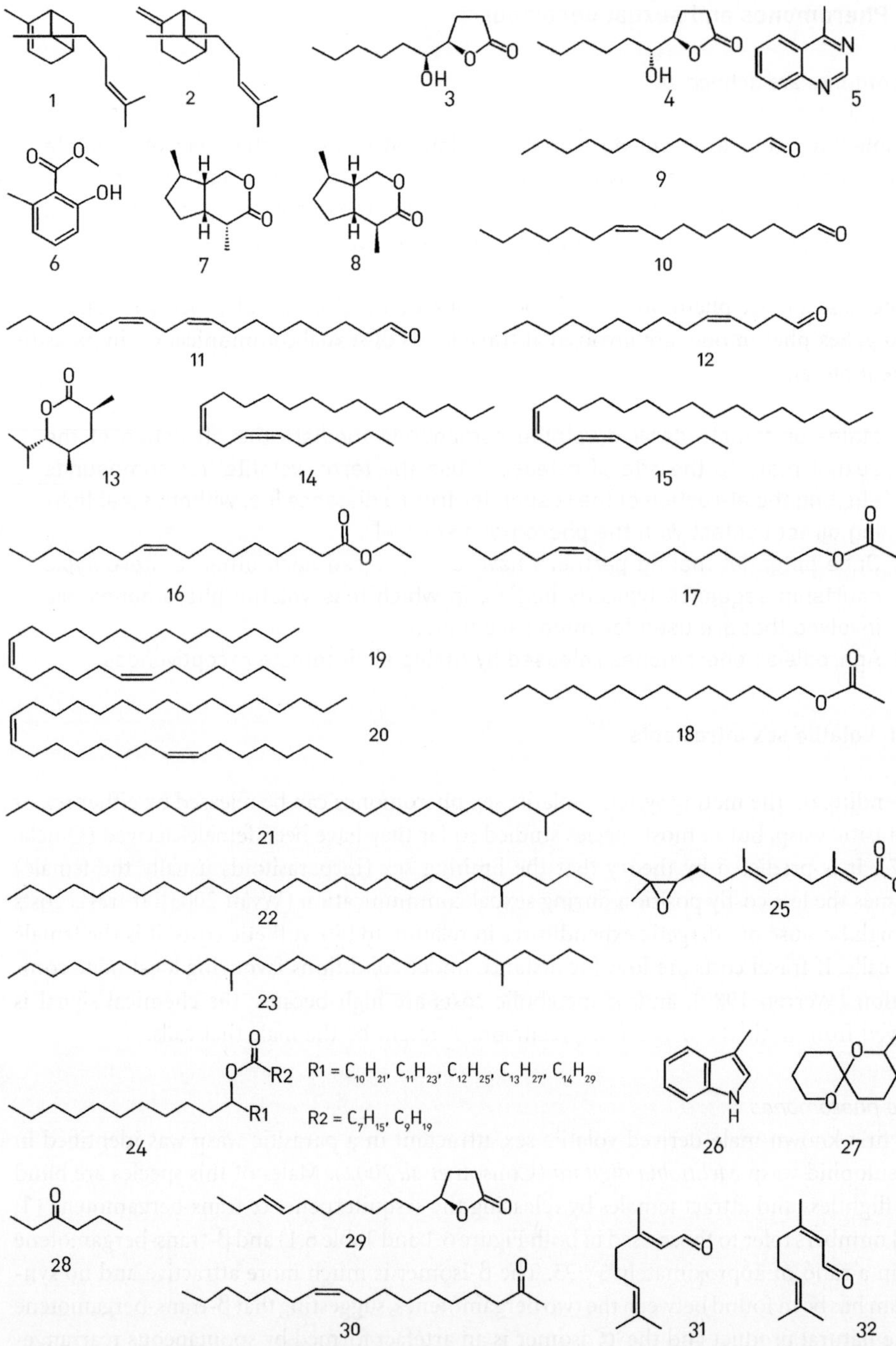
1
2
OH
3
OH
4
N
N
5
O
O
OH
6
H
H
O
O
7
H
H
O
O
8
O
9
O
10
O
11
O
12
O
O
13
14
15
O
O
16
O
O
17
19
O
O
18
20
21
22
O
O
O
25
23
O
O
R2
R1
R1 = $C_{10}H_{21}$, $C_{11}H_{23}$, $C_{12}H_{25}$, $C_{13}H_{27}$, $C_{14}H_{29}$
R2 = C_7H_{15}, C_9H_{19}
24
N
H
26
O
O
27
O
28
O
O
29
O
31
O
32
O
30

Males of the jewel wasp *Nasonia vitripennis* (Pteromalidae) release a mixture of (4*R*,5*S*)-(**3**) and (4*R*,5*R*)-5-hydroxy-4-decanolide (HDL) (**4**) to attract virgin females (Ruther *et al.* 2007), which is by far the most thoroughly investigated example of a parasitoid sex pheromone. The pheromone response depends on the female mating status (only virgins respond) (Ruther *et al.* 2007, 2010, Steiner & Ruther 2009a) and is synergized by the minor component 4-methylquinazoline (4-MQ) (**5**) (Ruther *et al.* 2008). The latter compound is responsible for the characteristic medicine-like odour of *Nasonia* males but is not attractive to females unless HDL is also present.

All pheromone components are synthesized in the rectal vesicle of the males. Gene expression analysis of the epoxide hydrolase gene, *Nasvi-EH1*, involved in HDL biosynthesis revealed the precise production site of HDL to be the male rectal papillae. These paired glandular structures are located adjacent to the rectal vesicle (Abdel-Latief *et al.* 2008). The pheromone is deposited on the ground by *N. vitripennis* males via the anus with dabbing movements of the abdominal tip (Steiner & Ruther 2009b). Quantitative analysis of pheromone deposits with regard to body size revealed amounts of total HDL between 250 ng (small males) and >700 ng (large males) (Blaul & Ruther 2012). Amounts as small as 10 ng can attract females in a bioassay. When given the choice between different doses, females were able to detect subtle differences, always preferring the higher dose (Ruther *et al.* 2009). Therefore, the *N. vitripennis* sex pheromone does not simply indicate the presence of a male but also allows females to discriminate against males of inferior quality, since pheromone titres and sexual attractiveness correlate with male quality parameters such as mating history, age (Ruther *et al.* 2009) and nutritional state (Blaul & Ruther 2011). Interestingly, *Nasonia* females prefer hosts for oviposition that contain more of the pheromone precursor linoleic acid, thus increasing the mating chances of their sons (Blaul & Ruther 2011). Marking behaviour is shown by males spontaneously, but clearly increases after mating or even after mere contact with a female (Steiner & Ruther 2009b). Pheromone markings are not only attractive to virgin females but also to the releasing males themselves (Ruther *et al.* 2011b). This site fidelity of males at pheromone marking is mediated by the minor component 4-MQ and, like the context-dependent pheromone release, can be interpreted as a kind of economical pheromone use. However, males are not able to distinguish their own markings from those of other males. The range of activity of the *N. vitripennis* pheromone is limited to approximately 5 cm around the markings, and is thus adapted to the natural habitat of *Nasonia* wasps (i.e. the nests of hole-breeding birds). Fresh pheromone markings remain attractive for approximately 2 hours (Steiner & Ruther 2009b). Males

Figure 6.1 Chemical structures of pheromones identified in parasitic wasps. **1** α-trans-bergamotene, **2** β-trans-bergamotene, **3** (4*R*,5*S*)-5-hydroxy-4-decanolide, **4** (4*R*,5*R*)-5-hydroxy-4-decanolide, **5** 4-methylquinazoline, **6** methyl 6-methylsalicylate, **7** (−)-iridomyrmecin, **8** (+)-isoiridomyrmecin, **9** dodecanal, **10** (*Z*)-9-hexadecenal, **11** (*Z*,*Z*)-9,12-octadecadienal, **12** (*Z*)-4-tridecenal, **13** (3*S*,5*R*,6*S*)-3,5-dimethyl-6-(methylethyl)-3,4,5,6-tetrahydropyran-2-one, **14** (6*Z*,9*Z*)-tricosadiene, **15** (6*Z*,9*Z*)-pentacosadiene, **16** ethyl (*Z*)-9-hexadecenoate, **17** (*Z*)-11-hexadecen-1-yl acetate, **18** tetradecyl acetate, **19** (*Z*,*Z*)-7,13-heptacosadiene, **20** (*Z*,*Z*)-7,15-hentriacontadiene, **21** 3-methylheptacosane, **22** 5-methylheptacosane, **23** 5,17-dimethylheptacosane, **24** series of 11-alkanol fatty acid esters, **25** juvenile hormone III, **26** skatol, **27** 2-methyl-1,7-dioxaspiro[5.5]undecane, **28** 3-hexanone, **29** (4*R*,11*E*)-tetradecen-4-olide, **30** (*Z*)-10-heptadecen-2-one, **31** geranial, **32** neral.

of the sister species *N. giraulti* and *N. longicornis*, as well as the closely related pteromalid wasp *Trichomalopsis sarcophagae*, use only (4R,5S)-HDL and the minor compound 4-MQ as pheromones (Niehuis *et al.* 2013; see Section 6.6).

Female pheromones

Females of the pteromalid wasp *Spalangia endius* produce methyl 6-methylsalicylate (MMS) (**6**) (Nichols *et al.* 2010). The compound and a second unknown chemical were found as a female-specific component in both the headspace and whole body washings. MMS and the unknown compound were electrophysiologically active, suggesting a pheromone function and, in fact, fractions containing MMS as well as synthetic MMS arrested males and elicited wing fanning in a few of them. The exact location of MMS in *S. endius* is unknown, but the authors suggest that it is distributed all over the cuticle (Nichols *et al.* 2010).

Males of *Leptopilina heterotoma* and *L. boulardi* (Figitidae), larval parasitoids of *Drosophila* fruit flies, have been attracted, both in the field and in wind tunnel bioassays, by volatiles released by virgin females of the respective species. In contrast, mated females were not attractive (Fauvergue *et al.* 1999). *Leptopilina heterotoma* females produce several iridoid compounds in a cephalic gland (most probably the mandibular gland). The two major compounds have been identified as (−)-iridomyrmecin (**7**) and (+)-isoiridomyrmecin (**8**), two iridomyrmecins not previously known from insects (Stökl *et al.* 2012). In contrast, males of *L. heterotoma* produce only (+)-isoiridomyrmecin, suggesting a sex pheromone function of the iridoids and, in fact, purified fractions containing the female-derived iridoids were highly attractive to males in a Y-tube bioassay and elicited wing fanning. Synthetic samples of (−)-iridomyrmecin and (+)-isoiridomyrmecin were only bioactive if the minor iridoids were also present and the females responded enantioselectively (I. Weiss, J. Hofferberth, J. Ruther and J. Stökl, unpublished data). Females of several species of hyperparasitoids from the subfamily Alloxystinae also produce iridoids in their mandibular gland (Völkl *et al.* 1994, Hübner *et al.* 2002). In *Alloxysta victrix*, the major compounds have been identified as unusual *trans*-fused iridomyrmecins (Hilgraf *et al.* 2012). However, although *A. victrix* females have been shown to produce a volatile sex pheromone, synthetic *trans*-fused iridomyrmecins were not attractive to males (Petersen 2000).

Males of the bethylid wasp *Cephalonomia tarsalis* are arrested by the odour of young virgin females and prefer it over the odour of older mated females (Collatz *et al.* 2009). Likewise, pieces of filter paper previously exposed to young virgin females arrested males, suggesting that the attractive chemicals are released actively or passively to the substrate. The chemical mediating male arrest has been extracted from the filter paper and identified as dodecanal (**9**), which was active at doses between 6 and 9 ng. The exact pheromone source and the release mechanism are still to be elucidated. The authors suggest that dodecanal functions as a substrate-borne trail sex pheromone guiding the males to females. In addition, females of the braconid wasp *Glyptapanteles flavicoxis* release a substrate-borne pheromone from their abdominal tip (Danci *et al.* 2006). Through a combination of behavioural bioassays, electrophysiological experiments and chemical analyses, the authors were able to show that the pheromone attracted males from distances below 10 cm and arrested them near the pheromone traces. Four electrophysiologically active components were detected in the pheromone extracts, which were all crucial for male attraction. However, structural elucidation failed, although 4500 female equivalents were pooled for GC-MS analysis.

Apart from these two examples, there are some further studies suggesting that female-derived substrate-borne sex pheromones are more common in parasitic wasps (Kainoh et al. 1991, Fauvergue et al. 1995, Pompanon et al. 1997, Bernal & Luck 2007). Interestingly, the only other identified trail sex pheromone, apart from dodecanal, is also an aldehyde: (Z)-9-hexadecenal (**10**) in the braconid wasp *Ascogaster reticulatus* (Kainoh et al. 1991). The pheromone is produced in the tibial gland of females (Kainoh & Oishi 1993) and elicits trail-following behaviour in males. Very similar compounds may also mediate long-range orientation. In *Ascogaster quadridentata*, for instance, (Z,Z)-9,12-octadecadienal (**11**) has been identified as a sex pheromone which attracted males in laboratory bioassays at doses of 1–10 ng (Delury et al. 1999). Unlike many other insects, virgin and mated females were equally attractive to males. Long-distance attractiveness of the pheromone was confirmed in field trials. Female *A. quadridentata* also produced (Z)-9-hexadecenal (**10**), the trail sex pheromone of the congeneric species *A. reticulatus*. However, although electrophysiologically active in males, the monounsaturated aldehyde was not attractive alone, nor did it synergize the dienal (Delury et al. 1999). In the braconid wasp *Macrocentris grandii*, females attract males with (Z)-4-tridecenal (**12**), which is formed by spontaneous oxidation from a series of female-specific long-chain 9,13-alkadienes (Swedenborg 1992). The attractiveness to males of the aldehydes is synergized by the more polar compound (3S,5R,6S)-3,5-dimethyl-6-(methylethyl)-3,4,5,6-tetrahydropyran-2-one (**13**), which is produced in the mandibular gland of either sex (Swedenborg et al. 1994). The formation of volatile bioactive compounds from oxidation-sensitive precursors such as the alkadienes in *M. grandii* might be a more common mechanism, since alkadienes are often produced specifically by one sex only. For instance, only females of the eurytomid wasp *Eurytoma amygdali* produce (6Z,9Z)-tricosadiene (**14**) and (6Z,9Z)-pentacosadiene (**15**), which were identified as sex pheromones eliciting landing responses in males both in laboratory bioassays and field trials (Krokos et al. 2001, Mazomenos et al. 2004). In this case, however, the alkadienes were volatile themselves, as indicated by headspace analysis using solid phase microextraction (SPME). Nevertheless, a systematic search for oxidation products of oxidizable sex-specific hydrocarbon precursors might reveal novel insights into the formation of female-derived parasitoid sex attractants.

Non-pheromonal semiochemicals involved in sexual behaviour
The host-finding process of parasitoid females can be divided into host habitat location, host location within the habitat, host recognition, and host acceptance (Vinson 1976, Steidle & van Loon 2002). It is well established that parasitoid females use a variety of chemical stimuli (kairomones and synomones) in all of these steps (see Chapters 3 and 4 in this volume). These stimuli may originate from the host itself (e.g. pheromones, faeces, or lipid footprints) or may be released from other sources that are present in the host habitat (e.g. plants being attacked by the hosts). Although it is self-evident that male parasitoids orienting towards the same chemical cues as foraging females might increase their chance of encountering receptive females, males have only rarely been studied with respect to their response to host-related cues (Metzger et al. 2010). Both sexes of the pteromalid wasp *Lariophagus distinguendus*, for instance, were innately attracted to volatiles released from the faeces of one of their hosts, the granary weevil *Sitophilus granarius* (Steiner et al. 2007). Males responding to host-associated volatiles are arrested in host patches with potentially high female density and thus increase their mating chances. There is increasing evidence that male parasitoids commonly use learnt non-pheromonal cues when searching

for mates. In fact, the examples listed in Section 6.4 suggest that males may associatively learn and subsequently respond to any chemical stimulus perceived during copulation.

Host-associated semiochemicals might also synergize the attractiveness of sex pheromones, as has been shown for the ichneumonid wasp *Venturia canescens* (Metzger *et al.* 2010). Males of this solitary larval parasitoid of pyralid moths responded 2.5 times more strongly to the female sex pheromone when it was offered in combination with host kairomones. In contrast, female and host odours were only slightly attractive when offered separately.

Another example of non-pheromonal semiochemicals used for mate finding has been reported in the solitary ichneumonid wasp *Pimpla disparis* (Hrabar *et al.* 2012). Males of this gypsy moth parasitoid are able to detect hosts containing potential mates that are about to emerge. For this purpose they use unidentified volatiles of an oral secretion which is released by the wasps when gnawing an emergence hole in the host. Although mediating an intraspecific interaction, the authors consider this an example of a mate-finding cue rather than a pheromonal signal because both emerging males and females release it. Thus, responding males are unable to recognize the sex of the emerging conspecific but may nevertheless benefit from the response by obtaining a better chance of encountering a virgin female. It is assumed that the primary function of the secretion is to soften the host's integument.

6.2.2 Female-derived courtship pheromones

After being attracted to each other by volatile pheromones or sexual kairomones (Ruther *et al.* 2002), almost all parasitic wasps show characteristic courtship behaviour. These stereotypic behavioural sequences have been studied in detail in a number of species (van den Assem 1989, Isidoro *et al.* 1999, Ruther *et al.* 2000, Battaglia *et al.* 2002, Romani *et al.* 2008) and sex pheromones are also involved in these short-range interactions. Female-derived chemicals are used by males for mate recognition and elicit courtship elements such as wing fanning, mounting, antennal movements/stroking or particular postures. Wing fanning is a very common courtship element which might facilitate mate finding by moving pheromone-laden air from front to rear, thus enabling directional orientation towards females (Danci *et al.* 2006). Alternatively, wing fanning might be performed to produce acoustic signals which may also be involved in the sexual communication of parasitic wasps (van den Assem & Putters 1980, Danci *et al.* 2010, Villagra *et al.* 2011, Benelli *et al.* 2012).

In some parasitic wasps, the same female-derived volatile pheromones that mediate long-distance attraction also trigger male courtship behaviour. Males of the ichneumonid *Syndipnus rubiginosus*, for instance, responded not only by positive anemotaxis to the female sex pheromone ethyl (*Z*)-9-hexadecenoate (**16**) but also displayed wing fanning or copulation attempts (Eller *et al.* 1984). Antennal drumming and mounting was elicited by tetradecyl acetate (**17**) and (*Z*)-11-hexadecenyl acetate (**18**) in males of the egg parasitoid *Trissolcus brochymenae* (Platygastridae) (Salerno *et al.* 2012). A possible long-range effect of these relatively volatile acetate esters, as known from many other insects, was not tested. The female-derived iridoids of *L. heterotoma* (I. Weiss, J. Hofferberth, J. Ruther and J. Stökl, unpublished data) and the unknown four-component close-range pheromone of *G. flavicoxis* (see above) also caused male wing fanning behaviour (Danci *et al.* 2006). Similarly, *E. amygdali* responded to the female sex pheromone components (6*Z*,9*Z*)-tricosadiene (**14**) and (6*Z*,9*Z*)-pentacosadiene (**15**) by anemotaxis and courtship (wing raising) (Krokos *et al.* 2001). Due to their twofold unsaturation, the latter two compounds, although having relatively high molecular mass, are volatile enough to cause anemotaxis. They belong to a class of chemicals which appear to be widely used as contact sex pheromones in parasitic wasps: the cuticular hydrocarbons (CHCs). CHCs cover the epicuticle of virtually any insect. They are composed of complex mixtures of straight-chain and methyl-branched alkanes and alkenes, and are involved in the communication of many insect species. The composition of CHC profiles is sex-specific in many parasitoids, and extracts from females and non-polar fractions of female extracts have both been shown to trigger courtship responses (Sullivan 2002, Steiner *et al.* 2005, 2006, Ruther *et al.* 2011a, Ablard *et al.* 2012, Kühbandner *et al.* 2012a). However, single components eliciting full behavioural responses have only rarely been identified (Ablard *et al.* 2012; Kühbandner *et al.* 2012a). This indicates a general problem which researchers face when studying CHCs with pheromone function: CHC profiles are often very complex and several compounds interact with each other for bioactivity. In particular, the role of more-polar lipids as modifiers of the behavioural function has apparently been ignored in the past. In the case of methyl-branched and unsaturated compounds, commercially available reference compounds are missing in most cases and have to be synthesized laboriously. In addition, methyl-branched CHCs are chiral and may be perceived enantioselectively. Despite these problems, some major advances have been made in recent years.

Males of the braconid wasp *Cardiochiles nigriceps* responded by displaying wing fanning, antennating, mounting, and copulation attempts when they came into contact with solvent extracts from females which had been applied to male dummies (Syvertsen *et al.* 1995). Chemical analyses revealed clear differences in the CHC profiles of males and females. In particular, males were lacking some long-chain alkadienes present on the female cuticle and in the Dufour's gland. Greatly decreased responses (antennating only) were recorded when two synthetic female-specific alkadienes – (Z,Z)-7,13-heptacosadiene (**19**) and (Z,Z)-7,15-hentriacontadiene (**20**) – were tested alone. This suggests that the compounds interact with other unidentified cuticular lipids.

The contact sex pheromones of some pteromalid wasps have also been studied extensively. Males of *Anisopteromalus calandrae*, *Roptrocerus xylophagorum* and *Lariophagus distinguendus* responded to models treated with female-derived extracts and/or non-polar CHC fractions by wing fanning, antennating, mounting, and copulation attempts (Yoshida 1978, Ruther *et al.* 2000, Sullivan 2002, Steiner *et al.* 2005). In *N. vitripennis*, males were arrested on filter paper discs treated with female CHCs but showed more complex behaviours such as the typical 'head nodding' (permitting the release of an aphrodisiac pheromone, see Section 6.2.3) or copulation attempts only when extracts were applied to solvent-washed male dummies (Steiner *et al.* 2006). In *Dibrachys cavus*, males were arrested on glass beads treated with female CHC fractions but were unresponsive when these were applied to filter paper (Ruther *et al.* 2011a). These results show that male responses to contact sex pheromones may depend on the physical properties of the models the pheromone is applied to. CHC-based sex pheromones are characterized by relatively low volatility and thus only have a narrow range of activity (Ruther *et al.* 2000), and therefore females in many species may remain attractive to males for several days after mating (Ruther *et al.* 2000, Sullivan 2002, Steiner *et al.* 2006). In *A. calandrae* and *L. distinguendus*, it seems that pheromone production has already started in the pupal stage, as males showed wing fanning towards pupae which had been excised from the hosts before emergence (Yoshida 1978, Steiner *et al.* 2005). The early presence of contact sex pheromones appears to be widespread in parasitic wasps, as males of several species have been observed to be arrested on hosts from which conspecifics were about to emerge (Yoshida 1978, Fauvergue *et al.* 1995, Steiner *et al.* 2005, Collatz *et al.* 2009, Goh & Morse 2010, Danci *et al.* 2011, Hrabar *et al.* 2012). Interestingly, male pupae also elicited wing fanning behaviour in *L. distinguendus* and it has been suggested that this might be a strategy of later-developing males to distract earlier-emerged competitors away from females, because the latter were equally arrested on host-infested wheat kernels containing male or female parasitoids (Steiner *et al.* 2005).

In some parasitoid species, newly emerged males also have the contact sex pheromone and are mistaken for females by older consexuals (Robacker *et al.* 1976, Steiner *et al.* 2005, Benelli and Canale 2012). In *L. distinguendus*, males deactivate the contact sex pheromone within the first 2 days after emergence. The mechanism involved is not yet understood, but in other insects the lipoprotein lipophorin has been shown to be involved in the internalization of topically applied hydrocarbons (Sevala *et al.* 2000). Pheromone deactivation in *L. distinguendus* requires males to be alive, because adult males that had been killed immediately after emergence remained attractive for many days (Steiner *et al.* 2005). Comparison of behaviourally inactive CHC profiles from older males with active ones from females and newly emerged males, respectively, allowed the identification of key components of the *L. distinguendus* contact sex pheromone. Deactivation was accompanied by the loss of 3-methylheptacosane (3-MeC27) (**21**) and some minor compounds from the male CHC profiles (Steiner *et al.* 2005). Removal of these components from bioactive female CHC profiles by size exclusion chromatography likewise resulted in a loss of bioactivity. This loss could be restored by adding synthetic 3-MeC27, pointing to a key function of this compound in the contact sex pheromone of *L. distinguendus*. 3-MeC27, however, was only behaviourally active when a chemical background of the other lipids was present. When applied to filter paper, only (*S*)-3-MeC27 was active, showing for the first time that high-boiling monomethylalkanes are perceived by an insect enantioselectively. When applied to dummy males, however, both enantiomers caused wing fanning (Kühbandner *et al.* 2012b). The stereoselective response of males to methyl-branched alkanes was impressively corroborated in a recent paper studying the courtship pheromone of the gypsy moth egg parasitoid *Ooencyrtus kuvanae* (Encyrtidae) (Ablard *et al.* 2012). CHC profiles of both sexes were qualitatively very similar but male-derived extracts that repelled consexuals contained more 5-methylheptacosane (5-MeC27) (**22**) and 5,17-dimethylheptacosane (5,17-DimeC27) (**23**). Behavioural bioassays using all possible combinations of synthetic chemicals revealed that only (*S*)-5-MeC27 + (5*S*,17*R*)-DimeC27 attracted males whereas (*R*)-5-MeC27 + (5*R*,17*R*)-DimeC27 repelled them, suggesting that the pheromone function is mediated by sex-specific differences in the stereochemistry of these two CHCs.

Remarkably, triacylglycerides (TAGs) were also necessary in order to elicit full behavioural responses by *L. distinguendus* males. This non-volatile class of polar lipids is well known as a dietary resource but they have long been ignored as possible semiochemicals. Since TAGs occur in high amounts in the body fat of insects, it was often not clear whether they actually occurred on the insect cuticle or were co-extracted from internal tissues. Selective analyses of the epicuticle of *L. distinguendus* males using solid phase micro extraction (SPME) with *in situ* transesterification revealed the presence of TAGs on the cuticle of *L. distinguendus* (Kühbandner *et al.* 2012b), corroborating recent findings in *Drosophila* flies (Yew *et al.* 2011).

More-polar lipids have also been shown to mediate courtship behaviour in the eulophid wasp *Diglyphus isaea* (Finidori-Logli *et al.* 1996). Extracts from females, as well as polar fractions thereof, elicited courtship in males. The polar fraction contained a series of

11-alkanol fatty acid esters (**24**), which were much more abundant in females. Also, the hydrocarbon fraction was slightly active. Experiments are now needed to address possible synergetic effects of the two lipid classes and to confirm the bioactivity of the esters by using synthetic reference compounds.

The studies described on *L. distinguendus* offer new insights into the contact sex pheromone function of cuticular lipids in parasitic wasps. They show that CHC profiles can be deactivated by removing key components which are perceived enantioselectively, and can interact with other CHCs and previously ignored TAGs. However, more species need to be studied before more general conclusions can be drawn. Strikingly, 3-methylalkanes are more abundant in female than male CHC profiles also in other parasitoids and sometimes even correlate with pheromonal activity (Howard 2001, Sullivan 2002, Darrouzet *et al.* 2010, Ruther *et al.* 2011a). In *D. cavus*, for instance, the proportions of 3-MeC29 and 3-MeC31 in the female CHC profile increase significantly when female wasps become sexually attractive for males within the first 1–2 days of their life (Ruther *et al.* 2011a). This suggests that 3-methylalkanes might have a key function not only in *L. distinguendus*.

6.2.3 Male-derived courtship pheromones

For almost three decades it has been well established that parasitoid antennae do not only house the most important sensory organs that are crucial for mate and host finding but also a number of glandular structures (Dahms 1984, Bin & Vinson 1986). In many species from various taxa, the male antennae are characterized by pores which have been found to be the external openings of integumentary glands (Bin & Vinson 1986, Isidoro & Bin 1995, Isidoro *et al.* 1996, 1999, Bin *et al.* 1999, Guerrieri *et al.* 2001, Battaglia *et al.* 2002, Romani *et al.* 2008). Depending on the taxon, these pores are typically restricted to particular antennomeres. They may be distributed inconspicuously but are often associated with spreading structures having the form of scales, plates, pegs or keels (Bin & Vinson 1986, Isidoro *et al.* 1996, Battaglia *et al.* 2002). The tyloids known from many ichneumonids, for instance, have been shown to be such release-and-spread structures rather than sensory organs (Bin *et al.* 1999, Klopfstein *et al.* 2010). In some parasitoids, the pores have been found to release viscous secretions of unknown composition (Isidoro *et al.* 1996, Battaglia *et al.* 2002).

During courtship, males of many parasitoid species establish antennal contact with the female and perform stroking movements along the female's antennae (van den Assem 1970, Isidoro & Bin 1995, Bin *et al.* 1999, Guerrieri *et al.* 2001, Battaglia *et al.* 2002, Romani *et al.* 2008). Often it is the antennomeres carrying release-and-spread structures that are particularly involved, suggesting that chemical substances are transferred. This has been corroborated by ablation experiments showing that antennal contact is crucial for the females to become receptive. Cross-ablation (i.e. removal of the right female and the left male antenna) prevented females from becoming receptive, whereas same-side ablation, allowing contact on at least one side, did not have this effect (Isidoro *et al.* 1999, Battaglia *et al.* 2002, Romani *et al.* 2008). Likewise, the targeted sealing of release-and-spread areas on male antennae with glue had the same interrupting effect (Bin *et al.* 1999, Romani *et al.* 2008). While much work has been done to describe the antennal glands involved in parasitoid courtship, to my knowledge not a single antennal pheromone has yet been identified. Even very basic experiments showing that antennal extracts may trigger female receptiveness are lacking. It appears that the antennal aphrodisiac pheromones are non-volatile and it has been suggested that the secretion released by males of the scelionid

Trissolcus basalis is proteinaceous (Bin & Vinson 1986), but this conclusion is rather tentative. Thus, it will be one of the future challenges of natural product chemists to characterize and identify putative antennal parasitoid pheromones.

Not only antennal glands have been found to produce male aphrodisiac pheromones. In *N. vitripennis*, the pheromone is released from an extruded oral gland (probably the mandibular gland) when they perform so-called 'head nodding', a characteristic courtship element which is shown by males after mounting the female. By sealing the mouthparts of courting males with glue, females can be prevented from becoming receptive, suggesting the involvement of an aphrodisiac pheromone (van den Assem *et al.* 1980, Ruther *et al.* 2010). Van den Assem *et al.* (1980) suggested that the pheromone was volatile. They puffed the headspace of courting couples into a chamber containing a constrained couple with a sealed male and reported that the female became receptive. However, the cited study is rather narrative, with many experimental details missing. We tried several times to repeat this experiment but were unsuccessful. Sealing the male prevented female receptiveness but we were never able to reverse this effect by applying the headspace of courting couples. A detailed video recording (available as supplemental online material in Ruther *et al.* 2010) demonstrates that the male oral gland establishes contact with the female antennae, suggesting that the compounds involved are non-volatile.

6.3 Other pheromones

6.3.1 Marking pheromones

Host organisms chosen by the mother are the only source of nutrients for the offspring of parasitic wasps. To make the optimal foraging decisions, it is therefore important for a parasitoid female to recognize whether or not a given host has already been parasitized by herself or by conspecific or heterospecific competitors. Therefore, ovipositing females of many parasitoid species release external or internal markings on – or close to – the host, which is subsequently less preferred for oviposition if alternative hosts are available (see reviews by Vinson 1976, van Lenteren 1981, Kainoh 1999, Nufio & Papaj 2001, Anderson 2002, Li 2006). In species with local mate competition (Werren 1980), the presence of previously parasitized hosts furthermore leads to a shift of the typically female-biased offspring sex ratios towards a higher proportion of males (King & Skinner 1991). It is likely that the same marking pheromones are involved here. The existence of marking pheromones has been demonstrated in parasitoids from numerous families and the sources have also been characterized (see reviews mentioned above) but, as with the antennal contact pheromones, information about the chemistry involved is scarce. In a number of species, there is evidence that the Dufour's gland is involved. Dufour's gland extracts applied to hosts have been shown to have an oviposition deterrent effect (Guillot & Vinson 1972, Mudd *et al.* 1982, Harrison *et al.* 1985, Rosi *et al.* 2001, Jaloux *et al.* 2005). The composition of the Dufour's gland often matches the cuticular hydrocarbon profiles of the respective species in both social and parasitic hymenopterans (Oldham *et al.* 1994, Ruther *et al.* 1998, Howard & Baker 2003, Jaloux *et al.* 2005), suggesting that these compounds do play a role. It has also been demonstrated that parasitoid-derived methyl-branched hydrocarbons are transferred to the host after oviposition. Whitefly nymphs parasitized by the parasitoid *Eretmocerus mundus* had four additional hydrocarbons in their CHC profile which are major components of the wasp's CHCs. However, although some kind of marking behaviour is shown by *E. mundus* after oviposition, it is not clear whether the additional compounds are deposited actively from a gland or released passively from the wasp's cuticle; it is also not clear whether they are bioactive at all (Buckner & Jones 2005). In the pteromalid wasp *Eupelmus viulleti*, artificial host containers (gelatine capsules) that had been treated with female CHCs by rubbing the abdomen of female wasps over the surface were less preferred for oviposition than controls. Treatment with male CHCs resulted in a shift of the offspring sex ratio in favour of females (Darrouzet *et al.* 2010). Furthermore, *E. viulleti* uses CHCs left by another pteromalid wasp, *Dinarmus basalis*, as a foraging kairomone. *Eupelmus viulleti* is a kind of kleptoparasite, preferring to deposit its eggs in hosts that have previously been parasitized by *D. basalis*. Hosts treated with both *D. basalis* Dufour's gland and cuticular extracts were preferred by *E. viulleti* females (Jaloux *et al.* 2005). Hence, CHCs released by a parasitoid externally can be exploited by natural enemies or competitors and thus become maladaptive, and it has been suggested that this might have favoured the evolution of internally released signals (Jaloux *et al.* 2005).

However, cuticular and glandular hydrocarbons are not the only source of oviposition marking pheromones. In some species, bioactive markings are extractable with ethanol or water, suggesting that the involved compounds are more polar (Kainoh 1999). For instance, females of the braconid wasp *Diachasma alloeum* parasitizing fruit-infesting *Rhagoletis* flies, mark hawthorn fruits after oviposition with a clear excretion-like fluid that is soluble in 50% water/ethanol and acts as an oviposition deterrent (Stelinski *et al.* 2006). Interestingly, the signal is also avoided by females of the hosts (Stelinski *et al.* 2009), showing that parasitoid oviposition markings are also involved in multitrophic interactions.

Juvenile hormone III (JH III) (**25**) has been suggested to be a host marking pheromone in the braconid hyperparasitoid *Dendrocerus carpenteri* (Höller *et al.* 1994). External application of JH III to aphid mummies containing the host modified the subsequent oviposition behaviour of *D. carpenteri* females and decelerated the development of the host *Aphidius uzbekistanicus*. The amounts of JH III necessary to cause these effects were much higher than those found in the ovaries, the putative source of the host marking pheromone in *D. carpenteri*. However, treatment of ovary extracts with JH esterase stopped bioactivity, supporting the pheromonal function of JH III.

6.3.2 Putative alarm and appeasement pheromones

Disturbed females of the bethylid wasp *Cephalonomia stephanoderis* release skatole (**26**) from their mandibular gland (Gomez *et al.* 2005). Undisturbed females responded to the headspace of disturbed females and synthetic skatole by agitated running and flight attempts, which have been interpreted as alarmed behaviour. Thus, skatole has been suggested to function as a kind of alarm pheromone in this species. Skatole has been described as a defence compound in lacewings (Blum *et al.* 1973). Furthermore, the fact that other parasitoids produce defence compounds in their mandibular glands (Völkl *et al.* 1994, Stökl *et al.* 2012) suggests that skatole is primarily released for defensive purposes and that alarmed wasps eavesdrop on the defensive responses of their conspecifics. Within the Bethylidae, skatole occurs in several other species of the subfamily Epyrinae, whereas the spiroacetal 2-methyl-1,7-dioxaspiro[5.5]undecane (**27**) occurs in the mandibular gland of wasps in the subfamily Bethylinae (Goubault *et al.* 2008). This compound is produced by both genders and has been shown to be released by contest losers in intraspecific aggressive female–female interactions in the bethylid wasp *Goniozus legneri* (Goubault *et al.* 2006). Since the authors observed a reduction in aggression following the release of the spiroacetal, the compound appears also to be a defensive compound with a secondarily evolved intraspecific function (appeasement pheromone).

6.3.3 Aggregation pheromones

The first parasitoid aggregation pheromone to be identified was found in the chalcidid wasp *Brachymeria intermedia* (Mohamed & Coppel 1987). Both males and females aggregated on filter paper which had been conditioned for several months with male and/or female wasps. The active compound has been identified as 3-hexanone (**28**). The congeneric species *B. lasus* also aggregated on *B. intermedia*-conditioned paper but did not respond to 3-hexanone, suggesting that other compounds are additionally involved in the aggregation of *Brachymeria* wasps. Only males of the braconid wasp *Spathius agrili* release a seven-component volatile mixture, three of which have been shown to function as an aggregation pheromone (Cossé *et al.* 2012). The bioactive compounds are (4R,11E)-tetradecen-4-olide (**29**), (Z)-10-heptadecen-2-one (**30**) and dodecanal (**9**). The volatile mixture elicited anemotaxis of male and female wasps, with males responding more weakly. As in the species using male sex pheromones, only virgin females were responsive. Interestingly, individual pheromone emission was lower if males were sampled in groups. This suggests that the pheromone has evolved as a male sex attractant but is exploited by opportunistic males to obtain access to females attracted by others. Therefore males might decrease their pheromone emission in the presence of competitors.

As mentioned above, both males and females of *N. vitripennis* are attracted and arrested by male pheromone markings (Ruther *et al.* 2007, 2011b). In this case, however, the pheromone should not be classified as an aggregation pheromone because the male response can be interpreted as site fidelity of the releasing males, enabling them to use their pheromone economically. However, it is not known whether opportunistic males respond to the pheromone markings of males in a natural environment in order to save their own resources.

An intriguing example of chemically mediated aggregation has been described in the gregarious eulophid wasp *M. digitata*, which parasitizes a number of hosts but especially mud-dauber wasps (Hymenptera: Sphecidae, Crabronidae). After development, female wasps collaborate to gnaw exit holes allowing them to leave the host and the subterranean mud nests. The cooperative chewing is mediated by an interaction of physical cues (chewing pits) and a gnawing pheromone which is present in the venom or Dufour's gland secretion and is released by stinging the surface to be gnawed on (Deyrup *et al.* 2005). However, the chemicals involved are unknown.

6.3.4 Anti-aggregation pheromones

For the same reasons listed for marking pheromones (avoidance of intraspecific competition) it might be beneficial for parasitoid females to detect from a distance host patches already being exploited by conspecific females. Females of *L. heterotoma*, for instance, avoid the odour of conspecific females when these had access to hosts feeding on stinkhorn (Janssen *et al.* 1995). The same effect was caused by the odour of congeneric *L. clavipes* females. However, *L. heterotoma* females responded indifferently when the same odour sources (parasitoid females + host larvae + stinkhorn) were offered but access of females to the hosts was prevented. It needs to be investigated whether the identified iridoid defence compounds with sex pheromone function (**7, 8**) (Stökl *et al.* 2012) are also involved in this interaction.

6.4 Variability in pheromone-mediated sexual behaviour

Sex pheromone communication in parasitic wasps can be influenced by a genetically fixed (innate) or learnt behavioural plasticity as well as by abiotic factors. These may affect either the sender of the signal, the receiver, or both. In order to carry out successful pheromone studies and to use parasitoids most effectively in pest management, it is essential to know as much as possible about these factors. A few recent studies that have addressed this aspect more thoroughly are reviewed here.

6.4.1 Innate plasticity of pheromone behaviour

Females of many parasitic wasps are monandrous and therefore it is predicted that they respond to male-derived sex attractants only as virgins and should stop their own pheromone production after mating. In contrast, males commonly mate several times and therefore both male pheromone production and response should be less constrained by their mating status. In fact, the females of several parasitoid species become unattractive to males after mating (Quicke 1997, Fauvergue *et al.* 1999). Furthermore, only virgin females of *M. digitata* and *N. vitripennis*, two parasitoids with male-derived sex attractants, respond to these chemical stimuli (Cônsoli *et al.* 2002, Ruther *et al.* 2007). In *N. vitripennis*, the mechanism of the mating-induced behavioural switch has been studied more thoroughly. Within minutes after mating, the olfactory preference of females switches from the male pheromone to host odours (Steiner & Ruther 2009a) and high doses of the male pheromone are even avoided after mating (Ruther *et al.* 2007). The behavioural switch is independent of the transfer of ejaculate and post-copulatory courtship (Ruther *et al.* 2010), which has been shown to be crucial for triggering resistance to remating in *Nasonia* (van den Assem & Visser 1976). Instead, it is the receptivity signal shown by females in response to the male oral aphrodisiac which makes them unresponsive to the abdominal sex attractant.

The impact of mating status, age and diel periodicity on pheromone communication has been studied in the aphid parasitoid *Aphidius ervi*, which uses two unknown female-derived sex pheromones for mate attraction and courtship (McClure *et al.* 2007). Females mated 2 hours or 24 hours before being tested in a bioassay were significantly less attractive to males than virgins were, while the male response was not affected by its mating history. In contrast, the factor 'age' decreased both the attractiveness of virgin females and male responsiveness when the wasps were more than 6 days old. Furthermore, both female attractiveness and male response were stronger in the morning than in the afternoon, indicating diel periodicity in pheromone communication. Similar results were obtained with the egg parasitoid *Anaphes listronoti*, where males responded most intensely to newly emerged virgin females 4–6 hours after the onset of the photophase. Older females (>1 day), however, appeared to release the pheromone earlier (Cormier *et al.* 1998).

6.4.2 Learnt plasticity of pheromone behaviour

Associative learning by parasitoid females has been studied extensively in the context of host finding (Vet & Groenewold 1990, Vet & Dicke 1992, Steidle & van Loon 2002, Hoedjes *et al.* 2011). However, the fact that males can also learn in the context of mate finding has been widely ignored in the past. Experienced males of *C. tarsalis*, for instance, were arrested by the odour of a grain-derived rearing medium infested by the sawtoothed grain beetle *Oryzaephilus surinamensis* (Collatz *et al.* 2009). This medium was also highly attractive to females. Likewise, experienced males and females of *L. distinguendus* responded to the pheromones of astigmatic mites co-occurring with granary weevil hosts in a stored product environment (Ruther & Steidle 2000). This suggests that any olfactory cue perceived during mating might be learnt by males and subsequently used for mate finding. This hypothesis has been corroborated by further experiments with *A. ervi*. Males exposed to vanilla odour during copulation with a female subsequently showed courtship behaviour in the absence of females when vanilla odour was present (Villagra *et al.* 2005). Relevant habitat-related odours can also be learnt associatively by *A. ervi*. Males exposed during copulation to different herbivore plant complexes (HPCs) preferred the odour of the respective HPC in a subsequent olfactometer bioassay and showed courtship behaviour even in the absence of females (Villagra *et al.* 2008). This response was specific for the respective HPC and no cross-reaction was observed. Males of the ichneumonid wasp *Itoplectis conquisitor* have even been shown to learn visual cues associatively. Males conditioned with pheromone dispensers containing female pheromone extracts subsequently responded to the dispensers even in the absence of the pheromone (Robacker *et al.* 1976).

6.4.3 Plasticity of pheromone behaviour caused by abiotic factors

The impact of the abiotic factors wind speed and atmospheric pressure on male pheromone responses have been studied in *Aphidius nigripes* and *A. ervi* (Marchand & McNeil 2000, McClure & McNeil 2009). In general, high wind speeds reduced or even stopped male responses to the female pheromones. Furthermore, results for *A. ervi* suggested that, at higher wind velocities (>50 cm/s), males increasingly forage for females by walking. Large variations (>5 millibars) in atmospheric pressure within 24 hours prior to the bioassay decreased male responsiveness significantly in *A. nigripes* (Marchand & McNeil 2000). In *A. ervi* no such effects were observed, but the fluctuations in atmospheric pressure were less pronounced in this study (McClure & McNeil 2009).

6.5 Pheromone biosynthesis

The biosynthesis of CHCs has been intensively studied in other insects (Blomquist 1993, 2010) and it is reasonable to assume that the same pathways are used by parasitic wasps. Studies on non-hydrocarbon pheromones have hitherto only been performed on the HDL stereoisomers in the genus *Nasonia*. Stable isotope labelling experiments using ^{13}C-labelled precursors revealed that linoleic and 12,13-epoxy-octadec-9Z-enoic acids are precursors of HDL in *N. vitripennis* (Abdel-Latief *et al.* 2008, Blaul & Ruther 2011). Furthermore, exposure of *N. vitripennis* males to [18]O_2 showed that atmospheric oxygen is incorporated into the pheromone (J. Ruther & L.-A. Garbe, unpublished data). RNA interference experiments revealed that the epoxide hydrolase gene *Nasvi-EH1* is functionally involved in the biosynthetic pathway (Abdel-Latief *et al.* 2008). Three putative alcohol dehydrogenases (ADHs) are involved in the transformation of (4R,5S)-HDL into (4R,5R)-HDL, thus being involved in the pheromone difference between *N. vitripennis* and its sister species (Niehuis *et al.* 2013; see Section 6.6).

6.6 Evolution of parasitoid sex pheromones

How chemical compounds evolve into sex pheromones and how these signals diversify during speciation is probably one of the most challenging questions that chemical ecologists are currently facing (Symonds & Elgar 2008). In order to make any progress, the mere identification of new pheromones is not sufficient. Rather, as many details as possible of a species' genetics, phylogeny, biochemistry, behaviour and ecology need to be considered in order to understand the process of chemosensory speciation (Symonds & Elgar 2008, Smadja & Butlin 2009). Often this information is available only for extensively studied model organisms. As can be deduced from the previous sections, some parasitoids such as *L. distinguendus* – and particularly *N. vitripennis* with its sequenced genome (Werren *et al.* 2010) – might be excellent candidates for studying pheromone evolution. Some of the most urgent questions are:

- What makes a natural product a sexual signal?
- Is it ecological factors that provide the selective forces for pheromone diversification or does it occur due to genetic drift?
- What genetic mechanisms and biochemical pathways underlie the modification of chemical signals during speciation?
- How do pheromone production and pheromone perception co-evolve?

There is increasing evidence that many parasitoid pheromones have evolved from chemicals with different primary functions such as chemical defence or physical protection. This has been referred to in the literature as 'semiochemical parsimony' (Blum 1996) and is not uncommon in insects. The iridoids, for instance, functioning as a female sex pheromone in *L. heterotoma* (see above) are effective repellents against ants (Stökl *et al.* 2012). Initially, both males and females could have been defended by one iridomyrmecin isomer only. Some presumably simple biosynthetic modifications might have caused the diversification of iridoids in *L. heterotoma* females, which is the prerequisite for their sex pheromone function. Other mandibular compounds with intraspecific functions such as skatole (**26**) also have defensive functions in other insects (Blum *et al.* 1973), which might support the secondary function hypothesis (see Section 6.3.2).

The primary function of CHCs is to form a hydrophobic layer on the insect cuticle, thus preventing desiccation (Gibbs 1998, Gibbs & Rajpurohit 2010). Through selection for certain structural parameters such as chain length, methyl branches or double bonds, the protective function of CHCs might have been optimized for each species because they influence physical properties such as melting point, viscosity and volatility (Gibbs & Rajpurohit 2010). The resulting species specificity of CHC profiles contains potential information which might subsequently have been exploited by male and female conspecifics following the evolution of the chemosensory prerequisites. Selective forces driving this process could have been fitness benefits such as improved host discrimination due to the early recognition of parasitized hosts by females or the early localization by males of virgin females within a host. This might have resulted in the secondary pheromone functions of CHCs as marking and contact sex pheromones, respectively (see Sections 6.2.2 and 6.3.1).

However, how does a species-specific CHC profile evolve into a sex-specific signal? The study on the *L. distinguendus* contact sex pheromone (Ruther & Steiner 2008) is one of the rare examples providing insights into the ecological factors driving the evolution of gender-specific CHC profiles: males localize wheat grains containing potential mating partners due to a CHC profile which is present on both female and male conspecifics. However, newly emerged males that smell like females are clearly disadvantaged by homosexual courtship activities and therefore males might have evolved the pheromone deactivation mechanism described previously (see Section 6.2.2), resulting in the removal of 3-MeC27 (**21**) from the CHC profile and converting the species-specific CHC profile to a gender-specific one (Steiner *et al.* 2005).

CHC profiles of polyphagous parasitoid species may vary with the host species (Howard 2001, Howard & Perez-Lachaud 2002, Kühbandner *et al.* 2012a). For instance, the CHC profiles of *L. distinguendus* females reared on different stored-product-infesting beetles were distinguishable by multivariate statistical methods. Wasps reared on the granary weevil *Sitophilus granarius* for many generations changed their CHC profile in a detectable manner after only one generation on the drugstore beetle *Stegobium paniceum* (Kühbandner *et al.* 2012a). Considering the pivotal role that CHCs play in sexual communication, these data suggest that host shifts in parasitic wasps might result in the reproductive isolation of host races due to a modification in the cuticular semiochemistry and might finally lead to speciation. However, the question of whether the observed modifications in CHCs actually influence the mating behaviour of the wasps needs to be studied in the future.

A well-explained example of pheromone diversification during insect speciation has been reported in the genus *Nasonia*. The *Nasonia* species complex consists of the four closely related species *N. vitripennis*, *N. giraulti*, *N. longicornis* and *N. oneida*. All species lay their eggs into the pupae of blow flies and flesh flies. *Nasonia vitripennis* is cosmopolitan, whereas the other species are restricted to North America. *Nasonia vitripennis* occurs sympatrically with *N. giraulti* in the eastern and with *N. longicornis* in the western part of the USA (Grillenberger *et al.* 2009). *N. oneida* has hitherto been found only in upstate New York where it is sympatric with *N. vitripennis* and *N. giraulti* (Raychoudhury *et al.* 2010). Interspecific mating is possible in *Nasonia* but hybrids do not normally occur, due to *Wolbachia*-mediated cytoplasmic incompatibility (Werren 1997). Therefore, interspecific mating imposes high fitness costs which might be avoided by pheromone-mediated prezygotic isolation. The sex pheromone of *N. vitripennis* consists of the three components (4*R*,5*S*)-HDL (**3**), (4*R*,5*R*)-HDL (**4**) and 4-MQ (**5**), whereas in the other species (4*R*,5*R*)-HDL is missing (Niehuis *et al.* 2013). This suggests that (4*R*,5*S*)-HDL plus 4-MQ is the ancestral pheromone phenotype and (4*R*,5*R*)-HDL has evolved when *N. vitripennis* diverged approximately a million years ago (Werren *et al.* 2010). The genetic basis responsible for this species difference in pheromone composition has been found, by quantitative trait locus (QTL) mapping, to be three highly homologous putative ADH genes localized on chromosome 1 (Niehuis *et al.* 2013). Knocking down these genes in *N. vitripennis* by RNA interference resulted in a loss of the *N. vitripennis* phenotype (i.e. these wasps were unable to synthesize (4*R*,5*R*)-HDL). ADHs are $NAD^+/NADH$-dependent oxidoreductases which may catalyse both the oxidation of alcohols to the respective carbonyls and vice versa (Tanner 2002, Kavanagh *et al.* 2008). Therefore, it has been suggested that the ADHs act as epimerases in *N. vitripennis*, catalysing the conversion of (4*R*,5*S*)-HDL into (4*R*,5*R*)-HDL. Apart from the pheromone diversification in the sender, novel signals have to be discriminated by the receiver. Since pheromone biosynthesis and perception should be inherited independently, novel phenotypes are predicted to be susceptible to stabilizing selection; that is, any modification in established communication channels should be selected against (Smadja & Butlin 2009). *N. vitripennis* females respond to the HDL stereoisomers stereoselectively. In olfactometer bioassays, (4R,5S)-HDL was attractive alone, whereas (4R,5R)-HDL was behaviourally inactive but had a strong synergetic effect in combination with (4R,5S)-HDL. In contrast, females of the sister species *N. giraulti* were unable to discriminate between their own and the *N. vitripennis* pheromone phenotype (Niehuis *et al.* 2013). This indicates that the evolution of the novel pheromone component (4R,5R)-HDL in *N. vitripennis* had no detrimental effects and might explain why it was not eliminated by stabilizing selection. Hence, *N. vitripennis* females had the opportunity to evolve the chemosensory adaptations to detect and to benefit from the derived pheromone phenotype. In conclusion, stereoselectivity in both pheromone biosynthesis and perception has resulted in a pheromone-mediated prezygotic isolation between *N. vitripennis* and the sympatric sister species.

6.7 Conclusions and outlook

Despite the significant advances made in recent years, hundreds of parasitoid pheromones remain to be discovered in the future. The recent studies described in this chapter should have opened the way for a more comprehensive review of chemical communication in these insects. The development of new analytical techniques for semiochemical identification and the availability of mass rearing methods for a number of species have eased the problems researchers had to face in the past when trying to identify parasitoid pheromones. Furthermore, some parasitoids appear to invest heavily in pheromone communication, resulting in surprisingly high amounts of the bioactive chemicals, which should facilitate structural elucidation of further pheromones. However, novel analytical approaches might be necessary to identify the first antennal or mandibular aphrodisiac pheromone, given that these are proteinaceous or composed of other non-volatile components. Gender-specific glands present not only on the antennae but also in other body parts of many parasitoids may facilitate the identification of candidate organisms for pheromone studies. Males of some Braconidae, for instance, possess an abdominal gland, the so-called Hagen's gland (Khoo & Lawrence 2002). This structure is responsible for the aromatic smell of these wasps and has been studied with respect to its chemical composition (Williams *et al.* 1988, Paddon-Jones *et al.* 1997, Mereyala *et al.* 2000) but, to my knowledge, its function has yet to be elucidated.

The interplay between pheromonal signals and non-pheromonal cues (sexual kairomones *sensu* Ruther *et al.* 2002), has probably been underestimated in the past. Males searching for mates may benefit from learning host-associated cues, just as foraging females do (Vet & Groenewold 1990, Vet & Dicke 1992). Future studies might reveal that this parsimonious strategy for increasing male mating chances is much more common than previously thought. In this context, it would also be interesting to study whether the mechanisms of memory formation, memory duration and memory structure in the brain of male and female parasitoids show any differences (Hoedjes *et al.* 2011, Schurmann *et al.* 2012).

The availability of the genomes of the parasitoid genus *Nasonia* represents a great opportunity to study not only the biosynthesis (Abdel-Latief *et al.* 2008) but also the perception of pheromones. In *Nasonia*, 225 putative olfactory receptor (OR) genes and 90 genes encoding odorant binding proteins (OBP) have been identified (Robertson *et al.* 2010, Vieira *et al.* 2012), which now await functional characterization. It will be interesting to study which of these ORs and OBPs are involved in the perception of sex pheromones and how they interact to maintain the observed stereoselectivity in the pheromone response.

Some recent studies have demonstrated that parasitoid pheromones offer great opportunities to test evolutionary hypotheses in the context of mate choice and sexual selection (Ruther *et al.* 2009, Steiner & Ruther 2009a, Blaul & Ruther 2011) which hitherto have mostly been studied using visual or acoustic signals (Steiger 2012). The benefits from using parasitoid pheromones to test those hypotheses are manifold: many potential study organisms have short generation times and can be reared in unlimited numbers. The sexual behaviour is often stereotypic and easy to quantify. Proximate parameters underlying parasitoid pheromone communication such as biosynthetic pathways and perception are accessible (see above), or will probably be accessible in the near future, enabling researchers to manipulate communication channels and study the effects of these modifications on mate choice decisions. For this purpose, both signal quality and strength can be accurately controlled using the well-established tools of analytical chemistry. Stable isotope labelling techniques have successfully been applied to track individual releases of these apparently invisible signals in real time (Goubault *et al.* 2006). Finally, the outcome of mate choice experiments can easily be analysed by using, for instance, eye colour mutants (Moynihan & Shuker 2011). Hence, a deeper study of parasitoid pheromones might lead to major advances in our general understanding of animal communication. The results of those studies might also be helpful for the optimal management of natural parasitoid populations in order to maximize their effectiveness as natural enemies and, thus, their contribution to ecosystem services (Suckling *et al.* 2002).

References

Abdel-Latief, M., Garbe, L.A., Koch, M. and Ruther, J. (2008) An epoxide hydrolase involved in the biosynthesis of an insect sex attractant and its use to localize the production site. *Proceedings of the National Academy of Sciences USA* **105**: 8914–19.

Ablard, K., Gries, R., Khaskin, G., Schaefer, P.W. and Gries, G. (2012) Does the stereochemistry of methylated cuticular hydrocarbons contribute to mate recognition in the egg parasitoid wasp *Ooencyrtus kuvanae*? *Journal of Chemical Ecology* **38**: 1306–17.

Anderson, P. (2002) Oviposition pheromones in herbivorous and carnivorous insects. In: Hilker, M. and Meiners, T. (eds) *Chemoecology of Insect Eggs and Egg Deposition*. Blackwell, Berlin, pp. 235–63.

Ayasse, M., Paxton, R.J. and Tengo, J. (2001) Mating behavior and chemical communication in the order Hymenoptera. *Annual Review of Entomology* **46**: 31–78.

Battaglia, D., Isidoro, N., Romani, R., Bin, F. and Pennacchio, F. (2002) Mating behaviour of *Aphidius ervi* (Hymenoptera: Braconidae): the role of antennae. *European Journal of Entomology* **99**: 451–6.

Benelli, G. and Canale, A. (2012) Do *Psyttalia concolor* (Hymenoptera: Braconidae) males gain in mating competitiveness from being courted by other males while still young? *Entomological Science* **15**: 257–60.

Benelli, G., Bonsignori, G., Stefanini, C. and Canale, A. (2012) Courtship and mating behaviour in the fruit fly parasitoid *Psyttalia concolor* (Szepligeti) (Hymenoptera: Braconidae): the role of wing fanning. *Journal of Pest Science* **85**: 55–63.

Bernal, J.S. and Luck, R.F. (2007) Mate finding via a trail sex pheromone by *Aphytis melinus* DeBach (Hymenoptera: Aphelinidae) males. *Journal of Insect Behavior* **20**: 515–25.

Bin, F. and Vinson, S.B. (1986) Morphology of the antennal sex gland in male *Trissolcus basalis* (Woll) (Hymenoptera, Scelionidae), an egg parasitoid of the green stink bug, *Nezara viridula* (Hemiptera, Pentatomidae). *International Journal of Insect Morphology and Embryology* **15**: 129–38.

Bin, F., Wäckers, F., Romani, R. and Isidoro, N. (1999) Tyloids in *Pimpla turionellae* (L.) are release structures of male antennal glands involved in courtship behaviour (Hymenoptera: Ichneumonidae). *International Journal of Insect Morphology and Embryology* **28**: 61–8.

Blaul, B. and Ruther, J. (2011) How parasitoid females produce sexy sons: a causal link between oviposition preference, dietary lipids and mate choice in *Nasonia*. *Proceedings of the Royal Society B – Biological Sciences* **278**: 3286–93.

Blaul, B. and Ruther, J. (2012) Body size influences male pheromone signals but not the outcome of mating contests in *Nasonia vitripennis*. *Animal Behaviour* **84**: 1557–63.

Blomquist, G.J. (1993) Hydrocarbon and hydrocarbon-derived sex pheromones in insects: biochemistry and endocrine regulation. In: Stanley-Samuelson, D.W. and Nelson, D.R. (eds) *Insect Lipids: Chemistry, Biochemistry, and Biology*. University of Nebraska Press, Lincoln, NE, pp. 317–51.

Blomquist, G.J. (2010) Biosynthesis of cuticular hydrocarbons. In: Blomquist, G.J. and Bagnères, A.G. (eds) *Insect Hydrocarbons*. Cambridge University Press, Cambridge, pp. 35–52.

Blum, M.S. (1996) Semiochemical parsimony in the Arthropoda. *Annual Review of Entomology* **41**: 353–74.

Blum, M.S., Wallace, J.B. and Fales, H.M. (1973) Skatole and tridecene: identification and possible role in a chrysopid secretion. *Insect Biochemistry* **3**: 353–7.

Buckner, J.S. and Jones, W.A. (2005) Transfer of methyl-branched hydrocarbons from the parasitoid, *Eretmocerus mundus*, to silverleaf whitefly nymphs during oviposition. *Comparative Biochemistry and Physiology A – Molecular & Integrative Physiology* **140**: 59–65.

Collatz, J., Tolasch, T. and Steidle, J.L.M. (2009) Mate finding in the parasitic wasp *Cephalonomia tarsalis* (Ashmead): more than one way to a female's heart. *Journal of Chemical Ecology* **35**: 761–8.

Cônsoli, F.L., Williams, H.J., Vinson, S.B., Matthews, R.W. and Cooperband, M.F. (2002) Transbergamotenes-male pheromone of the ectoparasitoid *Melittobia digitata*. *Journal of Chemical Ecology* **28**: 1675–89.

Cormier, D., Royer, L., Vigneault, C., Panneton, B. and Boivin, G. (1998) Effect of female age on daily cycle of sexual pheromone emission in gregarious egg parasitoid *Anaphes listronoti*. *Journal of Chemical Ecology* **24**: 1595–610.

Cossé, A.A., Petroski, R.J., Zilkowski, B.W., Vermillion, K., Lelito, J.P., Cooperband, M.F. and Gould, J.R. (2012) Male-produced pheromone of *Spathius agrili*, a parasitoid introduced for the biological control of the invasive emerald ash borer, *Agrilus planipennis*. *Journal of Chemical Ecology* **38**: 389–99.

Dahms, E.C. (1984) An interpretation of the structure and function of the antennal sense organs of *Mellittobia australica* (Hymenoptera: Eulophidae) with the discovery of a large dermal gland in the male scape. *Memoirs of the Queensland Musem* **21**: 361–77.

Danci, A., Gries, R., Schaefer, P.W. and Gries, G. (2006) Evidence for four-component close-range sex pheromone in the parasitic wasp *Glyptapanteles flavicoxis*. *Journal of Chemical Ecology* **32**: 1539–54.

Danci, A., Inducil, C., Schaefer, P.W. and Gries, G. (2011) Early detection of prospective mates by males of the parasitoid wasp *Pimpla disparis* Viereck (Hymenoptera: Ichneumonidae). *Environmental Entomology* **40**: 405–11.

Danci, A., Takacs, S., Schaefer, P.W. and Gries, G. (2010) Evidence for acoustic communication in the parasitoid wasp *Glyptapanteles flavicoxis*. *Entomologia Experimentalis et Applicata* **136**: 142–50.

Darrouzet, E., Lebreton, S., Gouix, N., Wipf, A. and Bagnères, A.G. (2010) Parasitoids modify their oviposition behavior according to the sexual origin of conspecific cuticular hydrocarbon traces. *Journal of Chemical Ecology* **36**: 1092–100.

Delury, N.C., Gries, G., Gries, R., Judd, G.J.R. and Brown, J.J. (1999) Sex pheromone of *Ascogaster quadridentata*, a parasitoid of *Cydia pomonella*. *Journal of Chemical Ecology* **25**: 2229–45.

Deyrup, L.D., Matthews, R.W. and Gonzalez, J.M. (2005) Cooperative chewing in a gregariously developing parasitoid wasp, *Melittobia digitata* Dahms, is stimulated by structural cues and a pheromone in crude venom extract. *Journal of Insect Behavior* **18**: 293–304.

Eller, F.J., Bartelt, R.J., Jones, R.L. and Kulman, H.M. (1984) Ethyl (*Z*)-9-hexadecenoate, a sex pheromone of *Syndipnus rubiginosus*, a sawfly parasitoid. *Journal of Chemical Ecology* **10**: 291–300.

Fauvergue, X., Fleury, F., Lemaitre, C. and Allemand, R. (1999) Parasitoid mating structures when hosts are patchily distributed: field and laboratory experiments with *Leptopilina boulardi* and *L. heterotoma*. *Oikos* **86**: 344–56.

Fauvergue, X., Hopper, K.R. and Antolin, M.F. (1995) Mate finding via a trail sex pheromone by a parasitoid wasp. *Proceedings of the National Academy of Sciences USA* **92**: 900–4.

Finidori-Logli, V., Bagnères, A.G., Erdmann, D., Francke, W. and Clément, J.L. (1996) Sex recognition in *Diglyphus isaea* Walker (Hymenoptera: Eulophidae): role of an uncommon family of behaviorally active compounds. *Journal of Chemical Ecology* **22**: 2063–79.

Gibbs, A.G. (1998) Waterproofing properties of cuticular lipids. *American Zoologist* **38**: 471–82.

Gibbs, A.G. and Rajpurohit, S. (2010) Cuticular lipids and water balance. In: Blomquist, G.J. and Bagnères, A.G. (eds) *Insect Hydrocarbons*. Cambridge University Press, Cambridge, pp. 100–20.

Godfray, H.C.J. (1994) *Parasitoids: Behavioral and Evolutionary Ecology*. Princeton University Press, Princeton, NJ.

Goh, M.Z. and Morse, D H. (2010) Male mate search for female emergence sites by a parasitic wasp. *Animal Behaviour* **80**: 391–8.

Gomez, J., Barrera, J.F., Rojas, J.C., Macias-Samano, J., Liedo, J.P. and Badii, M.H. (2005) Volatile compounds released by disturbed females of *Cephalonomia stephanoderis* (Hymenoptera: Bethylidae): a parasitoid of the coffee berry borer *Hypothenemus hampei* (Coleoptera: Scolytidae). *Florida Entomologist* **88**: 180–7.

Gonzalez, J.M., Matthews, R.W. and Matthews, J.R. (1985) A sex pheromone in males of *Melittobia australica* and *Melittobia femorata* (Hymenoptera, Eulophidae). *Florida Entomologist* **68**: 279–86.

Goubault, M., Batchelor, T.P., Linforth, R.S.T., Taylor, A.J. and Hardy, I.C.W. (2006) Volatile emission by contest losers revealed by real-time chemical analysis. *Proceedings of the Royal Society B – Biological Sciences* **273**: 2853–9.

Goubault, M., Batchelor, T.P., Romani, R., Linforth, R.S.T., Fritzsche, M., Francke, W. and Hardy, I.C.W. (2008) Volatile chemical release by bethylid wasps: identity, phylogeny, anatomy and behaviour. *Biological Journal of the Linnean Society* **94**: 837–52.

Grillenberger, B.K., van de Zande, L., Bijlsma, R., Gadau, J. and Beukeboom, L.W. (2009) Reproductive strategies under multiparasitism in natural populations of the parasitoid wasp *Nasonia* (Hymenoptera). *Journal of Evolutionary Biology* **22**: 460–70.

Guerrieri, E., Pedata, P.A., Romani, R., Isidoro, N. and Bin, F. (2001) Functional anatomy of male antennal glands in three species of Encyrtidae (Hymenoptera: Chalcidoidea). *Journal of Natural History* **35**: 41–54.

Guillot, F.S. and Vinson, S.B. (1972) Sources of substances which elicit a behavioral response from insect parasitoid, *Campoletis perdistinctus*. *Nature* **235**: 169–70.

Harrison, E.G., Fisher, R.C. and Ross, K.M. (1985) The temporal effects of Dufour's gland secretion in host discrimination by *Nemeritis canescens*. *Entomologia Experimentalis et Applicata* **38**: 215–20.

Hilgraf, R., Zimmermann, N., Lehmann, L., Tröger, A. and Francke, W. (2012) Stereoselective synthesis of *trans*-fused iridoid lactones and their identification in the parasitoid *Alloxysta victrix*. Part II. Iridomyrmecins. *Beilsteins Journal of Organic Chemistry* **8**: 1256–64.

Hoedjes, K.M., Kruidhof, H.M., Huigens, M.E., Dicke, M., Vet, L.E.M. and Smid, H.M. (2011) Natural variation in learning rate and memory dynamics in parasitoid wasps: opportunities for converging ecology and neuroscience. *Proceedings of the Royal Society B – Biological Sciences* **278**: 889–97.

Höller, C., Bargen, H., Vinson, S.B. and Witt, D. (1994) Evidence for the external use of juvenile hormone for host marking and regulation in a parasitic wasp, *Dendrocerus carpenteri*. *Journal of Insect Physiology* **40**: 317–22.

Howard, R.W. (2001) Cuticular hydrocarbons of adult *Pteromalus cerealellae* (Hymenoptera: Pteromalidae) and two larval hosts, angoumois grain moth (Lepidoptera: Gelechiidae) and cowpea weevil (Coleoptera: Bruchidae). *Annals of the Entomological Society of America* **94**: 152–8.

Howard, R.W. and Baker, J.E. (2003) Morphology and chemistry of Dufour glands in four ectoparasitoids: *Cephalonomia tarsalis*, *C. waterstoni* (Hymenoptera: Bethylidae), *Anisopteromalus calandrae*, and *Pteromalus cerealellae* (Hymenoptera: Pteromalidae). *Comparative Biochemistry and Physiology B – Biochemistry and Molecular Biology* **135**: 153–67.

Howard, R.W. and Perez-Lachaud, G. (2002) Cuticular hydrocarbons of the ectoparasitic wasp *Cephalonomia hyalinipennis* (Hymenoptera: Bethylidae) and its alternative host, the stored product pest *Caulophilus oryzae* (Coleoptera: Curculionidae). *Archives of Insect Biochemistry and Physiology* **50**: 75–84.

Hrabar, M., Danci, A., Schaefer, P.W. and Gries, G. (2012) In the nick of time: males of the parasitoid wasp *Pimpla disparis* respond to semiochemicals from emerging mates. *Journal of Chemical Ecology* **38**: 253–61.

Hübner, G., Volkl, W., Francke, W. and Dettner, K. (2002) Mandibular gland secretions in alloxystine wasps (Hymenoptera, Cynipoidea, Charipidae): do ecological or phylogenetical constraints influence occurrence or composition? *Biochemical Systematics and Ecology* **30**: 505–23.

Isidoro, N. and Bin, F. (1995) Male antennal gland of *Amitus spiniferus* (Brethes) (Hymenoptera, Platygastridae), likely involved in courtship behavior. *International Journal of Insect Morphology and Embryology* **24**: 365–73.

Isidoro, N., Bin, F., Colazza, S. and Vinson, S.B. (1996) Morphology of antennal gustatory sensilla and glands in some parasitoid Hymenoptera with hypothesis on their role in sex and host recognition. *Journal of Hymenopteran Research* **5**: 206–39.

Isidoro, N., Bin, F., Romani, R., Pujade-Villar, J. and Ros-Farre, P. (1999) Diversity and function of ale antennal glands in Cynipoidea (Hymenoptera). *Zoologica Scripta* **28**: 165–74.

Jaloux, B., Errard, C., Mondy, N., Vannier, F. and Monge, J.P. (2005) Sources of chemical signals which enhance multiparasitism preference by a cleptoparasitoid. *Journal of Chemical Ecology* **31**: 1325–37.

Janssen, A., van Alphen, J.J.M., Sabelis, M.W. and Bakker, K. (1995) Specificity of odor mediated avoidance of competition in *Drosophila* parasitoids. *Behavioral Ecology and Sociobiology* **36**: 229–35.

Kainoh, Y. (1999) Parasitoids. In: Hardie, J. and Minks, A.K. (eds) *Pheromones of Non-Lepidopteran Insects Associated with Agricultural Plants*. CABI Publishing, Wallingford, UK, pp. 383–404.

Kainoh, Y. and Oishi, Y. (1993) Source of sex pheromone of the egg-larval parasitoid, *Ascogaster reticulatus* Watanabe (Hymenoptera, Braconidae). *Journal of Chemical Ecology* **19**: 963–9.

Kainoh, Y., Nemoto, T., Shimizu, K., Tatsuki, S., Kusano, T. and Kuwahara, Y. (1991) Mating behavior of *Ascogaster reticulatus* Watanabe (Hymenoptera, Braconidae), an egg-larval parasitoid of the smaller tea tortrix, *Adoxophyes* sp. (Lepidoptera, Tortricidae). 3. Identification of a sex pheromone. *Applied Entomology and Zoology* **26**: 543–9.

Kavanagh, K., Jornvall, H., Persson, B. and Oppermann, U. (2008) The SDR superfamily: functional and structural diversity within a family of metabolic and regulatory enzymes. *Cellular and Molecular Life Sciences* **65**: 3895–906.

Keeling, C.I., Plettner, E. and Slessor, K.N. (2004) Hymenopteran semiochemicals. *Topics in Current Chemistry* **239**: 133–77.

Khoo, C.C.H. and Lawrence, P.O. (2002) Hagen's glands of the parasitic wasp *Diachasmimorpha longicaudata* (Hymenoptera: Braconidae): ultrastructure and the detection of entomopoxvirus and parasitism-specific proteins. *Arthropod Structure and Development* **31**: 121–30.

King, B.H. and Skinner, S.W. (1991) Proximal mechanisms of the sex ratio and clutch size responses of the wasp *Nasonia vitripennis* to parasitized hosts. *Animal Behaviour* **42**: 23–32.

Klopfstein, S., Quicke, D.L.J. and Kropf, C. (2010) The evolution of antennal courtship in diplazontine parasitoid wasps (Hymenoptera, Ichneumonidae, Diplazontinae). *BMC Evolutionary Biology* **10**: 218.

Krokos, F.D., Konstantopoulou, M.A. and Mazomenos, B.E. (2001) Alkadienes and alkenes, sex pheromone components of the almond seed wasp *Eurytoma amygdali*. *Journal of Chemical Ecology* **27**: 2169–81.

Kühbandner, S., Hacker, N., Niedermayer, S., Steidle, J.L.M. and Ruther, J. (2012a) Composition of cuticular lipids in the pteromalid wasp *Lariophagus distinguendus* is host dependent. *Bulletin of Entomological Research* **102**: 610–17.

Kühbandner, S., Sperling, S., Mori, K. and Ruther, J. (2012b) Deciphering the signature of cuticular lipids with contact sex pheromone function in a parasitic wasp. *Journal of Experimental Biology* **215**: 2471–8.

Li, G. (2006) Host-marking in hymenopterous parasitoids. *Acta Entomologica Sinica* **49**: 504–12.

Macfadyen, S., Craze, P.G., Polaszek, A., van Achterberg, K. and Memmott, J. (2011) Parasitoid diversity reduces the variability in pest control services across time on farms. *Proceedings of the Royal Society B – Biological Sciences* **278**: 3387–94.

Marchand, D. and McNeil, J.N. (2000) Effects of wind speed and atmospheric pressure on mate searching behavior in the aphid parasitoid *Aphidius nigripes* (Hymenoptera: Aphidiidae). *Journal of Insect Behavior* **13**: 187–99.

Mazomenos, B.E., Athanassiou, C.G., Kavallieratos, N. and Milonas, P. (2004) Evaluation of the major female *Eurytoma amygdali* sex pheromone components, (Z,Z)-6,9-tricosadiene and (Z,Z)-6,9-pentacosadiene for male attraction in field tests. *Journal of Chemical Ecology* **30**: 1245–55.

McClure, M. and McNeil, J.N. (2009) The effect of abiotic factors on the male mate searching behavior and the mating success of *Aphidius ervi* (Hymenoptera: Aphidiidae). *Journal of Insect Behavior* **22**: 101–10.

McClure, M., Whistlecraft, J. and McNeil, J.N. (2007) Courtship behavior in relation to the female sex pheromone in the parasitoid, *Aphidius ervi* (Hymenoptera: Braconidae). *Journal of Chemical Ecology* **33**: 1946–59.

Mereyala, H.B., Gadikota, R.R., Sunder, K.S. and Shailaja, S. (2000) Pd(II)Cl$_2$ mediated oxidative cyclisation of hydroxy-vinylfurans to lactols: synthesis of Hagen's gland lactones. *Tetrahedron* **56**: 3021–6.

Metzger, M., Fischbein, D., Auguste, A., Fauvergue, X., Bernstein, C. and Desouhant, E. (2010) Synergy in information use for mate finding: demonstration in a parasitoid wasp. *Animal Behaviour* **79**: 1307–15.

Mohamed, M.A. and Coppel, H.C. (1987) Pheromonal basis for aggregation behavior of parasitoids of the gypsy moth: *Brachymeria intermedia* (Nees) and *Brachymeria lasus* (Walker) (Hymenoptera: Chalcididae). *Journal of Chemical Ecology* **13**: 1385–93.

Moynihan, A.M. and Shuker, D.M. (2011) Sexual selection on male development time in the parasitoid wasp *Nasonia vitripennis*. *Journal of Evolutionary Biology* **24**: 2002–13.

Mudd, A., Fisher, R.C. and Smith, M.C. (1982) Volatile hydrocarbons in the Dufour's gland of the parasite *Nemeritis canescens* (Grav) (Hymenoptera, Ichneumonidae). *Journal of Chemical Ecology* **8**: 1035–42.

Nichols, W.J., Cosse, A.A., Bartelt, R.J. and King, B.H. (2010) Methyl 6-methylsalicylate: a female-produced pheromone component of the parasitoid wasp *Spalangia endius*. *Journal of Chemical Ecology* **36**: 1140–7.

Niehuis, O., Büllesbach, J., Gibson, J.D., Pothmann, D., Hanner, C., Judson, A.K., Navdeep, S., Gadau, J., Ruther, J. and Schmitt, T. (2013) Behavioural and genetic analysis on *Nasonia* shed light on pheromone evolution. *Nature* doi:10.1038/nature11838.

Nufio, C.R. and Papaj, D.R. (2001) Host marking behavior in phytophagous insects and parasitoids. *Entomologia Experimentalis et Applicata* **99**: 273–93.

Oldham, N.J., Billen, J. and Morgan, E.D. (1994) On the similarity of the Dufour gland secretion and the cuticular hydrocarbons of some bumblebees. *Physiological Entomology* **19**: 115–23.

Paddon-Jones, G.C., Moore, C.J., Brecknell, D.J., Konig, W.A. and Kitching, W. (1997) Synthesis and absolute stereochemistry of Hagen's-gland lactones in some parasitic wasps (Hymenoptera: Braconidae). *Tetrahedron Letters* **38**: 3479–82.

Petersen, G. (2000) *Signalstoffe der innerartlichen Kommunikation des Hyperparasitoiden* Alloxysta victrix *(Hymenoptera: Cynipidae) und ihre Wirkung auf den Primärparasitoiden* Aphidius uzbekistanicus *und die Grosse Getreideblattlaus* Sitobion avenae. PhD thesis, University of Kiel, Germany (in German).

Pompanon, F., DeSchepper, B., Mourer, Y., Fouillet, P. and Boulétreau, M. (1997) Evidence for a substrate-borne sex pheromone in the parasitoid wasp *Trichogramma brassicae. Journal of Chemical Ecology* **23**: 1349–60.

Quicke, D.L.J. (1997) *Parasitic Wasps*. Chapman & Hall, London.

Raychoudhury, R., Desjardins, C.A., Buellesbach, J., Loehlin, D.W., Grillenberger, B.K., Beukeboom, L., Schmitt, T. and Werren, J.H. (2010) Behavioral and genetic characteristics of a new species of *Nasonia. Heredity* **104**: 278–88.

Robacker, K.M. and Hendry, L.B. (1977) Neral and geranial: components of the sex pheromone of the parasitic wasp *Itoplectis conquisitor. Journal of Chemical Ecology* **3**: 563–77.

Robacker, D.C., Weaver, K.M. and Hendry, L.B. (1976) Sexual communication and associative learning in the parasitic wasp *Itoplectis conquisitor. Journal of Chemical Ecology* **2**: 39–48.

Robertson, H.M., Gadau, J. and Wanner, K.W. (2010) The insect chemoreceptor superfamily of the parasitoid jewel wasp *Nasonia vitripennis. Insect Molecular Biology* **19**: 121–36.

Romani, R., Rosi, M.C., Isidoro, N. and Bin, F. (2008) The role of the antennae during courtship behaviour in the parasitic wasp *Trichopria drosophilae. Journal of Experimental Biology* **211**: 2486–91.

Rosi, M.C., Isidoro, N., Colazza, S. and Bin, F. (2001) Source of the host marking pheromone in the egg parasitoid *Trissolcus basalis* (Hymenoptera: Scelionidae). *Journal of Insect Physiology* **47**: 989–95.

Ruther, J. and Steidle, J.L.M. (2000) Mites as matchmakers: semiochemicals from host-associated mites attract both sexes of the parasitoid *Lariophagus distinguendus. Journal of Chemical Ecology* **26**: 1205–17.

Ruther, J. and Steiner, S. (2008) Costs of female odour in males of the parasitic wasp *Lariophagus distinguendus* (Hymenoptera: Pteromalidae). *Naturwissenschaften* **95**: 547–52.

Ruther, J., Döring, M. and Steiner, S. (2011a) Cuticular hydrocarbons as contact sex pheromone in the parasitoid *Dibrachys cavus. Entomologia Experimentalis et Applicata* **140**: 59–68.

Ruther, J., Homann, M. and Steidle, J.L.M. (2000) Female-derived sex pheromone mediates courtship behaviour in the parasitoid *Lariophagus distinguendus. Entomologia Experimentalis et Applicata* **96**: 265–74.

Ruther, J., Matschke, M., Garbe, L.A. and Steiner, S. (2009) Quantity matters: male sex pheromone signals mate quality in the parasitic wasp *Nasonia vitripennis. Proceedings of the Royal Society B – Biological Sciences* **276**, 3303–10.

Ruther, J., Meiners, T. and Steidle, J. L. M. (2002) Rich in phenomena – lacking in terms: a classification of kairomones. *Chemoecology* **12**: 161–7.

Ruther, J., Sieben, S. and Schricker, B. (1998) Role of cuticular lipids in nestmate recognition of the European hornet *Vespa crabro* L. (Hymenoptera, Vespidae). *Insectes Sociaux* **45**: 169–79.

Ruther, J., Stahl, L.M., Steiner, S., Garble, L.A. and Tolasch, T. (2007) A male sex pheromone in a parasitic wasp and control of the behavioral response by the female's mating status. *Journal of Experimental Biology* **210**: 2163–9.

Ruther, J., Steiner, S. and Garbe, L.A. (2008) 4-methylquinazoline is a minor component of the male sex pheromone in *Nasonia vitripennis*. *Journal of Chemical Ecology* **34**: 99–102.

Ruther, J., Thal, K., Blaul, B. and Steiner, S. (2010) Behavioural switch in the sex pheromone response of *Nasonia vitripennis* females is linked to receptivity signalling. *Animal Behaviour* **80**: 1035–40.

Ruther, J., Thal, K. and Steiner, S. (2011b) Pheromone communication in *Nasonia vitripennis*: abdominal sex attractant mediates site fidelity of releasing males. *Journal of Chemical Ecology* **37**: 161–5.

Salerno, G., Iacovone, A., Carlin, S., Frati, F. Conti, E. and Anfora, G. (2012) Identification of sex pheromone components in *Trissolcus brochymenae* females. *Journal of Insect Physiology* **58**: 1635–42.

Schurmann, D., Sommer, C., Schinko, A.P.B., Greschista, M., Smid, H. and Steidle, J.L.M. (2012) Demonstration of long-term memory in the parasitic wasp *Nasonia vitripennis*. *Entomologia Experimentalis et Applicata* **143**: 199–206.

Sevala, V.L., Bagnères, A.G., Kuenzli, M., Blomquist, G.J. and Schal, C. (2000) Cuticular hydrocarbons of the dampwood termite, *Zootermopsis nevadensis*: caste differences and role of lipophorin in transport of hydrocarbons and hydrocarbon metabolites. *Journal of Chemical Ecology* **26**: 765–89.

Smadja, C. and Butlin, R.K. (2009) On the scent of speciation: the chemosensory system and its role in premating isolation. *Heredity* **102**: 77–97.

Steidle, J.L.M. and van Loon, J.J.A. (2002) Chemoecology of parasitoid and predator oviposition behaviour. In: Hilker, M. and Meiners, T. (eds) *Chemoecology of Insect Eggs and Egg Deposition.* Blackwell, Berlin, pp. 291–317.

Steiger, S. (2012) New synthesis – visual and chemical ornaments: what researchers of different signal modalities can learn from each other. *Journal of Chemical Ecology* **38**: 1.

Steiner, S. and Ruther, J. (2009a) How important is sex for females of a haplodiploid species under local mate competition? *Behavioral Ecology* **20**: 570–4.

Steiner, S. and Ruther, J. (2009b) Mechanism and behavioral context of male sex pheromone release in *Nasonia vitripennis*. *Journal of Chemical Ecology* **35**: 416–21.

Steiner, S., Hermann, N. and Ruther, J. (2006) Characterization of a female-produced courtship pheromone in the parasitoid *Nasonia vitripennis*. *Journal of Chemical Ecology* **32**: 1687–702.

Steiner, S., Steidle, J.L.M. and Ruther, J. (2005) Female sex pheromone in immature insect males: a case of pre-emergence chemical mimicry? *Behavioral Ecology and Sociobiology* **58**: 111–20.

Steiner, S., Steidle, J.L.M. and Ruther, J. (2007) Host-associated kairomones used for habitat orientation in the parasitoid *Lariophagus distinguendus* (Hymenoptera: Pteromalidae). *Journal of Stored Products Research* **43**: 587–93.

Stelinski, L.L., Pelz-Stelinski, K.S. and Gut, L.J. (2006) Male *Diachasma alloeum* parasitoids from two host species of tephritid fruit flies respond equally to female-produced sex pheromone. *Physiological Entomology* **31**: 178–83.

Stelinski, L.L., Rodriguez-Saona, C. and Meyer, W.L. (2009) Recognition of foreign oviposition-marking pheromone in a multi-trophic context. *Naturwissenschaften* **96**: 585–92.

Stökl, J., Hofferberth, J., Pritschet, M., Brummer, M. and Ruther, J. (2012) Stereoselective chemical defense in the *Drosophila* parasitoid *Leptopilina heterotoma* is mediated by (−)-iridomyrmecin and (+)-isoiridomyrmecin. *Journal of Chemical Ecology* **38**: 331–9.

Suckling, D.M., Gibb, A.R., Burnip, G.M. and Delury, N.C. (2002) Can parasitoid sex pheromones help in insect biocontrol? A case study of codling moth (Lepidoptera: Tortricidae) and its parasitoid *Ascogaster quadridentata* (Hymenoptera: Braconidae). *Environmental Entomology* **31**: 947–52.

Sullivan, B.T. (2002) Evidence for a sex pheromone in bark beetle parasitoid *Roptrocerus xylophagorum*. *Journal of Chemical Ecology* **28**: 1045–63.

Swedenborg, R.L.J. (1992) (*Z*)-4-Tridecenal, a pheromonally active air oxidation product from a series of (*Z,Z*)-9,13 dienes in *Macrocentrus grandii* Goidanich (Hymenoptera: Braconidae). *Journal of Chemical Ecology* **18**: 1913–31.

Swedenborg, P.D., Jones, R.L., Liu, H.W. and Krick, T.P. (1993) (3*R**,5*S**,6*R**)-3,5-Dimethyl-6-(methylethyl)-3,4,5,6-tetrahydropyran-2-one, a third sex pheromone component for *Macrocentrus grandii* (Goidanich) (Hymenoptera: Braconidae) and evidence for its utility at eclosion. *Journal of Chemical Ecology* **19**: 485–502.

Swedenborg, P.D., Jones, R.L., Zhou, H.-Q., Shin, I. and Liu, H.-W. (1994) Biological activity of (3*R**, 5*S**, 6*R**)-3,5-dimethyl-6-(methylethyl)-3,4,5,6-tetrahydropyran-2-one, a pheromone of *Macrocentrus grandii* (Goidanich) (Hymenoptera: Braconidae). *Journal of Chemical Ecology* **20**: 3373–80.

Symonds, M.R.E. and Elgar, M.A. (2008) The evolution of pheromone diversity. *Trends in Ecology and Evolution* **23**: 220–8.

Syvertsen, T.C., Jackson, L.L., Blomquist, G.J. and Vinson, S.B. (1995) Alkadienes mediating courtship in the parasitoid *Cardiochiles nigriceps* (Hymenoptera: Braconidae). *Journal of Chemical Ecology* **21**: 1971–89.

Tanner, M.E. (2002) Understanding nature's strategies for enzyme-catalyzed racemization and epimerization. *Accounts of Chemical Research* **35**: 237–46.

van den Assem, J. (1970) Courtship and mating in *Lariophagus distinguendus* (Först.) Kurdj. (Hymenoptera, Pteromalidae). *Netherlands Journal of Zoology* **20**: 329–52.

van den Assem, J. (1989) Mating behaviour in parasitic wasps. In: Waage, J. and Greathead, D. (eds) *Insect Parasitoids*. Academic Press, London, pp. 137–67.

van den Assem, J. and Putters, F.A. (1980) Patterns of sound produced by courting chalcidoid males and its biological significance. *Entomologia Experimentalis et Applicata* **27**: 293–302.

van den Assem, J. and Visser, J. (1976) Aspects of sexual receptivity in female *Nasonia vitripennis* (Hym., Pteromalidae). *Biology of Behaviour* **1**: 37–56.

van den Assem, J., Jachmann, F. and Simbolotti, P. (1980) Courtship behavior of *Nasonia vitripennis* (Hym., Pteromalidae): some qualitative, experimental evidence for the role of pheromones. *Behaviour* **75**: 301–7.

van Lenteren, J.C. (1981) Host discrimination by parasitoids. In: Nordlund, D.A., Jones, R.L. and Lewis, W.J. (eds) *Semiochemicals: Their Role in Pest Control*. John Wiley & Sons, New York, pp. 153–9.

Vet, L.E.M. and Dicke, M. (1992) Ecology of infochemical use by natural enemies in a tritrophic context. *Annual Review of Entomology* **37**: 141–72.

Vet, L.E.M. and Groenewold, A.W. (1990) Semiochemicals and learning in parasitoids. *Journal of Chemical Ecology* **16**: 3119–35.

Vieira, F.G., Forêt, S., He, X., Rozas, J., Field, L.M. and Zhou, J.-J. (2012) Unique features of odorant-binding proteins of the parasitoid wasp *Nasonia vitripennis* revealed by genome annotation and comparative analyses. *PLoS ONE* **7**(8): e43034.

Villagra, C., Pinto, C., Penna, M. and Niemeyer, H. (2011) Male wing fanning by the aphid parasitoid *Aphidius ervi* (Hymenoptera: Braconidae) produces a courtship song. *Bulletin of Entomological Research* **101**: 573–9.

Villagra, C.A., Vasquez, R.A. and Niemeyer, H.M. (2005) Associative odour learning affects mating behaviour in *Aphidius ervi* males (Hymenoptera : Braconidae). *European Journal of Entomology* **102**: 557–9.

Villagra, C.A., Vasquez, R.A. and Niemeyer, H.M. (2008) Olfactory conditioning in mate searching by the parasitold *Aphidius ervi* (Hymenoptera: Braconidae). *Bulletin of Entomological Research* **98**: 371–7.

Vinson, S.B. (1976) Host selection by insect parasitoids. *Annual Review of Entomology* **21**: 109–33.

Völkl, W., Hubner, G. and Dettner, K. (1994) Interactions between *Alloxysta brevis* (Hymenoptera, Cynipoidea, Alloxystidae) and honeydew collecting ants: how an aphid hyperparasitoid overcomes ant aggression by chemical defense. *Journal of Chemical Ecology* **20**: 2901–15.

Werren, J.H. (1980) Sex ratio adaptations to local mate competition in a parasitic wasp. *Science* **208**: 1157–9.

Werren, J.H. (1997) Biology of *Wolbachia*. *Annual Review of Entomology* **42**: 587–609.

Werren, J.H., Richards, S., Desjardins, C.A., Niehuis, O., Gadau, J., Colbourne, J.K. and the *Nasonia* Genome Working Group (2010) Functional and evolutionary insights from the genomes of three parasitoid *Nasonia* species. *Science* **327**: 343–8.

Williams, H.J., Wong, M., Wharton, R.A. and Vinson, S.B. (1988) Hagen's gland morphology and chemical content analysis for three species of parasitic wasps (Hymenoptera, Braconidae). *Journal of Chemical Ecology* **14**: 1727–36.

Wyatt, T.D. (2003) *Pheromones and Animal Behaviour: Communication by Smell and Taste*. Cambridge University Press, Cambridge.

Wyatt, T. (2010) Pheromones and signature mixtures: defining species-wide signals and variable cues for identity in both invertebrates and vertebrates. *Journal of Comparative Physiology A: Neuroethology, Sensory, Neural, and Behavioral Physiology* **196**: 685–700.

Yew, J.Y., Dreisewerd, K., de Oliveira, C.C. and Etges, W J. (2011) Male-specific transfer and fine scale spatial differences of newly identified cuticular hydrocarbons and triacylglycerides in a *Drosophila* species pair. *PLoS ONE* **6**: e16898.

Yoshida, S. (1978) Behaviour of males in relation to the female sex pheromone in the parasitoid wasp, *Anisopteromalus calandrae* (Hymenoptera: Pteromalidae). *Entomologia Experimentalis et Applicata* **23**: 152–62.

7

Chemical ecology of tachinid parasitoids

Satoshi Nakamura[1], Ryoko T. Ichiki[1] and Yooichi Kainoh[2]

[1] Japan International Research Centre for Agricultural Sciences, Owashi, Tsukuba, Ibaraki, Japan
[2] Faculty of Life and Environmental Sciences, University of Tsukuba, Tsukuba, Ibaraki, Japan

Abstract

Studies on the chemical ecology of insect parasitoids have attracted a great deal of attention in Hymenoptera during the past few decades. However, studies on this topic with regard to dipteran parasitoids are scarce compared with those of parasitic Hymenoptera, even though dipteran parasitoids are the second most dominant group of insect parasitoids after the parasitic Hymenoptera. In this chapter, we present information on the chemical ecology of dipteran parasitoids of the family Tachinidae in particular, which is the most dominant group of dipteran parasitoids. Among the few studies made on dipteran parasitoids, the chemical ecology of host foraging has been focused on Tachinidae. One of the interesting characteristics of Tachinidae species is that they show various oviposition strategies. In some groups, female tachinids must encounter a host and directly oviposit on or into it (the direct type) as most hymenopteran parasitoids do. On the other hand, females of some other groups do not necessarily need to encounter their hosts in order to parasitize them (the indirect type). In the latter case, larvae produced by a mother fly search for a host by themselves (the searching type) or wait for a passing host to arrive (the waiting type). Some other flies lay so-called microtype eggs on a food plant of the host, and these eggs should then be ingested by the host.

The direct type tachinids use odours emitted from healthy or host-infested plants as long-range olfactory cues to locate the food plants of their hosts. After female tachinids have located a potential host habitat, the females search for hosts using visual cues, mainly host movements. At this stage, the females also use chemicals derived from host frass as arrestants. After locating a potential host, the females inspect the texture, contrast, shape, firmness and curvature of the hosts with their front tarsi and finally lay an egg on or in the host. Some other direct type tachinids that attack stinkbugs exploit the volatile sexual pheromones of their hosts for host location. In the case of the indirect type tachinids, odours from host-infested plants are important as long-range olfactory cues for the location of host food plants. Then the searching type tachinids deposit their larvae on or near frass of their hosts in response to chemicals coming from the host frass. In the waiting-type tachinids, oviposition is stimulated by chemicals from host-infested plants or by chemicals from host frass. Plant leaf exudates elicit oviposition in the microtype tachinids. Tachinid parasitoids use various stimuli including the chemical stimulus associated with the host environment and hosts themselves during separate stages of the host-foraging process.

7.1 Introduction

The word 'parasitoid' probably gives most biologists an impression of parasitic Hymenoptera (Belshaw 1994). However, there is a great diversity of non-hymenopteran parasitoids, and roughly one-quarter of the described parasitoid species belong to either the Diptera or the Coleoptera (Eggleton & Belshaw 1992). Hymenopteran parasitoids have unique lifestyle features that differ from those of the other taxa (Feener & Brown 1997). For example, hymenopteran parasitoids are haplodiploid and possess an ovipositor with associated accessory glands. The parasitoid lifestyle probably evolved only once in the Hymenoptera, but may have evolved more than a hundred times in the Diptera (Eggleton & Belshaw 1992).

The family Tachinidae, with 10,000 described species, is one of the largest families in the Diptera (O'Hara 2008). All species in this family are endoparasitoids of other arthropods, primarily insects (Belshaw 1994, Feener & Brown 1997). Tachinids are often used as natural enemies of agricultural and forest pests (Grenier 1988), and are considered to be important biological control agents together with many hymenopteran families (Greathead 1986). Yet, despite their importance, surprisingly little information is available on this taxon or on its ecology and behaviour (Belshaw 1994, Stireman et al. 2006). Their chemical ecology is no exception, probably because rearing tachinids is relatively difficult in the laboratory (Greathead 1986, Stireman et al. 2006).

The tachinids are distinguished by a high level of host polyphagy (e.g. Askew & Shaw 1986, Belshaw 1994, Feener & Brown 1997, Stireman et al. 2006), even though endoparasitoids tend to have a narrow host range because they require a high degree of physiological adaptation to host immune systems (Belshaw 1994). This polyphagy indicates that female flies have to deal with various information from many host species in order to find and successfully parasitize them.

Studies of the chemical ecology of insects have focused on subjects such as sexual advertisement, social organization, defence, and finding and recognizing resources (Cardé & Millar 2004). However, we found little information on the chemical ecology of tachinids, except for a few studies about how they find and recognize their hosts. Successful parasitism requires a sequence of key steps, starting with host habitat location, continuing with host

Figure 7.1 An example of a tachinid, *Drino zonata*, that uses the direct strategy. A female is depositing an incubated egg on a larva of *Helicoverpa armigera*. Photograph by Takao Yoshida, under a Creative Commons Attribution-ShareAlike 3.0 Unported License.

location, and ending with host acceptance, before the insect can parasitize its host. Most hymenopteran parasitoids must find and directly contact their hosts for parasitism to occur, and the females of these species initiate the parasitism directly by ovipositing in the host. In contrast, some tachinid females take an indirect approach to parasitization, in which their larvae wait for, seek out, or are swallowed by the host.

One of the prominent features of tachinids is their wide variety of oviposition strategies and egg morphologies (Stireman *et al.* 2006). Based on the literature describing these oviposition strategies (e.g. Clausen 1940, Wood 1987, Belshaw 1994), we have defined two main groups to facilitate understanding of the chemical ecology of tachinid parasitoids. First, females of some species must contact their hosts directly, and oviposit on a host's cuticle (*direct-external*; Figs 7.1 and 7.3a) or inside a host's body (*direct-internal*; Fig. 7.3a). These females are oviparous or ovoviviparous: eggs are laid soon after fertilization, or are incubated in the uterus of females after fertilization until they are oviposited. The second main group is the *indirect* type: females of these species oviposit incubated eggs in the host's habitat, and it is the larvae rather than the females that find a host. In some species, the first-instar larvae search for the host by themselves (the *searching* type; Fig. 7.3b) or wait for a passing host (the *waiting* type; Figs 7.2 and 7.3c), then after contacting a host, they penetrate it. Another indirect type is referred to as '*microtype*' (Fig. 7.3d). Females of this type lay incubated eggs on the host's food plant, and the eggs ingested with the food hatch inside the host's gut. Soon after hatching, these larvae burrow into the host through the gut wall to reach host organs such as a silk gland or a ganglion. Table 7.1 summarizes the tachinid species known to use each of these different host-finding strategies.

The indirect strategies are diverse, and are used by nearly half of the tachinid species (Eggleton & Gaston 1992). The advantage of an indirect strategy is that it permits many species to parasitize concealed or nocturnal hosts that would otherwise be inaccessible to the female (Belshaw 1994). Moreover, this approach offers an advantage for females because

Figure 7.2 An example of a tachinid, *Linnaemya longirostris*, that uses the indirect strategy; (a) a female and (b) a larva on the lead of a mechanical pencil. Photographs by Takao Yoshida, under a Creative Commons Attribution-ShareAlike 3.0 Unported License.

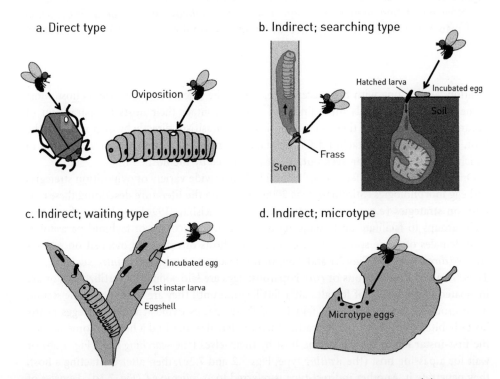

Figure 7.3 Oviposition strategies of tachinid flies. In the direct strategy (a), females oviposit on or inside a suitable host. In the indirect strategies (b, c, d), females oviposit incubated eggs in the host's habitat. The eggs hatch soon after oviposition and the hatched larvae must either seek a host (b) or wait for a passing host (c). The microtype eggs must be ingested by a host (d).

Table 7.1 Cues used during host searching behaviour of tachinid flies.

Cue	Oviposition strategy	Egg type	Species	Host species	Plant[a]	References
Volatile: long-range orientation cue						
Host-infested plant volatiles	Direct-external	Unincubated	*Exorista japonica*	*Mythimna separata* (Noctuidae)	Maize	Kainoh *et al.* (1999), Ichiki *et al.* (2008), Hanyu *et al.* (2009, 2011), Ichiki *et al.* (2011)
	Direct-external	Unincubated	*Bessa harveyi*	*Pristiphora erichsonii* (Tenthredinidae)	Pine	Monteith (1964)
	Direct-external	Incubated	*Drino bohemica*	*Neodiprion lecontei* (Diprionidae)	Pine	Monteith (1955, 1964)
	Indirect, microtype	Incubated	*Cyzenis albicans*	*Operophtera brumata* (Geometridae)	Oak	Roland *et al.* (1989, 1995)
	Indirect, microtype	Incubated	*Leschenaultia exul*	*Malacosoma disstria* (Lasiocampidae)	Aspen poplar	Mondor & Roland (1997)
	Indirect, microtype	Incubated	*Patelloa pachypyga*	*Malacosoma disstria* (Lasiocampidae)	Aspen poplar	Mondor & Roland (1997)
	Indirect, microtype	Incubated	*Pales pavida*	*Mythimna separata*	Maize	R.T. Ichiki (pers. obs.)
	Indirect, microtype	Incubated	*Zenillia dolosa*	*Mythimna separata*	Maize	R.T. Ichiki (pers. obs.)
	Indirect, searching	Incubated	*Lixophaga diatraeae*	*Diatraea saccharalis* (Crambidae)	Sugarcane	Roth *et al.* (1982)
	Indirect, waiting	Incubated	*Linnaemya (=Bonnetia) comta*	*Agrotis ipsilon* (Noctuidae)	Maize	Clement *et al.* (1986)

(cont'd)

Table 7.1 (cont'd)

Cue	Oviposition strategy	Egg type	Species	Host species	Plant[a]	References
Artificially damaged and healthy plant volatiles	Direct-external	Unincubated	Exorista mella	Grammia geneura (Arctiidae)	Weakleaf bur ragweed	Stireman (2002)
	Direct-internal	Incubated	Eucelatoria bryani	Heliothis virescens (Noctuidae)	Okra, cotton, etc.	Nettles (1979, 1980), Martin et al. (1990)
	Indirect, searching	Incubated	Lydella grisescens	Ostrinia nubilalis (Crambidae)	Maize	Franklin & Holdaway (1966)
Aggregation pheromones of host	Direct-external	Unincubated	Trichopoda pennipes	Nezara viridula (Pentatomidae)	–	Mitchell & Mau (1971), Harris & Todd (1980), Aldrich et al. (1987)
	Direct-external	Unincubated	Hemyda aurata	Podisus maculiventris (Pentatomidae)	–	Aldrich et al. (1984), Aldrich (1995a, 1995b)
	Direct-external	Unincubated	Euclytia flava	Podisus maculiventris, Euschistus spp. (Pentatomidae)	–	Aldrich et al. (1984, 1991), Aldrich (1995a, 1995b)
	Direct-external	Incubated	Euthera tentatrix	Podisus maculiventris	–	Aldrich et al. (2007)
	Direct-internal	Unincubated	Cylindromyia fumipennis	Podisus maculiventris	–	Aldrich et al. (2007)
	Direct-external	Unincubated	Gymnosoma rotundatum	Plautia stali (Pentatomidae)	–	Moriya & Shiga (1984), Mishiro & Ohira (2002), Adachi et al. (2007), Jang & Park (2010), Higaki & Adachi (2011), Jang et al. (2011)
	Direct-external	Unincubated	Gymnosoma par	Thyanta spp. (Pentatomidae)	–	Aldrich et al. (2007)

Contact chemical: short-range orientation cue

Plant					
Direct-external	Unincubated	*Exorista japonica*	*Mythimna separata*	Maize	Hanyu *et al.* (2011)
Direct-internal	Incubated	*Eucelatoria bryani*	*Heliothis virescens*	Okra	Nettles (1982)
Indirect, microtype	Incubated	*Blepharipa pratensis*	*Lymantria dispar* (Lymantriidae)	Oak	Odell & Godwin (1984)
Indirect, microtype	Incubated	*Cyzenis albicans*	*Operophtera brumata* (Geometridae)	Oak	Hassell (1968), Roland (1986)
Indirect, microtype	Incubated	*Leschenaultia exul*	*Malacosoma disstria*	Aspen poplar	Mondor & Roland (1998)
Indirect, microtype	Incubated	*Pales pavida*	*Mythimna separata*	Maize	Ichiki (pers. obs.)
Indirect, microtype	Incubated	*Zenillia dolosa*	*Mythimna separata*	Maize	Ichiki (pers. obs.)
Indirect, waiting	Incubated	*Linnaemya (=Bonnetia) comta*	*Agrotis ipsilon*	Maize	Clement *et al.* (1986)

(cont'd)

Table 7.1 (cont'd)

Cue	Oviposition strategy	Egg type	Species	Host species	Plant[a]	References
Host frass	Direct-external	Unincubated	*Exorista japonica*	*Mythimna separata*	Maize	Tanaka *et al.* (2001)
	Direct-internal	Incubated	*Eucelatoria bryani*	*Heliothis virescens*	Artificial diets	Nettles (1982)
	Indirect, searching	Incubated	*Lixophaga diatraeae*	*Diatrea saccharali, Heliothis zea* (Noctuidae)	Sugarcane, cotton boll	Roth *et al.* (1978), Thompson *et al.* (1983)
	Indirect, searching	Incubated	*Lydella grisescens*	Two natural hosts: *Ostrinia nubilalis, Papaipema nebris* (Noctuidae), and other two non-hosts	Maize, cat-tail weed	Hsiao *et al.* (1966)
	Indirect, searching	Incubated	*Triarthria setipennis*	*Forficula auricularia* (Forficulidae)	Artificial diets	Kuhlmann (1995)
	Indirect, searching	Incubated	*Doleschella* sp.	*Pantorhytes szentivanyi* (Curculionidae)	Cacao pod	Baker (1978)
	Indirect, waiting	Incubated	*Linnaemya (=Bonnetia) comta*	*Agrotis ipsilon*	Maize, chickweed, artificial diet	Rubink & Clement (1982), Clement *et al.* (1986)

Host body	Indirect, waiting	Incubated	Archytas marmoratus	Heliothis virescens, other four noctuids and two pyralids	H. virescens: artificial diets, others: unknown	Nettles & Burks (1975)
Host cuticular	Direct-internal	Incubated	Eucelatoria bryani	Heliothis virescens	Artificial diets	Burks & Nettles (1978)
Host vomit and hemolymph	Direct-internal	Incubated	Eucelatoria bryani	Heliothis virescens	Artificial diets	Nettles (1982)

Physical cue (tactile and visual cues): long-range or short-range orientation cues

Plant colour	Direct-external	Unincubated	Exorista japonica	Mythimna separata	Maize	Ichiki et al. (2011)
Firmness of host	Direct-external	Incubated	Drino inconspicua	Gilpinia hercyniae (Diprionidae)	–	Dippel & Hilker (1998)
Texture, curvature and contrast of host	Direct-external	Unincubated	Exorista japonica	Mythimna separata	–	Tanaka et al. (1999), Yamawaki & Kainoh (2005)

(cont'd)

Table 7.1 (cont'd)

Cue	Oviposition strategy	Egg type	Species	Host species	Plant[a]	References
Host shape	Direct-internal	Incubated	Eucelatoria bryani	Heliothis virescens	–	Burks & Nettles (1978)
Host motion	Direct-external	Unincubated	Exorista japonica	Mythimna separata	–	Yamawaki et al. (2002), Yamawaki & Kainoh (2005)
	Direct-external	Unincubated	Exorista mella	Grammia geneura	–	Stireman (2002)
	Direct-external	Unincubated	Bessa parallela	Pryeria sinica [Zygaenidae]	–	Ichiki et al. (2006)
	Direct-external	Incubated	Drino inconspicua	Gilpinia hercyniae	–	Dippel & Hilker (1998)
	Direct-external	Incubated	Drino bohemica	Neodiprion lecontei	–	Monteith (1956)
	Direct-internal	Incubated	Compsilura concinnata	Lymantria dispar	–	Weseloh (1980)

it reduces the risk of injury caused by host defences such as beating, tumbling or biting, and because it decreases handling time (Stireman *et al.* 2006). Tachinids that use the indirect type of parasitism tend to have a higher fecundity than the direct type, presumably due to increased mortality of eggs or the first-instar larvae (Belshaw 1994). Although these strategies for contacting the host exist in hymenopteran parasitoids, they are very rare (Belshaw 1994).

In the direct strategy, successful parasitization requires that female flies first locate a habitat where suitable hosts are likely to exist, then locate hosts within this habitat, and finally determine whether to accept a host. In contrast, females that adopt the indirect strategy do not have to find and contact hosts, and therefore only need to locate a suitable habitat or a combination of suitable habitat and hosts present within that habitat. Subsequently, they can oviposit in the host's environment. Our literature review found no study of the host-seeking or host acceptance behaviour of the searching and waiting types of larvae in the indirect category. For the microtype strategy, there is no solid information of how the first-instar larvae detect and reach the specific organ where they will live for a while after they hatch in the host's gut. It seems likely that some infochemicals are responsible, but these chemicals have not yet been found.

In this chapter, we discuss aspects of chemical ecology related to the host location behaviour of tachinid parasitoids. We have divided this discussion into two sections: long-range orientation, which involves the location of habitat suitable for the host; and short-range orientation, which involves the location of hosts. We also discuss the visual and tactile cues that tachinids use to find suitable hosts. For each component of the searching behaviour, we focus separately on the direct and indirect oviposition strategies. Although some tachinids (e.g. the tribe Ormiini) that parasitize orthopteran species can hear and use the sexual calls of their hosts as acoustic cues for host location (e.g. Cade 1975, Robert *et al.* 1992, Walker 1993), their host location behaviour is not discussed in this chapter.

7.2 Long-range orientation

Chemical evidence for herbivores inducing plants to actively attract the herbivores' natural enemies was first studied in the context of plant–mite interactions (Sabelis & van de Baan 1983), but the majority of subsequent studies have shown that herbivore-induced plant volatiles (HIPVs) can attract parasitoids (Turlings & Wäckers 2004). We further categorize the plant volatiles that can affect tachinid behaviour into HIPVs produced by an infested plant, and green leaf volatiles (GLVs) that are released by both healthy and artificially damaged plants. Another category of long-range attraction is referred to as 'chemical eavesdropping' (Stowe *et al.* 1995), in which the parasitoid is attracted to the pheromones used by its host insect (i.e. kairomones).

7.2.1 Long-range orientation by direct type parasitoids

Based on both behavioural observations and chemical analyses, Turlings & Wäckers (2004) suggested that HIPVs play a key role in host and prey location. In early research, Monteith (1955, 1964) reported that two direct type tachinids, *Drino bohemica* Mesnil and *Bessa harveyi* (Meigen), were attracted to herbivore-damaged foliage, especially to that of unhealthy plants. We started our research on the tritrophic tachinid–armyworm–maize system in 1997, and soon demonstrated that females of the tachinid *Exorista japonica* were attracted to maize plants infested by larvae of the rice armyworm *Mythimna separata* (Walker) (Lepidoptera: Noctuidae) (Kainoh *et al.* 1999). We studied their behaviour in relation to the condition of the maize plants and their chemicals in wind tunnel bioassays (Ichiki *et al.* 2008, 2011, Hanyu *et al.* 2009, 2011). When filter paper impregnated with the head-space volatiles collected from plants infested by the parasitoid's host (henceforth, 'host-infested plants') was attached to intact plants, females landed on these plants at a significantly higher rate than on intact plants without the filter paper. They also responded to a synthetic blend of nine chemicals that had been identified from host-infested plants. Of the nine chemicals identified previously by Takabayashi *et al.* (1995), four ((*E*)-4,8-dimethyl-1,3,7-nonatriene, indole, 3-hydroxy-2-butanone and 2-methyl-1-propanol), released only by host-infested plants, were together classified as a host-induced blend. The other five compounds ((*Z*)-3-hexen-1-yl acetate, (*E*)-2-hexenal, hexanal, (*Z*)-3-hexen-1-ol and linalool), released not only by infested plants but also by intact or artificially damaged plants, were together classified as a non-specific blend. The females showed a significantly higher response (Fig. 7.4) to a mixture of the non-specific and host-induced blend than to the control (Ichiki *et al.* 2008).

Another interesting aspect of long-range orientation is the timing of the initial release of HIPVs and how long they continue to attract tachinids. Hanyu *et al.* (2009) reported that *E. japonica* females can continue to detect HIPVs for at least 24 hours after the hosts stop feeding on a plant. When a maize plant was infested by 20 last-instar *M. separata* for 1 hour and the host larvae were then removed from the plant, the attraction of female flies to the plant in a wind tunnel assay remained high (57–73%) for 5 hours and then decreased gradually to 48% after 24 hours. When a maize plant was continuously infested with five

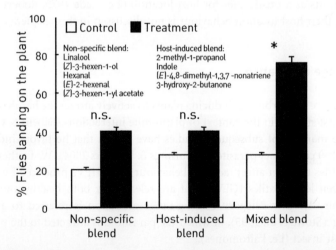

Figure 7.4 Flight response (%±SE) of female *Exorista japonica* to a non-specific blend of chemicals released by plants, by a host-induced blend, and by a mixture of the two blends in a wind tunnel assay. The *asterisk* indicates a significant difference between the treatment and a control leaf without the chemical blend added (Fisher's exact probability test, *P* < 0.05). From Ichiki, R.T., Kainoh, Y., Kugimiya, S., Takabayashi, J. and Nakamura, S. (2008) Attraction to herbivore-induced plant volatiles by the host-foraging parasitoid fly *Exorista japonica*. *Journal of Chemical Ecology* **34**: 614–621, Fig. 4. © Springer Science + Business Media, LLC 2008. With kind permission from Springer Science and Business Media.

host larvae for the whole 24-hour period, the attraction of females to the plants remained high (between 60% and 70%) for 4 days. In contrast, the attraction to artificially damaged maize plants was high (85%) when the plants were tested soon after damage, but decreased to 40% within 1 hour after damage. However, the release of HIPVs from an infested plant is confined to the herbivore-damaged portion of the plant (i.e. it is not systemic). Uninfested leaves of infested plants did not attract the females (Hanyu *et al.* 2009). In addition to being attracted by the odour of target plants, females of *E. japonica* probably employ plant colour to locate host-infested plants: females showed a significantly higher landing rate on the green paper plant model (84.6%) than on the yellow (53.8%), blue (38.5%) or red (30.8%) models when odours of the host-infested plants were present in the wind tunnel (Ichiki *et al.* 2011).

It has long been known that predators, and especially parasitoids of herbivores, are attracted to infested plants (Turlings & Wäckers 2004). In previous research, predators and parasitoids were found more often on certain plant species, and differences in plant volatiles may have been responsible for this differential attractiveness. There are several reports of tachinid behaviour indicating the attraction of flies to a healthy plant (e.g. Nettles 1979, 1980, Martin *et al.* 1990), but we do not know whether the response is to specific plant volatiles or to general GLVs.

Females of the tachinid *Eucelatoria bryani* Sabrosky were attracted to volatiles from cotton and okra plants (Nettles 1979, 1980), and subsequent research confirmed that they were attracted to fresh leaves of healthy plants (maize, okra, pigeon pea, sorghum, tomato, velvetleaf) in an olfactometer (Martin *et al.* 1990). *Exorista mella* (Walker) females responded to GLVs associated with cut sprigs of the food plant of their host in a simple flask olfactometer (Stireman 2002), but details of how the plant had been treated (healthy or artificially damaged) were not clearly described. *Exorista japonica* was attracted to volatiles produced by artificially damaged maize plants in a wind tunnel bioassay (Hanyu *et al.* 2011), and flies were attracted to GLVs soon after the host began to damage the plant.

7.2.2 Long-range orientation by indirect type parasitoids

Microtype
Microtype tachinids oviposit on foliage rather than on the host insects. Many species that use this strategy are attracted to host-infested plants. The infested plants can attract flies from a distance (Roth *et al.* 1982, Roland *et al.* 1989, 1995, Mondor & Roland 1997). The microtype tachinid *Cyzenis albicans* (Fallen) is attracted to volatiles from oak trees infested by the winter moth *Operophtera brumata* (L.) (Lepidoptera: Geometridae) (Roland *et al.* 1989, 1995). This tachinid had a high rate of response to borneol among the 10 volatile compounds identified from a crude extract of oak foliage (Roland *et al.* 1995). In a field experiment (Roland 1990), artificial dispensers of oak leaf extracts made with different solvents (water, hexane, chloroform, ethyl acetate) were placed in apple trees, and the number of *C. albicans* eggs deposited was significantly higher than in the control with the hexane extract only. This result showed that the hexane-soluble attractive odour was released from oak leaves.

In another microtype tachinid, *Leschenaultia exul* (Townsend), the flies were attracted to the host–trembling aspen complex in a wind tunnel (Mondor & Roland 1997). In a wind tunnel test, *L. exul* and *Patelloa pachypyga* (Aldrich & Webber), two microtype tachinids of the forest tent caterpillar *Malacosoma disstria* (Lepidoptera: Lasiocampidae), responded differently depending on the tree species on which the host fed (Mondor & Roland 1997).

Female *L. exul* were preferentially attracted to the host–trembling aspen complex rather than to the host–balsam poplar complex.

We compared the attraction of two microtype tachinid flies, *Pales pavida* (Meigen) and *Zenillia dolosa* (Meigen), to odours from maize leaves infested by larvae of their common host, *M. separata*, in a wind tunnel. Naïve *P. pavida* females showed a higher rate of landing on host-infested maize plants than on artificially damaged or intact maize plants, whereas *Z. dolosa* landed on both the infested and the artificially damaged plants at a higher rate than on the intact maize plants (Ichiki *et al.* 2012).

Searching type

Many tachinid species that use this strategy attack hosts concealed in stems, fruits and seeds by ovipositing near the entrance holes made by the hosts (Clausen 1940). A few of these species are attracted to plant odour. The tachinid *Lixophaga diatraeae* (Townsend), which is attracted to sugarcane infested with larvae of the sugarcane borer *Diatraea saccharalis* (Fabricius) (Lepidoptera: Crambidae), is attracted to HIPVs (Roth *et al.* 1982). Another example of the searching strategy by a tachinid that is attracted to a healthy plant is *Lydella grisescens* Robineau-Desvoidy, a parasitoid of the European corn borer *Ostrinia nubilalis* (Hübner) (Lepidoptera: Crambidae). The females were attracted to both healthy and infested maize plants (Franklin & Holdaway 1966). The field experiment was done with two different maize hybrids. One hybrid attracted the tachinid females but the other did not, and the parasitoid consequently oviposited in larger numbers on the first one because it released a greater amount of GLVs or HIPVs.

Waiting type

Tachinid flies that use this strategy ensure that their larvae are positioned in the vicinity of their hosts. The larvae then wait for a passing host and attach themselves to, and burrow into, a suitable one. The tachinid *Linnaemya* (=*Bonnetia*) *comta* (Fallen) appears to respond to host-infested plants. Female *L. comta* oviposited in significantly greater numbers on maize seedlings infested by the black cutworm *Agrotis ipsilon* (Hufnagel) (Lepidoptera: Noctuidae) than on artificially damaged maize seedlings (Clement *et al.* 1986). An olfactometer was not used in this experiment, but attraction to the odour of the infested plant might explain the increased oviposition rate.

7.2.3 Host pheromones used by direct type parasitoids

Pheromones have evolved as chemical signals between individuals of the same species, so they are usually highly species-specific. This makes them reliable semiochemical cues for parasitoids that must seek particular hosts (Powell 1999). For example, several species of tachinid parasitoids use the aggregation pheromone produced by pentatomid bugs as a host-finding kairomone (Aldrich 1995b), a chemical signal which is adaptively favourable to the receiver but not to the emitter. The behaviour can be categorized as both host habitat location and host location behaviour. We have chosen to describe this in our discussion of long-range cues because aggregation pheromones generally function as long-range communication signals (Powell 1999). In this sense, they are similar to the plant volatiles that lead parasitoids to their host's habitat.

In this category, all the tachinids that we found in the literature followed the direct strategy. Female *Trichopoda pennipes* (F.) were caught in cages containing males of the southern green stink bug *Nezara viridula* (L.) (Heteroptera: Pentatomidae), in which the male-emitted aggregation pheromone acts as a kairomone for *T. pennipes* females (Mitchell & Mau 1971, Harris & Todd 1980, Aldrich *et al.* 1987). *Hemyda aurata* Robineau-Desvoidy and *Euclytia flava* (Townsend) parasitoids were caught in a trap baited with synthetic aggregation pheromone of the spined soldier bug *Podisus maculiventris* (Say) (Heteroptera: Pentatomidae) (Aldrich *et al.* 1984, Aldrich 1995a, 1995b). In addition, *E. flava*, *Gymnosoma* spp. and *Euthera* spp., which are parasitoids of *Euschistus* stink bugs, were caught in traps baited with synthetic pheromone (Aldrich *et al.*, 1991). Traps baited with methyl (*E,Z,Z*)-2,4,6-decatrienoate, or combinations of other compounds released by the stink bugs, attracted *E. flava*, *Gymnosoma par* Walker, *Euthera tentatrix* Loew, *H. aurata* and *Cylindromyia fumipennis* (Bigot) (Aldrich *et al.* 2007). Females of the tachinid *Gymnosoma rotundatum* (L.) are attracted to male-produced aggregation pheromones of the brown-winged green bug *Plautia stali* Scott (Moriya & Shiga 1984, Mishiro & Ohira 2002, Adachi *et al.* 2007, Jang & Park 2010, Higaki & Adachi 2011, Jang *et al.* 2011). The pheromone attracted both female and male tachinid flies, since the males may increase their chance of mating with pheromone-attracted females by waiting near an appropriate pheromone source (Higaki & Adachi 2011).

7.3 Short-range orientation

After parasitoid females have located a potential host habitat, they show an arrestment response when they detect low-volatility kairomones (sometimes referred to as 'contact chemicals') deposited by their hosts on the substrate (van Alphen & Jervis 1996). These short-range chemical cues produced by host activity strongly indicate the presence of a host and serve as reliable information for female parasitoids to detect the host (Vet & Dicke 1992). Several studies have shown that host location behaviour by female tachinids can be elicited by tactile-chemosensory cues associated with the host, and by the host's food plants (Stireman *et al.* 2006). In these cases, females use chemosensors on their front tarsi, which may function similarly to the chemosensors on the long antennae of many hymenopteran parasitoids (Stireman *et al.* 2006). In addition to tactile-chemosensory cues, visual cues are also important in the final stages of the tachinid host location process, especially for species that use the direct strategy.

7.3.1 Short-range orientation by direct type parasitoids

Among tachinids that directly attack their hosts, frass produced by the host is known to cause females to exhibit an arrestment response when they enter areas that either contain or are likely to contain hosts. The kairomone seems to be perceived by a female's front tarsi. Tanaka *et al.* (2001) demonstrated that females of the tachinid *E. japonica* are arrested in response to chemicals derived from the fresh frass of larvae of *M. separata*. The females showed obvious behavioural changes when they touched host frass with their front tarsi; their walking speed decreased and they began intensive exploration of the frass-containing patch by tapping the frass with their legs and walking or turning within the patch (Tanaka *et al.* 2001). The arrestant chemicals appear to include polar compounds, as they were extracted by methanol (Fig. 7.5) but not by acetone, ether or hexane. Females of the tachinid *E. bryani* exhibit a similar response to the fresh frass of *Heliothis virescens* larvae as well

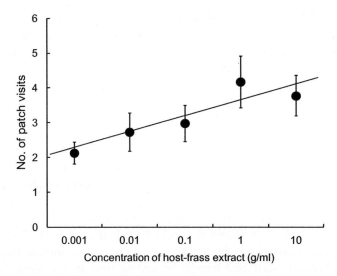

Figure 7.5 Dose-dependent mean (±SE) frequency of visits to a patch by *Exorista japonica* females in response to methanol extracts of *Mythimna separata* larval frass. Regression for the number of patch visits: $y = 0.20 \ln(x) + 3.6$, $r^2 = 0.83$, $P < 0.05$. From Tanaka, C., Kainoh, T. and Honda, H. (2001) Host frass as arrestant chemicals in locating host *Mythimna separata* by the tachinid fly *Exorista japonica*. *Entomologia Experimentalis et Applicata* **100**: 173–178.

as to stale dried frass that had been stored at 25°C for 1 week (Nettles 1982). In the case of *E. bryani*, however, the arrestant chemicals in the frass were extracted by hexane only and not by acetone, ethanol or water, suggesting that apolar compounds were included.

Contact chemicals from the host's food plants may also affect the behaviour of tachinids at short range. After *E. japonica* females are attracted to volatile chemicals emitted by maize plants damaged by host larvae (Ichiki *et al.* 2008, Hanyu *et al.* 2009, 2011, Ichiki *et al.* 2011), they orient towards the damaged part of the maize leaves (Hanyu *et al.* 2009). They then explore the leaf extensively by tapping the leaf surface with their legs while walking. The females spent significantly more time exploring infested plants than exploring artificially damaged and undamaged plants (Hanyu *et al.* 2011). Nettles (1982) reported that *E. bryani* actively seeks contact with paper infused with a dichloromethane extract of okra leaves, suggesting that contact chemicals from okra may affect this tachinid's host location behaviour.

Visual cues may be a key factor in host location by the direct type tachinids. When *E. japonica* females encounter a host, they turn towards it ('fixation'), walk to within 1 cm of the host ('approach'), and then pursue the crawling larva to guide it (Yamawaki *et al.* 2002). Once the fly has approached the host, it begins its 'examination' behaviour, which consists of facing and touching the host with its front tarsi (Nakamura 1997). Yamawaki *et al.* (2002) found a correlation between movements of *E. japonica* during host pursuit and the visual stimuli provided by the host, which suggests that pursuit of the host was controlled mainly by visual cues at short range. The females show fixation, approach and examination behaviours in response to a moving freeze-dried larva of *M. separata* and to a moving black rubber tube (Yamawaki & Kainoh 2005), suggesting that larval movement attracts the female. Females of the tachinid *Drino inconspicua* Meigen walk directly towards a moving object over distances of 9–10 cm (Dippel & Hilker 1998). They prefer to contact a dummy moving larva made of filter paper rather than a stationary dummy or a stationary freshly killed larva of the European spruce sawfly *Gilpinia hercyniae* (Hartig) (Hymenoptera: Diprionidae). A similar behavioural preference for moving objects was found in the tachinids *Compsilura concinnata* (Meigen) (Weseloh 1980) and *E. mella* (Stireman 2002). The tachinid *D. bohemica* prefers to enter the arm of a Y-tube olfactometer that contains moving feathers rather than the arm without feathers, even when the odours of larvae of the red-headed pine sawfly *Neodiprion lecontei* (Fitch) (Hymenoptera: Diprionidae) flow from both arms (Monteith 1956). Field observations have shown that the tachinid *Bessa parallela* (Meigen) prefers to deposit its eggs on late final-instar larvae of the Euonymus leaf notcher *Pryeria sinica* Moore (Lepidoptera: Zygaenidae) that were actively moving around on the ground before pupation (Ichiki *et al.* 2006).

Chemical components of the host's surface may also be important elicitors of oviposition for the direct type tachinids. Oviposition of *E. bryani* is stimulated by cuticular substances from the larvae of the host, *H. virescens* (Burks & Nettles 1978). When *E. bryani* females were provided with a host cuticle coated on one half with Tween 80, flexible collodion or a Cab-o-sil slurry, they deposited 10 times more larvae on the uncoated than on the coated half. The chemicals that stimulated oviposition were extracted only with a chloroform–methanol mixture (2:1) and not with hexane, chloroform, dichloromethane, acetone, methanol or water used either singly or serially.

Contact chemicals on the host's cuticle play a minor role in oviposition by *D. inconspicua* (Dippel & Hilker 1998). Female *D. inconspicua* prefer to oviposit on soft dummy rather than on hard dummy hosts, suggesting that the softness of the dummy may be crucial for stimulating oviposition by this tachinid. Female *E. japonica* check the texture and curvature of the host by means of tarsal examination before oviposition. They prefer to oviposit on a cylindrical shape rather than on a flat board or a cube, and prefer to oviposit on a surface with a rubbery texture rather than on a surface with a paper or silicone texture (Tanaka *et al.* 1999). Female *E. bryani* prefer to oviposit on a cylindrical shape with a 4 mm diameter (similar to the natural shape of their host), and will not oviposit on a variety of flattened or semi-flattened shapes, suggesting that the correct host shape may be essential for oviposition by *E. bryani* (Burks & Nettles 1978).

7.3.2 Short-range orientation by indirect type parasitoids

Microtype
Plant leaf exudates are known to elicit arrestment behaviour of females and stimulate them to oviposition in areas that either contain or are likely to contain hosts. The chemical components of the leaf exudates are perceived by the female's front tarsi. When females of the tachinid *Blepharipa pratensis* (Meigen) contact a recently damaged plant edge, they orient perpendicular to the edge and move back and forth with their front tarsi grasping the damaged edge (Odell & Godwin 1984). Leaf exudates appear to arrest the fly on the leaf and to increase the intensity of tarsal examination behaviour. If flies come into contact with the edge of a damaged leaf, oviposition usually follows. Female *B. pratensis* laid significantly more eggs on northern red oak leaves infested by larvae of the gypsy moth *Lymantria dispar* (L.) (Lepidoptera: Lymantriidae) than on artificially damaged or undamaged leaves (Odell & Godwin 1984). Tarsal contact with damaged leaves has also been implicated in oviposition in two other microtype tachinids: *C. albicans* (Hassell 1968,

Roland 1986) and *L. exul* (Mondor & Roland 1997). Hassell (1968) demonstrated that sugars at the edge of feeding-damaged oak leaves act as an oviposition stimulant for *C. albicans*. Our recent observations (Ichiki *et al.* 2012) demonstrated that two microtype tachinids have different oviposition preferences for *M. separata*-infested maize plants: female *P. pavida* scattered their eggs throughout a damaged leaf, whereas *Z. dolosa* concentrated their oviposition around an area of host infestation or an artificially damaged spot. Reflecting this difference in oviposition preferences, eggs of *P. pavida* laid on maize leaves survived for about 2 weeks, whereas most of the eggs of *Z. dolosa* died within a few days.

Searching type
Tachinid females often help their larval progeny to locate hosts by depositing them where there is evidence of host activity (Godfray 1994). Several studies reported that female flies deposit their larvae on or near frass of their hosts. Tarsal contact with host frass appears to be the primary stimulus for oviposition by searching type tachinids such as *L. diatraeae*, which attacks the sugarcane borer *D. saccharalis* (Roth *et al.* 1978, Thompson *et al.* 1983); by *L. grisescens* (Hsiao *et al.* 1966), which attacks the European corn borer *O. nubilalis*; by *Doleschalla* sp., which attacks the cacao weevil borer *Pantorhytes szentivanyi* Marsh (Coleoptera: Curculionidae) (Baker 1978); and by *Triarthria setipennis* (Fallen), which attacks the European earwig *Forficula auricularia* L. (Dermaptera: Forficulidae) (Kuhlmann 1995). Roth *et al.* (1978) found that oviposition preference of *L. diatraeae* was governed in part by the host's food source, and that the oviposition stimulant could be extracted in significant amounts only from the host's alimentary canal, suggesting that the active substance originates from the host's food source and that the amount present varies with the source. The contact chemicals in the host frass that elicit oviposition by *L. diatraeae* can be extracted with methanol and seem to be polar non-volatile compounds (Thompson *et al.* 1983). In *L. grisescens*, contact chemicals in the host frass are soluble in ethanol and water, and seem to be a metabolic product of the caterpillars rather than a product of plant fermentation (Hsiao *et al.* 1966). Oviposition on or near the frass should result in a high degree of host-finding success by the searching tachinid larvae. Although it is known that members of the tribe *Dexiini* oviposit on the ground and that their first instars find hosts such as scarab beetle larvae dwelling in the soil (O'Hara 2008), the cues related to the host location behaviour of these tachinids are unknown.

Waiting type
Oviposition of the tachinid *L. comta* appears to be stimulated by chemicals from host-infested plants and by host frass. Females prefer to oviposit on maize seedlings infested by the black cutworm rather than on artificially damaged maize seedlings (Clement *et al.* 1986). The females also actively oviposit in patches with fresh or old (8 days) frass, but not in patches with oven-dried frass or black India ink dots (1.5–2.0 mm diameter) (Clement *et al.* 1986). The oviposition by females was positively correlated with the quantity of host frass (Rubink & Clement 1982). *L. comta* may discriminate between the frass of suitable and unsuitable hosts. For example, females prefer the frass of *A. ipsilon*, which has frequently been recorded as a host, to the frass of the armyworm *Pseudaletia unipuncta*, which has sometimes been recorded as a host, or to the frass of the pearly underwing *Peridroma saucia*, which has never been recorded as a host (Rubink & Clement 1982). Clement *et al.* (1986) found no oviposition preference by *L. comta* among the frass of *A. ipsilon* fed on

fresh maize plants, fresh chickweed plants, or an artificial diet. Nettles & Burks (1975) reported that the oviposition stimulant that induces the tachinid *Archytas marmoratus* (Townsend) to oviposit is most probably a metabolite of the host (*H. virescens*) and seems to be a protein with a molecular weight of 30,000 ± 5000. They found that the oviposition stimulant was present throughout the larval, pupal and early adult stages of *H. virescens*, and was also present in larvae of five noctuid species and two pyralid species. Low volatility of an oviposition stimulant would be advantageous for the survival of *A. marmoratus* larvae, which are sessile and do not ordinarily leave the substrate until they are touched by a host or other moving object. If the oviposition stimulant were volatile, flies would deposit larvae over a much wider area. Consequently many larvae would not survive, because their chances of encountering a host would decrease (Nettles & Burks 1975).

7.4 Conclusions

In this chapter we reviewed studies of the chemical ecology of tachinid parasitoids and have demonstrated the importance of chemical cues in their host location behaviour, although there is relatively little information on tachinids compared with the larger body of research on hymenopteran parasitoids. In the long-range host location phase, females that use direct and indirect strategies both use volatile chemical cues elicited from the host's food plant. After arriving in the host's habitat, tactile and chemosensory cues from the host's body or frass and from the host's food plant are important host location cues during short-range host orientation behaviour. Visual stimuli are particularly impor-tant for tachinids that use a direct strategy, which seems to be different from the mecha-nisms of host selection used by hymenopteran parasitoids. Some tachinids that use a direct strategy to attack heteropteran species use host pheromones as chemical cues for host location.

The tachinids have a high level of host polyphagy. It is unknown how female flies find a host habitat by using or choosing different chemical cues derived from diverse host species during the long-distance orientation phase. The levels of polyphagy might affect how tachinid females receive, use, learn and remember chemical cues from various resources related to their preferred hosts. As far as we know, there has been no study of the host location behaviour of tachinid larvae that use an indirect strategy, or how searching- and waiting-type larvae successfully find and attach to their hosts. We also do not know how microtype larvae find the specific host organs they require for development after they hatch in the host's gut. We believe that more information on, and more intensive study of, the host location behaviour of tachinid species with different oviposition strategies will lead not only to a better understanding of their ecology and behaviour but will also help when applying attractants of tachinids in the target field for effective biological control.

Acknowledgements

We are grateful to Drs Hiroshi Shima, Takuji Tachi and Jun Tabata for their useful com-ments to improve the manuscript. We would also like to thank Dr Takao Yoshida for giving us permission to use photographs of tachinid flies.

References

Adachi, I., Uchino, K. and Mochizuki, F. (2007) Development of a pyramidal trap for monitoring fruit-piercing stink bugs baited with *Plautia crossota stali* (Hemiptera: Pentatomidae) aggregation pheromone. *Applied Entomology and Zoology* **42**: 425–31.

Aldrich, J.R. (1995a) Testing the 'new associations' biological control concept with a tachinid parasitoid (*Euclytia flava*). *Journal of Chemical Ecology* **21**: 1031–42.

Aldrich, J.R. (1995b) Chemical communication in the true bugs and parasitoid exploitation. In: Cardé, R.T. and Bell, W.J. (eds) *Chemical Ecology of Insects*, 2nd edn. Chapman & Hall, London, pp. 318–63.

Aldrich, J.R., Hoffmann, M.P., Kochansky, J.P., Lusby, W.R., Eger, J.E. and Payne, J.A. (1991) Identification and attractiveness of a major pheromone component for nearctic *Euschistus* spp. stink bugs (Heteroptera: Pentatomidae). *Environmental Entomology* **20**: 477–83.

Aldrich, J.R., Khrimian, A. and Camp, M.J. (2007) Methyl 2,4,6-decatrienoates attract stink bugs and tachinid parasitoids. *Journal of Chemical Ecology* **33**: 801–15.

Aldrich, J.R., Kochansky, J.P. and Abrams, C.B. (1984) Attractant for a beneficial insect and its parasitoids: pheromone of the predatory spined soldier bug, *Podisus maculiventris* (Hemiptera: Pentatomidae). *Environmental Entomology* **13**: 1031–6.

Aldrich, J.R., Oliver, J.E., Lusby, W.R., Kochansky, J.P. and Lockwood, J.A. (1987) Pheromone strains of the cosmopolitan pest, *Nezara viridula* (Heteroptera: Pentatomidae). *Journal of Experimental Zoology* **244**: 171–5.

Askew, R.R. and Shaw, M.R. (1986) Parasitoid communities: their size, structure, and development. In: Waage, J.K. and Greathead, D. (eds) *Insect Parasitoids*. Academic Press, London, pp. 225–64.

Baker, G.L. (1978) The biology of a species of *Doleschalla* (Diptera: Tachinidae), a parasite of *Pantorhytes szentivanyi* (Coleoptera: Curculionidae). *Pacific Insects* **19**: 53–64.

Belshaw, R. (1994) Life history characteristics of Tachinidae (Diptera) and their effect on polyphagy. In: Hawkins, B.A. and Sheehan, W. (eds) *Parasitoid Community Ecology*. Oxford University Press, Oxford, pp. 145–62.

Burks, M.L. and Nettles, W.C., Jr (1978) *Eucelatoria* sp.: effects of cuticular extracts from *Heliothis virescens* and other factors on oviposition. *Environmental Entomology* **7**: 897–900.

Cade, W. (1975) Acoustically orienting parasitoids: fly phonotaxis to cricket song. *Science* **190**: 1312–13.

Cardé, R.T. and Millar, J.G. (2004) *Advances in Insect Chemical Ecology*. Cambridge University Press, Cambridge.

Clausen, C.P. (1940) *Entomophagous Insects*. Hafner, New York.

Clement, S.L., Rubink, W.L. and McCartney, D.A. (1986) Larviposition response of *Bonnetia comta* (Dipt.: Tachinidae) to a kairomone of *Agrotis ipsilon* (Lep.: Noctuidae). *Entomophaga* **31**: 277–84.

Dippel, C. and Hilker, M. (1998) Effects of physical and chemical signals on host foraging behavior of *Drino inconspicua* (Diptera: Tachinidae), a generalist parasitoid. *Environmental Entomology* **27**: 682–7.

Eggleton, P. and Belshaw, R. (1992) Insect parasitoids: an evolutionary overview. *Philosophical Transactions of the Royal Society of London (Series B)* **337**: 1–20.

Eggleton, P. and Gaston, K.J. (1992) Tachinid host ranges: a reappraisal (Diptera: Tachinidae). *Entomologist's Gazette* **43**: 139–43.

Feener, D.H., Jr and Brown, B.V. (1997) Diptera as parasitoids. *Annual Review of Entomology* **42**: 73–97.

Franklin, R.T. and Holdaway, F.G. (1966) A relationship of the plant to parasitism of European corn borer by the tachinid parasite *Lydella grisescens*. *Journal of Economic Entomology* **59**: 440–1.

Godfray, H.C.J. (1994) *Parasitoids: Behavioural and Evolutionary Ecology*. Princeton University Press, Princeton, NJ.

Greathead, D. (1986) Parasitoids in classical biological control. In: Waage, J.K. and Greathead, D. (eds) *Insect Parasitoids*. Academic Press, London, pp. 289–318.

Grenier, S. (1988) Applied biological control with tachinid flies (Diptera, Tachinidae): a review. *Anzeiger für Schädlingskunde, Pflanzenschutz, Umweltschutz* **51**: 49–56.

Hanyu, K., Ichiki, R.T., Nakamura, S. and Kainoh, Y. (2009) Duration and location of attraction to herbivore-damaged plants in the tachinid parasitoid *Exorista japonica*. *Applied Entomology and Zoology* **44**: 371–8.

Hanyu, K., Ichiki, R.T., Nakamura, S. and Kainoh, Y. (2011) Behavior of the tachinid parasitoid *Exorista japonica* (Diptera: Tachinidae) on herbivore-infested plants. *Applied Entomology and Zoology* **46**: 565–71.

Harris, V.E. and Todd, J.W. (1980) Male-mediated aggregation of male, female and 5th-instar southern green stink bugs and concomitant attraction of a tachinid parasite, *Trichopoda pennipes*. *Entomologia Experimentalis et Applicata* **27**: 117–26.

Hassell, M.P. (1968) The behavioural response of a tachinid fly (*Cyzenis albicans* (Fall.)) to its host, the winter moth (*Operophtera brumata* (L.)). *Journal of Animal Ecology* **37**: 627–39.

Higaki, M. and Adachi, I. (2011) Response of a parasitoid fly, *Gymnosoma rotundatum* (Linnaeus) (Diptera: Tachinidae), to the aggregation pheromone of *Plautia stali* Scott (Hemiptera: Pentatomidae) and its parasitism of hosts under field conditions. *Biological Control* **58**: 215–21.

Hsiao, T.H., Holdaway, F.G. and Chiang, H.C. (1966) Ecological and physiological adaptations in insect parasitism. *Entomologia Experimentalis et Applicata* **9**: 113–23.

Ichiki, R.T., Ho, G.T., Wajnberg, E., Kainoh, Y., Tabata, J. and Nakamura, S. (2012) Different uses of plant semiochemicals in host location strategies of the two tachinid parasitoids. *Naturwissenschaften* **99**: 687–94.

Ichiki, R.T., Kainoh, Y., Kugimiya, S., Takabayashi, J. and Nakamura, S. (2008) Attraction to herbivore-induced plant volatiles by the host-foraging parasitoid fly *Exorista japonica*. *Journal of Chemical Ecology* **34**: 614–21.

Ichiki, R.T., Kainoh, Y., Yamawaki, Y. and Nakamura, S. (2011) The parasitoid fly *Exorista japonica* uses visual and olfactory cues to locate herbivore-infested plants. *Entomologia Experimentalis et Applicata* **138**: 175–83.

Ichiki, R., Nakamura, S., Takasu, K. and Shima, H. (2006) Oviposition behaviour of the parasitic fly *Bessa parallela* (Meigen) (Diptera: Tachinidae) in the field. *Applied Entomology and Zoology* **41**: 659–65.

Jang, S.A. and Park, C.G. (2010) *Gymnosoma rotundatum* (Diptera: Tachinidae) attracted to the aggregation pheromone of *Plautia stali* (Hemiptera: Pentatomidae). *Journal of Asia-Pacific Entomology* **13**: 73–5.

Jang, S.A., Cho, J.H., Park, G.M., Choo, H.Y. and Park, C.G. (2011) Attraction of *Gymnosoma rotundatum* (Diptera: Tachinidae) to different amounts of *Plautia stali* (Hemiptera: Pentatomidae) aggregation pheromone and the effect of different pheromone dispensers. *Journal of Asia-Pacific Entomology* **14**: 119–21.

Kainoh, Y., Tanaka, C. and Nakamura, S. (1999) Odor from herbivore-damaged plant attracts a parasitoid fly, *Exorista japonica* Townsend (Diptera: Tachinidae). *Applied Entomology and Zoology* **34**: 463–7.

Kuhlmann, U. (1995) Biology of *Triarthria setipennis* (Fallen) (Diptera: Tachinidae), a native parasitoid of the European earwig, *Forficula auricularia* L. (Dermaptera: Forficulidae), in Europe. *Canadian Entomologist* **127**: 507–17.

Martin, W.R., Jr, Nordlund, D.A. and Nettles, W.C., Jr (1990) Response of parasitoid *Eucelatoria bryani* to selected plant material in an olfactometer. *Journal of Chemical Ecology* **16**: 499–508.

Mishiro, K. and Ohira, Y. (2002) Attraction of a synthetic aggregation pheromone of the brown-winged green bug, *Plautia crossota stali* Scott to its parasitoids, *Gymnosoma rotundata* and *Trissolcus plautiae*. *Kyushu Plant Protection Research* **48**: 76–80 (in Japanese with English summary).

Mitchell, W.C. and Mau, R.F.L. (1971) Response of the female southern green stink bug and its parasite, *Trichopoda pennipes*, to male stink bug pheromones. *Journal of Economic Entomology* **64**: 856–9.

Mondor, E.B. and Roland, J. (1997) Host locating behaviour of *Leschenaultia exul* and *Patelloa pachypyga*, two tachinid parasitoids of the forest tent caterpillar, *Malacosoma disstria*. *Entomologia Experimentalis et Applicata* **85**: 161–8.

Mondor, E.B. and Roland, J. (1998) Host searching and oviposition by *Leschenaultia exul*, a tachinid parasitoid of the forest tent caterpillar, *Malacosoma disstria*. *Journal of Insect Behavior* **11**: 583–92.

Monteith, L.G. (1955) Host preferences of *Drino bohemica* Mesn. (Diptera: Tachinidae), with particular reference to olfactory responses. *Canadian Entomologist* **87**: 509–30.

Monteith, L.G. (1956) Influence of host movement on selection of hosts by *Drino bohemica* Mesn (Diptera: Tachinidae) as determined in an olfactometer. *Canadian Entomologist* **88**: 583–6.

Monteith, L.G. (1964) Influence of the health of the food plant of the host on host-finding by tachinid parasites. *Canadian Entomologist* **96**: 1477–82.

Moriya, S. and Shiga, M. (1984) Attraction of the male brown-winged green bug, *Plautia stali* Scott. (Heteroptera: Pentatomidae) for males and females of the same species. *Applied Entomology and Zoology* **19**: 317–22.

Nakamura, S. (1997) Ovipositional behaviour of the parasitoid fly, *Exorista japonica* (Diptera: Tachinidae), in the laboratory: diel periodicity and egg distribution on a host. *Applied Entomology and Zoology* **32**: 189–95.

Nettles, W.C., Jr (1979) *Eucelatoria* sp. females: factors influencing response to cotton and okra plants. *Environmental Entomology* **8**: 619–23.

Nettles, W.C., Jr (1980) Adult *Eucelatoria* sp.: response to volatiles from cotton and okra plants and from larvae of *Heliothis virescens*, *Spodoptera eridania*, and *Estigmene acrea*. *Environmental Entomology* **9**: 759–63.

Nettles, W.C., Jr (1982) Contact stimulants from *Heliothis virescens* that influence the behavior of females of the tachinid, *Eucelatoria bryani*. *Journal of Chemical Ecology* **8**: 1183–91.

Nettles, W.C., Jr and Burks, M.L. (1975) A substance from *Heliothis virescens* larvae stimulating larviposition by females of the tachinid, *Archytas marmoratus*. *Journal of Insect Physiology* **21**: 965–78.

Odell, T.M. and Godwin, P.A. (1984) Host selection by *Blepharipa pratensis* (Meigen), a tachinid parasite of the gypsy moth, *Lymantria dispar* L. *Journal of Chemical Ecology* **10**: 311–20.

O'Hara, J.E. (2008) Tachinid flies (Diptera: Tachinidae). In Capinera, J.L. (ed) *Encyclopedia of Entomology*, 2nd edn. Springer, Dordrecht, pp. 3675–86.

Powell, W. (1999) Parasitoid hosts. In: Hardie, J. and Minks, A.K. (eds) *Pheromones of Non-Lepidopteran Insects Associated with Agricultural Plants*. CABI Publishing, Wallingford, UK, pp. 405–27.

Robert, D., Amoroso, J. and Hoy, R.R. (1992) The evolutionary convergence of hearing in a parasitoid fly and its cricket host. *Science* **258**: 1135–7.

Roland, J. (1986) Parasitism of winter moth in British Columbia during build-up of its parasitoid *Cyzenis albicans*: attack rate on oak v. apple. *Journal of Animal Ecology* **55**: 215–34.

Roland, J. (1990) Parasitoid aggregation: chemical ecology and population dynamics. In: Mackauer, M., Ehler, L.E. and Roland, J. (eds) *Critical Issues in Biological Control*. Intercept, Andover, UK, pp. 185–211.

Roland, J., Denford, K.E. and Jimenez, L. (1995) Borneol as an attractant for *Cyzenis albicans*, a tachinid parasitoid of the winter moth, *Operophtera brumata* L. (Lepidoptera, Geometridae). *Canadian Entomologist* **127**: 413–21.

Roland, J., Evans, W.G. and Myers, J.H. (1989) Manipulation of oviposition patterns of the parasitoid *Cyzenis albicans* (Tachinidae) in the field using plant extracts. *Journal of Insect Behavior* **2**: 487–503.

Roth, J.P., King, E.G. and Thompson, A.C. (1978) Host location behavior by the tachinid, *Lixophaga diatraeae*. *Environmental Entomology* **7**: 794–8.

Roth, J.P., King, E.G. and Hensley, S.D. (1982) Plant, host, and parasite interactions in the host selection sequence of the tachinid *Lixophaga diatraeae*. *Environmental Entomology* **11**: 273–7.

Rubink, W.L. and Clement, S.L. (1982) Reproductive biology of *Bonnetia comta* (Fallen) (Diptera: Tachinidae), a parasitoid of the black cutworm, *Agrotis ipsilon* Hufnagel (Lepidoptera: Noctuidae). *Environmental Entomology* **11**: 981–5.

Sabelis, M.W. and van de Baan, H.E. (1983) Location of distant spider mite colonies by phytoseiid predators: demonstration of specific kairomones emitted by *Tetranychus urticae* and *Panonychus ulmi*. *Experimental and Applied Acarology* **33**: 303–14.

Stireman, J.O., III (2002) Host location and acceptance in a polyphagous tachinid parasitoid. *Entomologia Experimentalis et Applicata* **103**: 23–34.

Stireman, J.O., III, O'Hara, J.E. and Wood, D.M. (2006) Tachinidae: evolution, behavior, and ecology. *Annual Review of Entomology* **51**: 525–55.

Stowe, M.K., Turlings, T.C.J., Loughrin, J.H., Lewis, W.J. and Tumlinson, J.H. (1995) The chemistry of eavesdropping, alarm, and deceit. *Proceedings of the National Academy of Sciences USA* **92**: 23–8.

Takabayashi, J., Takahashi, S., Dicke, M. and Posthumus, M.A. (1995) Developmental stage of the herbivore *Pseudaletia separata* affects production of herbivore-induced synomone by corn plants. *Journal of Chemical Ecology* **21**: 273–87.

Tanaka, C., Kainoh, Y. and Honda, H. (1999) Physical factors in host selection of the parasitoid fly, *Exorista japonica* Townsend (Diptera: Tachinidae). *Applied Entomology and Zoology* **34**: 91–7.

Tanaka, C., Kainoh, T. and Honda, H. (2001) Host frass as arrestant chemicals in locating host *Mythimna separata* by the tachinid fly *Exorista japonica*. *Entomologia Experimentalis et Applicata* **100**: 173–8.

Thompson, A.C., Roth, J.P. and King, E.G. (1983) Larviposition kairomone of the tachinid *Lixophaga diatraeae*. *Environmental Entomology* **12**: 1312–14.

Turlings, T.C.J. and Wäckers, F. (2004) Recruitment of predators and parasitoids by herbivore-injured plants. In Cardé, R.T. and Millar, J.G. (eds) *Advances in Insect Chemical Ecology*. Cambridge University Press, Cambridge.

van Alphen, J.J.M. and Jervis, M.A. (1996) Foraging behaviour. In: Jervis, M. and Kidd, N. (eds) *Insect Natural Enemies: Practical Approaches to Their Study and Evolution*. Chapman and Hall, London, pp. 1–62.

Vet, L.E.M. and Dicke, M. (1992) Ecology of infochemical use by natural enemies in a tritrophic context. *Annual Review of Entomology* **37**: 141–72.

Walker, T.J. (1993) Phonotaxis in female *Ormia ochracea* (Diptera: Tachinidae), a parasitoid of field crickets. *Journal of Insect Behavior* **6**: 389–410.

Weseloh, R.M. (1980) Host recognition behavior of the tachinid parasitoid, *Compsilura concinnata*. *Annals of the Entomological Society of America* **73**: 593–601.

Wood, D.M. (1987) Tachinidae. In: McAlpine, J.F.M. (ed) *Manual of the Nearctic Diptera, Vol. 2.* Agriculture Canada, Ottawa, pp. 1193–269.

Yamawaki, Y. and Kainoh, Y. (2005) Visual recognition of the host in the parasitoid fly *Exorista japonica*. *Zoological Science* **22**: 563–70.

Yamawaki, Y., Kainoh, Y. and Honda, H. (2002) Visual control of host pursuit in the parasitoid fly *Exorista japonica*. *Journal of Experimental Biology* **205**: 485–92.

8

Climate change and its effects on the chemical ecology of insect parasitoids

Jarmo K. Holopainen[1], Sari J. Himanen[2] and Guy M. Poppy[3]

[1]Department of Environmental Science, University of Eastern Finland, Kuopio, Finland
[2]MTT Agrifood Research Finland, Plant Production Research, Mikkeli, Finland
[3]Centre for Biological Sciences, University of Southampton, UK

Abstract

Since parasitoids represent one of the higher trophic levels in a food chain, they are likely to be affected by many abiotic factors related to climate change. Climate not only directly affects the behaviour and performance of insect parasitoids, but also has indirect bottom-up affects via the behaviour and survival of their host species. These hosts, in turn, are also impacted by any response to the climate taking place in their host plant. Parasitoids of plant-feeding insects will face many direct impacts of climate changes, for example through temperature and humidity changes as well as the effects of various atmospheric pollutants. In addition, the indirect effects caused by altered plant chemistry and subsequent nutrition mediating host insect survival may have even more drastic impacts than the direct effects of environmental changes itself. In this chapter we examine how factors involved in climate change have direct and indirect effects on the chemical defences of plants. We look at how these changes are reflected in the chemical communication between trophic levels and how the changes at lower levels might be accumulated and reflected in the ecology of species at higher trophic levels. We also present examples from tritrophic systems, which are frequently studied and used as model systems, to estimate the impacts of environmental disturbance on the components of chemical tritrophic communication between plants, herbivores and parasitoids.

Chemical Ecology of Insect Parasitoids, First Edition. Eric Wajnberg and Stefano Colazza.
© 2013 John Wiley & Sons, Ltd. Published 2013 by John Wiley & Sons, Ltd.

We especially focus on the impacts of oxidative pollutants such as ozone, elevated CO_2, and rising temperature on the atmospheric behaviour of volatile signalling compounds, which are particularly important in parasitoid host location – a significant stage for determining parasitoid success, reproduction, and population development.

8.1 On climate change and chemical ecology

The Earth's climate has been highly variable throughout history and there have been numerous periods with colder and warmer annual mean temperatures over geological time. The scientific community is largely unanimous in proposing that a major reason for the recent global warming and consequent alterations in climatic conditions, in all continents, is due to anthropogenic impacts on the atmosphere (IPCC 2007). Global temperature deviations since the 1970s are now large enough to exceed the bounds of natural variability (Karl & Trenberth 2003). Raising of the global mean temperature is caused by increased forcing of solar radiation by greenhouse gases, mostly carbon dioxide (CO_2), related to the burning of fossil fuels (IPCC 2007). Global warming is affecting flooding and drought risks in many regions of the world, largely through changing the timing and intensity of dry and rainy periods. The decrease of snow cover in warming polar areas will result in the thawing of permafrost and the rapid release of another important greenhouse gas, methane, from the frozen soil layer (Mackelprang *et al.* 2011), possibly magnifying the greenhouse effect even further. Formation of tropospheric ozone (O_3) in the lower atmosphere is related to the release of oxides of nitrogen (NOx) in the burning process of fuels. O_3 is a phytotoxic greenhouse gas which is increasing in concentration in the lower troposphere and could significantly reduce CO_2 uptake by vegetation (Sitch *et al.* 2007). A decline in the stratospheric O_3 layer is linked to global climate change, and its capacity to screen solar UV-B radiation has decreased. Trophic interactions (herbivory, decomposition, etc.) in terrestrial ecosystems are considered to be sensitive to variations in UV-B irradiance (Ballare *et al.* 2011). In future, variations in UV-B and UV-A radiation resulting from changes in climate and land use are expected to have even more important consequences for terrestrial ecosystems than the changes in UV caused by O_3 depletion (Ballare *et al.* 2011).

There is substantial evidence from recent years that ecosystems are responding to the rapid change in climatic conditions (Walther *et al.* 2002). A change of climate may affect nature in the form of increased frequency of extreme weather episodes or a gradual extension of the period when the weather deviates from the 'normal' conditions to which local populations of organisms have adapted. An increasing variability of climate itself may affect specialist and generalist species, which differ in their adaptation capabilities. For example, an impaired ability of specialized Hymenoptera parasitoid wasps to track lepidopteran host populations may occur, while tachinid flies, which possess a wider host range, seem to be better buffered against climatic fluctuations (Stireman *et al.* 2005). In most cases, elevated temperatures will cause long-term drought and salinity stress to vegetation and, at the other extreme, occasional hurricanes, heavy rains and flooding can also lead to drastic changes in ecosystem structure and function. Changes in precipitation may amplify fluctuations in a range of populations, leading to rapid extinctions of plant species and their specialist herbivores, including their parasitoids (McLaughlin *et al.* 2002). The most distinctive and large-scale impact in ecosystems is caused by the warming of climate, which results in, for example, a poleward shift in the distribution of herbivorous insect species in Europe (Parmesan *et al.* 1999). At the ecosystem level, this will often lead to spatial and temporal asynchrony between herbivores and their natural enemies, including parasitoids (Parmesan 2006, Hance *et al.* 2007). This host–parasitoid synchrony influences subsequent parasitoid population size and the rate of colonization in previously uninhabited host populations (van Nouhuys & Lei 2004).

The functionality of ecosystems is based on the success of primary producers such as green plants or planktonic microalgae. Therefore, any changes in the environmental conditions that affect vegetation are amplified through the higher trophic levels. Chemical ecology is defined as ecological interactions mediated by the chemicals that organisms produce (Jones 1988). Thus, any alteration in the capacity of a particular organism to produce a chemical compound which is involved in species interaction, or in the ability of a receiving organism to perceive this compound, or even a tendency of the molecules of mediating chemical compounds to resist or become vulnerable to environmental degradation, will affect the chemical ecology of parasitoids. In particular, parasitoids are affected due to their position at the top trophic level of food chains.

The impacts of climate change on plant secondary chemistry are multifaceted, since the synthesis of these chemicals occurs via multiple metabolic routes, some of these being dependent on primary metabolism and many being induced by environmental stresses or variations. Chemical ecology regarding climate change is closely linked to plant secondary chemistry, as the upper trophic level interactions are often sensitive to plant responses to abiotic changes. Biotic and various abiotic stresses, singly or in combination, are known to induce general physiological responses in plants, for example by inducing oxidative stress, which results in reactive oxygen species (ROS) bursts (Vickers *et al.* 2009). ROS are important first-step signalling molecules in eliciting defence responses in plants, and also activate several other signalling pathways (see Ode, Chapter 2, this volume). The interaction and balance of these pathways determines whether the cell will stay alive or whether ROS signalling will lead to cell death (Overmyer *et al.* 2003). Localized cell death is an important defence response in plants to limit the damage caused for instance by ozone, piercing and sucking insects, or plant pathogens. Signal transduction activated by ROS is responsible for induced plant defence via three major pathways:

1 *the octadecanoid pathway*, which involves jasmonic acid (JA) signalling leading, for example, to the production of 'green leaf volatiles' and terpenoids (Holopainen & Gershenzon 2010);
2 *the shikimate pathway*, leading to salicylic acid (SA) signalling and the production of phenylpropanoids and benzenoids (Pinto *et al.* 2010);
3 *the ethylene pathway*, controlling ethylene production at different stages of plant development and during recovery from stress (Overmyer *et al.* 2003).

Currently, a simplified view has been put forward (Thaler *et al.* 2012), which suggests that SA signalling mainly induces resistance against biotrophic pathogens and some phloem-feeding insects, and JA signalling induces resistance against necrotrophic pathogens, some phloem-feeding insects and chewing herbivores (see Ode, Chapter 2, this volume). Most of the major groups of plant secondary compounds induced by both abiotic and biotic stress and controlled by JA and SA signalling are known to act as antioxidants. These include phenolic compounds such as flavonoids (Hernandez *et al.* 2009), pro-anthocyanidins, flavonols and catechins (Martz *et al.* 2010), alkaloids (Boga *et al.* 2011), and various terpenoids with an isoprenoid skeleton (Giuliano *et al.* 2008).

As summarized above, a typical plant response to oxidative stress is increased antioxidant production, including the biosynthesis of various isoprenoids (Vickers *et al.* 2009). Volatile isoprenoids have an important ecological role in indirect plant defence by attracting preda-tors and parasitoids to herbivore-attacked plants (Dicke & Baldwin 2010). Research on the ecological functions and environmental stability of isoprenoids has been very active recently.

Among climate change parameters, temperature is considered to be the major abiotic influence on the functioning of multitrophic communication and indirect plant defence through the attraction of parasitoids and other natural enemies by volatile organic compounds (VOCs) (Peñuelas & Staudt 2010). In addition to signal production in plants and signal perception in parasitoids, another important component of indirect defence is the lifetime of chemical signals in the atmosphere (Yuan *et al.* 2009). Oxidative air pollut-ants, related to climate warming, may significantly shorten the atmospheric lifetimes of herbivore-induced VOCs and thus influence their signal strength, quality and duration. In addition to plant-produced volatile compounds, host-related volatiles such as alarm phe-romones produced by aphids (Cui *et al.* 2012) are important attractants to parasitoids. The impacts of air pollutants and temperature changes may thus play a large role in determin-ing the efficiency of host–parasitoid communication by volatiles through bottom-up (first and second trophic level originating) effects.

As well as the volatile cues, parasitoids can use visual cues to locate host herbivores. For example, the generalist tachinid *Exorista japonica* (Diptera: Tachinidae) is attracted to odours derived from maize plants infested with the larvae of the noctuid moth *Mythimna separata*, but the coloration of the moth's host plant can affect the host preference of the moth parasitoid (Ichiki *et al.* 2011; see Nakamura, Ichiki & Kainoh, Chapter 7, this volume). Host plant coloration, such as the synthesis of foliar pigments or symptoms of leaf senes-cence, could be very sensitive to changes in environmental factors and may affect the visual sensing of the environment by parasitoids. The great variety of changes and the potential interactions between change factors estimated to be induced by climate change could have a multitude of effects where the impact on parasitoid chemical ecology is significant. Figure 8.1 gives an overview of direct and indirect effects for the currently recognized main factors to be affected by climate change. Multiple abiotic factors and a plethora of ecological interactions between and within trophic levels determine the overall responses of our ecosystems and food webs to changing climate.

8.2 Direct climate change impacts on parasitoids

Changes in temperature have a significant effect on the active and dormant stages of para-sitoids as well as their distribution and synchronicity with their host species. Direct physi-ological responses of parasitoids to changing environmental conditions such as temperature extremes are covered comprehensively in a recent review by Hance *et al.* (2007).

External disturbances such as more frequent heat waves, drought periods, extreme precipitation events and air pollutants may affect parasitoids' physiological capacity to

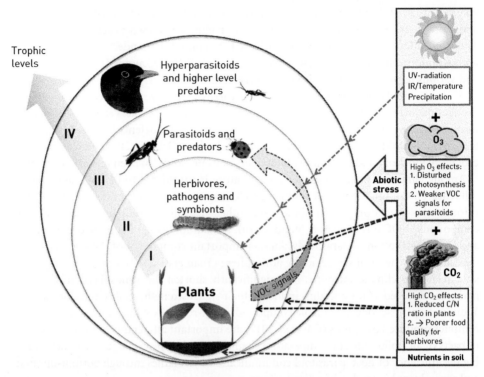

Figure 8.1 Schematic overview of effects and possible interactions in ecosystems affected by changing climate. General stress by combination of abiotic factors related to climate change may have a unified impact in oxidative defence of organisms. Each type of single stress has a specific impact on organisms and their interactions between different trophic levels. VOC: volatile organic compounds.

perceive chemical and visual signals from their environment. These factors could have serious detrimental effects on the chemical ecology and behaviour of parasitoids. As this chapter focuses on climate change impacts on the chemical ecology of parasitoids, in the following sections we review these direct effects and how they impact on volatile sensing by parasitoids, together with the indirect effects of climate change via other interacting organisms.

8.3 Climate change and bottom-up impacts on parasitoids: herbivore host and plant host quality

Tritrophic interactions established between plants, herbivores and their parasitoids and predators are presumed to have a long history of coevolution. The presence of three species from different trophic levels, with their specific contribution and function in the food chain and relations between co-occurring individuals/species, both at one's own or down- or upward trophic level, can make such multitrophic interactions extremely sensitive to external disturbances. In fact, Whittaker (2001) considered between-species interactions on

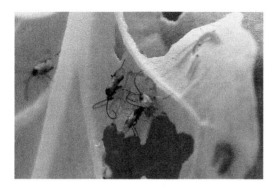

Figure 8.2 Approaching individuals of the parasitoid *Cotesia vestalis* landing on a freshly damaged broccoli plant leaf area to sense the emanating herbivore-induced volatiles with their antennae. No attention is given at this stage to the actual host (diamondback moth, *Plutella xylostella*) larvae on the same leaf (right). Video frame capture by J.K. Holopainen.

plants to be the most sensitive indicator of environmental change, especially those relating to air pollutants in the atmosphere. For parasitoids that attack hosts and emerge from their host herbivores, the effects of environmental factors on the food plant of the host herbivore also has a crucial role for the survival of hosts, parasitoids and the parasitoids' offspring.

Most parasitoids have been adapted to locate and identify their host by the composition of volatiles emitted by herbivore-damaged plants. Egg parasitoids of herbivorous insects can, for example, orientate on the basis of specific plant VOCs, for example from conifer needles, induced by their ovipositing hosts (Köpke *et al.* 2010). Parasitoids of leaf-feeding larvae, in particular, use feeding site-specific signalling cues such as the VOCs emitted from the margins of the damaged leaf area (Connor *et al.* 2007), but also cues from frass and silk on leaves (Mattiacci & Dicke 1995). For example, when arriving on herbivore-damaged plants, *Cotesia vestalis*, a braconid parasitoid of the diamondback moth (*Plutella xylostella*), first land on the damaged area of the leaf and intensively investigate the leaf surface, paying little attention to nearby host larvae (Fig. 8.2). The importance of herbivore-induced attractive plant volatiles to successful parasitization rate has been shown in studies where the production of herbivore-induced volatiles was manipulated in plants. As summarized by Holopainen (2011), the parasitization rate on manipulated plants in those studies was 37–180% higher than on control plants. Volatile communication between herbivore-damaged plants and parasitoids has recently been studied with the aim of improving biological pest control by releasing the most attractive plant-based volatile compounds from dispensers (Kaplan 2012).

Any modification in global climate change factors that affect the plant's capacity to emit herbivore-inducible VOCs, such as change in growth, maturation, stomatal function, substrate availability and changes VOC biosynthesis pathways, are expected to affect parasitoid performance (Yuan *et al.* 2009). Furthermore, oxidative pollutants which affect the atmospheric lifetime of herbivore-inducible VOCs could disrupt the orientation of parasitoids when the concentration of the signalling compounds is diluted as a result of degradation. Laboratory experiments with controlled ozone concentrations have shown that ozone exposure decreases the concentrations of herbivore-inducible mono-, homo- and sesquiterpenes as well as C_6 green leaf volatiles originating from the oxylipin pathway. Lower concentrations of these signal compounds in the atmosphere are directly reflected in the behaviour of parasitoids such as *C. vestalis* (Himanen *et al.* 2009a).

The performance of parasitoids is not only dependent on the physiological status of their host larvae, but also on the risk of becoming parasitized by hyperparasitoids or being eaten by predators. It is commonly known that plant secondary metabolites sequestered by herbivorous insect larvae will give efficient protection against many generalist predators (see Baden *et al.* 2011, Petschenka *et al.* 2011). In the presence of generalist predatory bugs, lepidopteran larvae can even increase the accumulation of plant defence compounds in their body in order to improve their chemical defence at the pupal stage (Bowers & Stamp 1997). For generalist predators, detoxification of defence chemicals is a great challenge because the cost of detoxification increases when plant-derived toxins of several pathways are concerned. Therefore, generalist predators often avoid herbivorous larvae rich in plant-derived toxins and, at the same time, endoparasitoids of the herbivores may take advantage of the lower predation rate (van Nouhuys *et al.* 2012).

As the larvae of endoparasitoids of herbivorous insects have a very intimate relationship with their hosts, they have to be able to detoxify plant-derived compounds in their host larvae (van Nouhuys *et al.* 2012). There is evidence that the content of defence chemicals in the host plant can affect the performance of butterfly larvae, but parasitoid responses could be variable. A change in climatic conditions, such as increase in daytime temperature, may have a significant effect on the concentration of defence compounds in plants (Tamura 2001). For example, the concentration of the iridoid glycoside catalpol in *Plantago lanceolata* plants is found to increase in a temperature-dependent manner (Tamura 2001). This compound and other iridoid glycosides affect the performance of specialist (Nieminen *et al.* 2003) and generalist (Harvey *et al.* 2005, Reudler *et al.* 2011) lepidopteran defoliators and their parasitoid wasps. Reudler *et al.* (2011) reported that larval performance of the specialist butterfly *Melitaea cinxia* and the parasitoid *Cotesia melitaearum* was unaffected by the total iridoid glycoside content variation on *P. lanceolata*. The authors expected such a response from these highly specialized species. In contrast, the generalist moth herbivores *Spodoptera exigua* and *Chrysodeixis chalcites* showed poorer performance on plants with high iridoid glycoside content, indicating that plant-specific defence compounds are most efficient against generalist herbivores. Interestingly, the development time responses of the generalist parasitoids *Hyposoter didymator* and *Cotesia marginiventris* differed from those of their generalist hosts. The development time of *H. didymator* was unaffected by iridoid glycoside content in the diet of its host, while *C. marginiventris* actually developed faster on hosts grown on high iridoid glycoside genotypes, although the adult longevity of this parasitoid became shorter.

These examples show how sensitive the performance of a single parasitoid species to variation in host plant chemical quality can be. Therefore, the impact of climate change factors on herbivores and their parasitoids as a result of changes in host plant defence chemistry could be species-specific and not easy to generalize under multiple climate drivers (Tylianakis *et al.* 2008). In more complex systems, the response of hyperparasitoids to variations in plant-derived toxins may also play a role in parasitoid performance. Hyperparasitoids are secondary parasitoids that have a strong impact on the performance of primary parasitoids of their hosts. A recent study (van Nouhuys *et al.* 2012) showed that both specialist and generalist hyperparasitoid wasp species can cope with various levels of plant defence compounds in their host larvae which enable them to face a potentially diverse array of chemical challenges. This may also suggest that hyperparasitoids adapt well to any climate-induced change in plant-derived toxins in their hosts (Tamura 2001) if they are not directly harmed by climate change-related abiotic stresses.

8.4 Impacts of climate change-related abiotic stresses on parasitoid ecology and behaviour

8.4.1 Impacts of elevated temperature

Parasitoid behaviour is strongly dependent on how they perceive the information from their environment. Typical of the parasitoids of herbivorous, carnivorous and detritivorous invertebrates is that the odour of the host's food items is usually important for efficient searching behaviour. The release rate of VOCs from living and dead organic material is affected by their synthesis (Peñuelas & Staudt 2010) or by the emission rates in the breakdown process of decaying tissues (Holopainen *et al.* 2010). Emission rates, in turn, are also affected by physicochemical characteristics of VOCs, mainly by their solubility, volatility and diffusivity. Individual volatile chemical compounds typically have a high vapour pressure; in other words, their evaporation rate from a liquid phase to a gas phase is high and increases with rising temperature. Therefore, ecological interactions based on volatile signals can be very sensitive to changes in ambient temperature (Holopainen & Gershenzon 2010, Peñuelas & Staudt 2010).

Herbivore-induced plant volatiles (HIPVs) have been found to play an important role in attracting the natural enemies of herbivorous insects under laboratory conditions, and evidence exists for volatile attraction to function also in nature (Dicke & Baldwin 2010). Parasitoid wasps are attracted by many volatile and semivolatile compounds such as sesquiterpenes and homoterpenes, which are emitted at higher rates by plants after they have been damaged by host herbivores. Diurnal and nocturnal variation in ambient temperature has been shown to affect the volatility and emission of herbivore-induced VOCs. During colder night conditions, semivolatile VOCs released by damaged plants are condensed on plant surfaces. In the morning, some of the emissions of semivolatile VOCs are composed of depositions from the plant surface, where warming temperature releases the condensed compounds back to the atmosphere (Schaub *et al.* 2010). Part of the emissions of semivolatile compounds may even originate from neighbouring plants (Himanen *et al.* 2010). Therefore, the composition of a VOC blend in the morning may differ depending both on night temperature and on the ratio of directly synthesized and emitted VOCs to those semivolatile VOCs synthesized earlier and released from condensation storage on the plant surface. Variation in night temperature may particularly affect parasitoids which use VOCs released during nocturnal induction, such as indole and homoterpene DMNT (4,8-dimethylnona-1,3,7-triene), as orientation cues (Signoretti *et al.* 2012).

Herbivore-induced VOCs are also released from the below-ground parts of plants and are reported to attract, for example, insect pathogenic nematodes of root-feeding beetle larvae (Rasmann *et al.* 2005). However, the specialist parasitoid *Trybliographa rapae* (Hymenoptera: Figitidae) of cabbage root fly, *Delia radicum*, larvae was not attracted by the VOCs released from the shoots and root system of turnip plants suffering root damage by fly larvae (Pierre *et al.* 2011). The efficiency of VOCs released from a damaged plant root system in attracting parasitoids is not well understood. The luring of flying parasitoids to root emissions could be complicated, as soil particles may condense part of the VOC blend on their surfaces in a temperature-dependent way. A higher temperature could have the potential to improve the atmospheric emission rates of root volatiles. In forest ecosystems, the situation is even more complex and the detection of root VOCs and those released from plant litter might become more difficult to distinguish by parasitoids. C_6 'green leaf volatiles', known to be important attractants of egg and larval parasitoids of leaf herbivores (see Reddy *et al.* 2002), are released from decomposing leaf litter (Holopainen *et al.* 2010). Furthermore, conifer litter releases parasitoid-attracting terpenes such as β-ocimene (Isidorov *et al.* 2010). Indirect evidence suggests that VOCs emitted by needle litter or roots could attract parasitoids of the common pine sawfly (*Diprion pini*) pupae in the soil beneath needle litter in poor understorey vegetation sites (Herz & Heitland 2005). This result suggests that the increasing complexity of ground flora, such as may occur under a warming climate, is capable of impeding host location by pupal parasitoids if they use litter VOCs as signals during searching.

Climatic warming has the potential to affect life-cycle synchronization between Diptera larvae feeding in animal faeces and carrion and their parasitoids. Optimal temperature ranges for the emergence success of some cosmopolitan fly parasitoids could be much more limited than those of their carrion-feeding hosts (Ferreira de Almeida *et al.* 2002). Furthermore, temperature changes may also affect the orientation behaviour of parasitoids of carrion-feeding insects, as the changing pattern of VOC emissions is a reliable cue for parasitoids to find suitable hosts during the succession of decomposition (Voss *et al.* 2009). In the typically scattered distribution of rotting animal material, parasitoids need reliable volatile cues to identify the appropriate stage of the decaying process of the carcass in order to maximize the detection of suitable host larvae. The cosmopolitan blowfly encryptid parasitoid *Tachinaephagus zealandicus* relies on the odour of the decaying substrate of flesh eaten by blow fly larvae in their search for hosts (Voss *et al.* 2009). At this stage of decomposition, mammal corpses emit a wide variety of volatiles which are dominated by several sulphur-rich VOCs, such as dimethyl disulphide, but also several less volatile alcohols and esters (Statheropoulos *et al.* 2011). Temperature dependence of the ratio of these compounds is important; for instance, when the carcass is at the later dried-out stages of decomposition, a higher proportion of benzyl butyrate in the VOC emission plume indicates to parasitoids that blowflies have been replaced with necrophagous beetles (von Hoermann *et al.* 2011).

8.4.2 Precipitation and drought

Variability in precipitation was found to be the most important factor in explaining caterpillar–parasitoid interactions across a broad gradient of climatic variability in an assessment of combined data from 15 geographically dispersed databases (Stireman *et al.* 2005). Parasitism by specialist hymenopterans decreased with increasing variability in precipitation, but tachinid flies – as generalists – were not significantly affected. Deviation from the optimal precipitation levels may affect parasitoids at different stages of development. For example, larval parasitoids most obviously suffer from the poor performance of herbivorous host insects on drought-stressed trees, whereas heavy rains and flooding may have detrimental effects on parasitoids of the pupae in soil.

Drought was found to have negative effects on phloem- and mesophyll-feeding sucking insects in an extensive meta-analysis (Huberty & Denno 2004). Recently, there has been an observation of reduced parasitization rates in aphid populations on drought-stressed plants (Johnson *et al.* 2011). Plants suffering from drought in the summer had lower numbers of the aphid *Rhopalosiphum padi* and, in addition, drought had a negative impact on the abundance of the aphid parasitoid *Aphidius ervi*, beyond its impacts on aphid density (Johnson *et al.* 2011). This result provides only correlative evidence between the herbivore and parasitoid response to drought, and there need to be more mechanistic studies to elucidate whether there is drought-induced alteration in host plant chemistry (see Turtola *et al.* 2003), which may result in reduced host quality for parasitoids. Forest tent caterpillar (*Malacosoma disstria*) population dynamics were found to be strongly dependent on parasitoid fecundity and mortality and variation in environmental factors such as drought (Babin-Fenske & Anand 2011). The authors predicted that forests affected by drought and other abiotic stressors will have more severe and frequent defoliation from these insects than surrounding unaffected forests.

Increased flooding as a result of land use and climate change has serious impacts on ecosystems and ecosystem services, especially close to river areas that are at risk (Verburg *et al.* 2012). Terrestrial ecosystems flushed and covered by water will suffer drastic changes in the composition of communities of soil organisms. Waterlogging will have a particularly strong impact on the root systems of perennial plants such as trees, causing root anoxia and also changes in the physiology of tree foliage (Copolovici & Niinemets 2010). Flooding is capable of inducing emissions of volatile signalling compounds (Copolovici & Niinemets 2010) and it is obvious that herbivore–parasitoid systems are strongly dependent on these compounds and could therefore be disturbed.

The negative impact of flood-induced root anoxia on foliage quality may reduce the performance of herbivorous species and extend the period when herbivores are exposed to parasitoids. On the other hand, aphids are able to take advantage of flooding stress when deciduous trees reduce their leaf area by translocation of nutrients from older senescent foliage to new growth, leading to higher densities of aphids on trees responding strongly to flooding (Holopainen *et al.* 2009). Such rapid aggregation of aphids on foliage showing symptoms of senescence provides a 'hot-spot' for aphid parasitoids and predators (Chacon & Heimpel 2010). Thus, changes in host plant quality as a result of flooding could lead to a temporary increase in aphid density, with a subsequent increase in the parasitization rate of aphids, but also to a higher mortality rate of parasitoids by intraguild predators (Chacon & Heimpel 2010).

8.4.3 Gaseous reactive air pollutants

Sulphur oxide (SO_2) and oxides of nitrogen (NOx), including nitric oxide (NO) and nitrogen dioxide (NO_2), are air pollutants which are particularly formed during the burning of fossil fuels, the major cause of anthropogenic climate warming, with the highest concentrations in the vicinity of urban and industrial areas. When plants are exposed to SO_2 and NO_2, there is often increased growth and increased reproduction rates of aphids (see Dohmen *et al.* 1984, Bolsinger & Flückiger 1987, Holopainen *et al.* 1991). Higher population densities of aphids were observed in most cases to be a result of improved food quality, such as a higher concentration of amino acids in phloem sap, rather than reduced parasitization rate. Gate *et al.* (1995) demonstrated in experiments with a fruit fly (*Drosophila subobscura*), which feeds on decaying plant and fungal material, that the braconid parasitoid *Asobara tabida* successfully found its host larvae in chamber exposures to elevated SO_2 and NO_2 concentrations. The proportion of hosts parasitized was equal in filtered air and in air with elevated SO_2 and NO_2, but the searching efficiency of parasitoids was significantly reduced in ozone-rich air, resulting in approximately 10% lower parasitism. It has been suggested that pollutants, particularly ozone (O_3), may interfere with the olfactory responses of parasitoids (Gate *et al.* 1995).

Ozone formation is strongly related to anthropogenic NOx emissions. In addition, VOC emissions are related to anthropogenic and biogenic sources. Furthermore, climate warming can lead to increased emissions of biogenic VOCs from plants and emissions of NO from soil, resulting in increased formation of O_3, HONO and NO_3 radicals in the atmosphere (Sitch *et al.* 2007, Jacobson & Streets 2009). Increased flashing in a warming climate is substantially increasing this trend (Jacobson & Streets 2009). Rising temperature is, in particular, increasing the emission of semivolatile compounds from vegetation and promoting their reactivity with atmospheric oxidants (Holopainen & Gershenzon 2010).

Ozone has been observed to degrade and shorten the atmospheric lifetime of many of the volatile compounds that attract parasitoids (reviewed in McFrederick *et al.* 2009, Holopainen & Gershenzon 2010, Pinto *et al.* 2010, Holopainen 2011). The ecological impacts of air pollution and ozone include disturbances in the behaviour of pollinating insects due to shortened signalling distances. Flower scent trails of the most reactive compounds such as linalool, β-myrcene and β-ocimene were detectable several kilometres downwind during pre-industrial times but nowadays the same compounds are detectable only at a distance of less than 200 metres in the most polluted environments (McFrederick *et al.* 2008). Many herbivore-induced VOC emissions could be more reactive than floral emissions in the presence of O_3, OH and other atmospheric radicals such as NO_3 in polluted air (Atkinson & Arey 2003), because the proportion of highly reactive sesquiterpenes and homoterpenes is higher in the herbivore-induced odour blend than in flower emissions (Pinto *et al.* 2007a).

Laboratory experiments with lima bean, cabbage (Pinto *et al.* 2008, 2007b) and oilseed rape (Himanen *et al.* 2009a) have clearly shown that O_3 is an efficient oxidant which can substantially reduce herbivore-induced volatiles in the atmosphere. However, these experiments also showed that the diamondback moth larvae parasitoid *Cotesia vestalis* is still able to locate the damaged plant when comparing plumes from intact and damaged plants in substantially elevated O_3 concentrations (60 and 120 ppb). This observation, although made in the laboratory using an olfactometer, indicates that hymenopterous parasitoids have possibly adapted to use, at least to some extent, the reaction products of plant-emitted VOCs. In nature, there is absolutely no 'clean' air without the atmospheric reaction products of biogenic VOCs. In particular, the ratio of concentrations of original plant-emitted volatile compounds and the reaction products of these compounds in the same plume could be an efficient cue for the proximity of a plant releasing 'cry for help' signals. Observations by Himanen *et al.* (2009a), assessing the impact of O_3 on herbivore-induced VOCs of oilseed rape, showed that when herbivore damage and the consequent induced VOC emissions are at very low level, O_3 may demolish the signal and *C. vestalis* is not able to discriminate between intact and herbivore-damaged plants. On the other hand, an open field experiment using an O_3 exposure system with ambient or moderately elevated O_3 did not show a significant difference in the parasitization rates of diamondback moth larvae by *C. vestalis* (Pinto *et al.* 2008). Elevated O_3 was approximately 50% higher than in natural air plots and peaked above 80 ppb, which has the potential to degrade herbivore-induced VOCs in the atmosphere. It can be concluded that moderately increased O_3 levels do not significantly disturb herbivore-induced volatile signalling compounds, the orientation behaviour of parasitoid wasps, or the rate of parasitism. However, in the field experiment, the host plant of herbivores was exposed to elevated O_3 for only a few days and obviously was not physiologically disturbed. Based both on laboratory experiments with elevated O_3 and CO_2 exposures (Vuorinen *et al.* 2004, Pinto *et al.* 2007a, Himanen *et al.* 2009a) and on these field observations, it can be concluded that any disturbance in the host plant physiology, which will reduce the capacity of the plant to produce the attractive volatile compounds when under herbivore attack, may lead to a lower host-finding efficiency of parasitoid wasps.

8.4.4 Atmospheric CO_2 concentration

Current climate warming is linked to a continuous increase in atmospheric carbon dioxide (CO_2) concentration (IPCC 2007). The annual average atmospheric background CO_2 concentration at Mauna Loa, Hawaii has continuously risen from 316 ppm in 1959 to 394 ppm in 2012 (see updated statistics at http://www.esrl.noaa.gov/gmd/ccgg/trends/). An increase of 25% in atmospheric concentration of CO_2 during 53 years is extremely rapid in evolutionary terms.

Improved carbon availability has substantial effects on plant growth and chemical composition (Peñuelas & Estiarte 1998). Pollinating (Goyret *et al.* 2008) and blood-feeding insects (Erdelyan *et al.* 2012) are able to perceive changes in atmospheric CO_2 concentration, as CO_2 is an important volatile orientation cue for these species. Night-blooming *Datura wrightii* can emit up to 200 ppm above natural CO_2 during peak nectar production, and the pollinating hawk moth *Manduca sexta* is able to detect variation at a resolution of 0.5 ppm in atmospheric CO_2 concentrations (Goyret *et al.* 2008). Less is known about the direct responses of parasitoids of herbivorous insects to changes in atmospheric CO_2 concentrations.

Studies on herbivorous insects have shown that the effects of elevated CO_2, using concentrations 50–100% above ambient (Bezemer *et al.* 1998, Chen *et al.* 2007, Himanen *et al.* 2009a) as projected for the early 2100s, are mostly mediated by an increased C/N ratio of the host plant. Defoliating chewing insects are particularly affected by nitrogen dilution in host plants, as occurs under elevated CO_2 atmospheres (Docherty *et al.* 1997, Stiling & Cornelissen 2007), showing poorer performance and greater feeding damage. Sap-feeding insects may also respond to elevated CO_2, although research on aphid populations has generated results which have been variable and less consistent (Holopainen 2002). A meta-analysis (Stiling & Cornelissen 2007) revealed that elevated CO_2 significantly decreased herbivore abundance, increased their relative consumption rates, development time and total consumption, and significantly decreased their relative growth rate, conversion efficiency and pupal weight. In general, there are smaller and fewer herbivore hosts at elevated CO_2 atmospheres, which could easily result in reduced parasitoid performance. On the other hand, an extended feeding period of herbivore larvae in elevated CO_2 will expose larvae to a longer searching and parasitizing period by parasitoids. This may increase the probability of higher rates of parasitism and increased death rates of herbivores from natural enemies (Stiling & Cornelissen 2007). However, evidence from CO_2 exposure experiments with parasitoids of gypsy moths suggests that the effects of CO_2 on third trophic levels are minor (Lindroth 1996). Experiments with deciduous trees in an open field CO_2 and O_3 exposure system supported this observation (Percy *et al.* 2002).

Although the impacts of elevated CO_2 on herbivorous insects has been extensively studied on woody plants and crop plants (Lindroth, 1996, Docherty *et al.* 1997, Stiling & Cornelissen 2007), there is less information on CO_2 effects on parasitoids. Bezemer *et al.* (1998) found, in a long-term study with the addition of 200 ppb to ambient CO_2 concentration and a 2°C increase in temperature, that aphid abundance was enhanced by both the CO_2 and the temperature treatment. Parasitism rates on potato aphids (*Myzus persicae*) by the parasitoid *Aphidius matricariae* remained unchanged in elevated CO_2, but showed an increasing trend in conditions of elevated temperature. Chen *et al.* (2007) observed that the abundance of the parasitized individuals of the aphid *Sitobion avenae* by the parasitoid *Aphidius picipes* significantly increased in 550 and 750 ppm CO_2 compared with natural CO_2 concentration, but parasitoid emergence rate at 750 ppm was reduced. The parasitoid substantially suppressed aphid abundance in an elevated CO_2 atmosphere, but the authors did not propose any mechanism to explain how the CO_2-affected parasitoid's behaviour would lead to a higher parasitism rate.

Vuorinen *et al.* (2004) demonstrated that the specialist parasitoid of diamondback moth, *Cotesia vestalis*, preferred the scent of damaged plants of two white cabbage cultivars grown at ambient CO_2 but did not differentiate between volatiles from intact and herbivore-damaged cv. Lennox plants under elevated CO_2. These results suggest that elevated atmospheric CO_2 concentration could weaken the plant response induced by insect herbivore feeding and thereby lead to a disturbance of signalling to the third trophic level. A study with oilseed rape (Himanen *et al.* 2009a) indicated that lower herbivory reduced herbivore-inducible emissions from transgenic, insect-resistant Bt (*Bacillus thuringiensis* insecticidal toxin produced) plants. As well as the study on cabbage plants by Vuorinen *et al.* (2004), the study by Himanen *et al.* (2009b) demonstrated that when plants were grown under elevated CO_2 there were increased emissions of most herbivore-inducible terpenoids, and *C. vestalis* always orientated to host-damaged plants independently of plant herbivore resistance or CO_2 concentration.

8.4.5 Parasitoid response to combined abiotic stresses

To understand how abiotic environmental stresses relating to climate change affect trophic interactions in ecosystems and the role of parasitoids in the food chain, experiments should include both laboratory and field studies and analysis for numerous environmental factors in combination to simulate the variation and dynamics that organisms face in nature. Large open field exposure systems have been established to create changes in abiotic factors in a controlled fashion while also maintaining the original microclimatic conditions such as natural diurnal changes of temperature and humidity (Percy *et al.* 2002, Karnosky *et al.* 2007). An advantage of such exposure systems is that they allow studies of, for example, mature trees instead of tree seedlings. These facilities also allow monitoring of the natural community of organisms related to ecosystems dominated by a particular tree species (Karnosky *et al.* 2007). The most serious obstacle for this type of research approach is the enormous cost of establishment of a field set-up with sufficient instrumentation and replication. In addition, the associated running costs are high, even in combined exposures of only two factors such as ozone and CO_2 (Karnosky *et al.* 2007).

The studies described so far have concentrated on assessing the climate change impacts with respect to one factor only. Running these experiments in greenhouses or exposure chambers with seedlings or clonal plantlets will give a very limited simulation of expected environmental changes and totally ignore possible interactive effects of abiotic factors. However, meta-analyses (e.g. Zvereva & Kozlov 2010) of published works using one-factor exposures will improve the reliability of data significantly for modelling and forecasting climate change effects on species interactions and the chemical ecology of parasitoids. Greenhouse and growth chamber experiments cannot throw light on certain types of ecosystem interaction which are affected, for instance, by elevated atmospheric ozone resulting ultimately in the lower efficiency of aphid predators and parasitoids in controlling aphid populations (Percy *et al.* 2002). Since climate change causes a combination of direct effects on orientation signals of parasitoids, and indirect impacts on parasitoids via plant defences against herbivores, it is difficult to predict the overall outcome based on data obtained from simplistic experiments conducted in growth chambers and greenhouses.

8.5 Climate change impacts on biological control

Biological control is based on the introduction of biocontrol species or the augmentation of natural populations of parasitoids and other natural enemies of key pests of crop plants. Therefore, any effect of climate change on the chemical and/or behavioural ecology of parasitoids will have a substantial effect on the success of biocontrol programmes. Thomson *et al.* (2010) reviewed the effects of climate change on natural enemies of agricultural pests. They concluded that there are more negative than positive effects on the efficiency of biological pest control by parasitoids and predators. Crop plant growth-related effects include loss of synchrony between crop plant and host herbivore development, resulting in altered timing of suitable instars and thus decreased fitness of parasitoids. A lower chemical quality of crop plants reduces the fitness both of pests and of their biocontrol species. An increase in crop plant biomass could reduce host density, resulting in a longer search time on the part of parasitoids. The altered distribution of crop plants may reduce the efficiency of biological control by natural enemies when parasitoids do not follow the range expansion of herbivore population equally fast. If biocontrol organisms are more sensitive to thermal and humidity fluctuations than their host herbivorous species, host location could have a lower rate of success. The efficiency of biocontrol organisms is improved when the reduced quality of host plants leads to increased development time and reduced size of the host herbivores. A longer exposure time results in higher host encounter and attack rate, and decreased herbivore size increases predation rate to satiate the predators. Some of the management options to improve biocontrol efficiency and to maintain predator and parasitoid populations under a changing climate are water conservation through mulching, reduced tillage, cover crops, higher-tolerance cultivars, and various remnant vegetation areas and shelter belts.

Genetically modified (GM) crops that are better able to tolerate biotic and abiotic stresses may affect the efficiency of biocontrol agents, and climate change factors could modify this effect. A GM cultivar may lead to lower biocontrol success, as in the case when there is an impact on the behaviour of parasitoids such as reduced orientation rate to moth-damaged GM cultivars because of a stronger negative effect of ozone on VOC signalling on these cultivars (Himanen *et al.* 2009a). Other risks are that the protective component of an insect-resistant GM cultivar could also be harmful to susceptible biocontrol organisms. Using an artificial diet system containing GNA lectin, Couty *et al.* (2001) were able to quantify the movement of GNA lectin from the diet into the aphid and then into the parasitoid. This compound was transferred through the trophic levels and had a dose-dependent effect on parasitoid development. Parasitoid larvae excreted most of the ingested lectin as a meconial pellet prior to pupation, but a small amount was still detected in the pupae. This transfer process of a toxin from aphids through to parasitoids and excretion by parasitoid larvae could be disturbed under the stress caused by climate change, which might lead to some effects on parasitoid performance.

Schuler *et al.* (1999) used Bt-resistant *Plutella xylostella* pests on Bt oilseed rape to evaluate the direct effects of Bt toxins on parasitoid biology, since the *Plutella xylostella* larvae were not affected by the Bt. They found that there was no effect on the survival or host-seeking ability of the pest's natural parasitoid enemy, *Cotesia vestalis*, when the larvae were not affected by the Bt. This observation indicates that Bt plants may have an environmental advantage over broad-spectrum insecticides, since the parasitoid and GM plant do not work antagonistically, as frequently occurs when using synthetic insecticides. However, it remains uncertain as to whether climate change may affect the risks related to Bt crops on non-targets such as parasitoids. A high dose of ozone exposure was found to increase the Bt toxin concentration in Bt oilseed rape (Himanen *et al.* 2009b), which may cause a threshold to be exceeded, as there could be a dose-dependent response to Bt which most GM plants currently do not exceed.

8.6 Ecosystem services provided by parasitoids: impact of changing climate

The Millennium Ecosystem Assessment (2005) was a landmark publication which has had significant impact on both science and government policy. One of the ecosystem services identified as an important regulatory service is pest control using natural enemies. This service is also an essential component of the sustainable intensification of agriculture, suggested as a way to address global food security (Royal Society 2009) alongside ecological security (United Nations Environmental Programme 2011). The United Nations report highlighted how agroecosystems require many ecosystem services in order for them to function, and how they also generate many services/disservices which require them to be appropriately managed. It is argued that agroecosystems need to increase the cultural, regulation and supporting services alongside the provisioning services, in order to create more balance between the services as occurs in 'natural' ecosystems. Thus, the management of parasitoid natural enemies is an important part of sustainably intensifying agriculture, and understanding how climate change affects this service is very important when considering how to manage agroecosystems in the future.

How climate change will affect farmed and natural landscapes will have considerable impact on parasitoids and their role as pest regulators. Habitat/landscape heterogeneity is a key factor in promoting biodiversity in agricultural landscape (Benton *et al.* 2003, Holzschuh *et al.* 2007), and thus the influence of climate change on habitat fragmentation, crop type, landscape use and management might have considerable effects on parasitoid biodiversity and potentially on their function as a pest regulation service. While some studies have addressed the effects of farm management techniques on enhancing the abundance of natural enemies, few have shown how this relates to the function of pest regulation, thus protecting the crop without the need for synthetic pesticides or resulting in losses/gains to productivity. In fact, in a recent study involving organic farms, there was no link found between parasitoid richness and pest regulation services (Macfadyen *et al.* 2009). A recent meta-analysis of 46 studies on crop pest and natural enemy responses to landscape complexity found that natural enemies have a strong positive response to landscape complexity, with generalists responding at all scales while specialists responded more strongly at smaller scales (Chaplin-Kramer *et al.*, 2011). However, there was no link to pest control and the authors argued the case for more population dynamics studies to characterize the relationship between landscape complexity and pest control services using natural enemies. As we can be sure that climate change will affect the complexity of landscapes, the effects on parasitoid–pest dynamics will need to be studied at a range of scales over appropriate time periods and involving a complex of parasitoid food webs (Memmott 2009). It is important that we adopt the landscape perspective both for understanding the negative and positive effects of agricultural land use and for enhancing biodiversity and thus ecosystem services, which are underpinned by biodiversity (Tscharntke *et al.* 2005), but we still lack a good understanding of the biodiversity–function relationship which will be important in predicting the effects of climate change on particular agroecosystems.

The current interest in differences in pest control by natural enemies on conventional versus organic farms is not only useful in understanding how best to manage these agroecosystems, but can also provide insights into how to manage the agroecosystems of the future in a changing climate. Interestingly, one can achieve 'organic-level' biodiversity by carefully managing conventional agriculture (Gibson *et al.* 2007). Organic farms are often more heterogeneous, and it is the heterogeneity which provides the complexity for enhanced biodiversity (Rundlöf *et al.* 2008). Thus, how future climate change affects heterogeneity will become important in parasitoid biodiversity. According to theory, agricultural intensification results in habitat homogeneity, and the use of agrochemicals across large spatial scales will decrease natural enemy diversity and will thus affect pest outbreaks (Wilby & Thomas 2002). However, the theory has rarely been empirically investigated until a recent study, focusing on organic/conventional farms, provided some important insights into the issues in terms of the function of biodiversity in the service of pest regulation (Macfadyen *et al.* 2011). The authors found a positive relationship between parasitoid species richness and temporal stability in parasitism rates – higher richness caused less variation in parasitism rate, which is still useful for farmers as it could be seen to offer insurance that it is not better or worse but is more consistent. They also showed that there are more species per functional group on organic farms compared with conventional farms. The increased biodiversity of parasitoids on the organic farms did not increase the resilience and/or robustness of the pest regulation service. This was demonstrated by removing species from organic and conventional systems through manipulation, and then measuring the pest regulation service. The service of pest regulation by parasitoids did not respond differently to such species loss, even though the organic systems had more biodiversity before the manipulations. Hence there was no difference in resilience to shock, which becomes even more important considering ecosystems in future climates. However, it is usually the functional diversity (amount of different functional groups) and especially response diversity (diversity in responses to change) possessed by the system that has most effect on system resilience to change (Elmqvist *et al.* 2003). Finally, the type of habitat was important, and hedgerows substantially increased richness, especially on organic farms, and thus such landscape features will be important under future cropping regimes/agronomic management brought about in adapting to future climates.

8.7 Future research directions and conclusions

We consider that climate change will include a rapid increase in atmospheric CO_2 concentration, leading to a warmer climate, more extreme weather episodes such as drought, hurricanes and flooding, and rapid changes in atmospheric quality due to increased methane, NO and biogenic VOC emissions. This, in turn, would lead to the more extensive formation of oxidative atmospheric air pollutants such as ozone, OH and NO_3 radicals, and would ultimately promote the formation of secondary aerosols as natural biosphere–atmosphere feedback systems (IPCC 2007). Thus, climate change will have a substantial effects on the whole biosphere and will have many expected and unexpected impacts on ecological interactions mediated by plant secondary compounds and affecting chemical ecology at higher trophic levels. In the future, food production using crop plants should be sustainable and all methods of maintaining biodiversity and natural ecosystem services provided by parasitoids should be exploited. Improved cultivation techniques in crop maintenance and management of agroecosystems and their environment, through providing diversity-supporting shelter belts and ecological corridors, will help to keep parasitoid population densities and diversity at sufficiently high levels. To reduce the use of synthetic pesticides, GM cultivars, which have a higher resistance against herbivorous insect pests, should be bred, but they should not pose a risk of toxicity for parasitoids of target or non-target pests or for human health. Furthermore, these cultivars should also possess properties that improve their resistance against increasing fluctuations in temperature and precipitation. Resistance against oxidative pollutants could possibly be improved by seeking or creating cultivars that produce reactive VOCs when faced with abiotic stress. Cultivars may also be found with improved production of herbivore-induced VOCs after herbivore attack, which might help parasitoids orientate more efficiently to attacked plants.

To conclude, it is evident that, so far, we have too little knowledge about how climate change factors affect the chemical ecology of parasitoids. However, analyses of experiments concerning plant exposure to climate change factors, singly and in combination, have created enough evidence to state that changes in the physical and chemical environment will affect the primary producers in an ecosystem with an intensity that will affect the chemical ecology of herbivores and their parasitoids. Time series of ecological observations from natural environments will become an important tool to assess how parasitoids respond and adapt to climate change. The increase in mean temperature and concentration of atmospheric pollutants seems to occur so rapidly that substantial changes can be detected in insect population densities and range shifts become visible within a decade or even less. To understand how climate change will affect particular ecosystem functions and multitrophic interactions such as parasitism, we need more studies on the dynamics and comparative performance of species from different trophic levels under a combination of varying climate change factors, and analyses at the genomic, metabolomic, phenological and ecological levels. By monitoring these changes with modern technology and using them to help model interactions under many potential future climate scenarios, we should gain an understanding which will help in predicting and protecting ecosystem functioning under future climates. Empirical studies should assess both bottom-up and top-down effects simultaneously to understand the key mechanisms affecting the chemical ecology, biological control and performance of parasitoids in future climates. Specialist and generalist species may also react very differently, with large differences in their adaptation capabilities: a topic which deserves increasing attention in terms of sustaining species and biodiversity conservation.

References

Atkinson, R. and Arey, J. (2003) Gas-phase tropospheric chemistry of biogenic volatile organic compounds: a review. *Atmospheric Environment* **37**: S197–S219.

Babin-Fenske, J. and Anand, M. (2011) Agent-based simulation of effects of stress on forest tent caterpillar (*Malacosoma disstria* Hübner) population dynamics. *Ecological Modelling* **222**: 2561–9.

Baden, C.U., Geier, T., Franke, S. and Dobler, S. (2011) Sequestered iridoid glycosides: highly effective deterrents against ant predators? *Biochemical Systematics and Ecology* **39**: 897–901.

Ballare, C.L., Caldwell, M.M., Flint, S.D., Robinson, A. and Bornman, J.F. (2011) Effects of solar ultraviolet radiation on terrestrial ecosystems patterns, mechanisms, and interactions with climate change. *Photochemical and Photobiological Sciences* **10**: 226–41.

Benton, T.G., Vickery, J.A. and Wilson, J.D. (2003) Farmland biodiversity: is habitat heterogeneity the key? *Trends in Ecology and Evolution* **18**: 182–8.

Bezemer, T., Jones, T. and Knight, K. (1998) Long-term effects of elevated CO_2 and temperature on populations of the peach potato aphid *Myzus persicae* and its parasitoid *Aphidius matricariae*. *Oecologia* **116**: 128–35.

Boga, M., Kolak, U., Topcu, G., Bahadori, F., Kartal, M. and Farnsworth, N.R. (2011) Two new indole alkaloids from *Vinca herbacea* L. *Phytochemistry Letters* **4**: 399–403.

Bolsinger, M. and Flückiger, W. (1987) Enhanced aphid infestation at motorways: the role of ambient air-pollution. *Entomologia Experimentalis et Applicata* **45**: 237–43.

Bowers, M. and Stamp, N. (1997) Effect of hostplant genotype and predators on iridoid glycoside content of pupae of a specialist insect herbivore, *Junonia coenia* (Nymphalidae). *Biochemical Systematics and Ecology* **25**: 571–80.

Chacon, J.M. and Heimpel, G.E. (2010) Density-dependent intraguild predation of an aphid parasitoid. *Oecologia* **164**: 213–20.

Chaplin-Kramer, R., O'Rourke, M.E., Blitzer, E.J. and Kremen C. (2011) A meta-analysis of crop pest and natural enemy response to landscape complexity. *Ecology Letters* **14**: 922–32.

Chen, F.J., Wu, G., Parajulee, M.N. and Ge, F. (2007) Impact of elevated CO_2 on the third trophic level: a predator *Harmonia axyridis* and a parasitoid *Aphidius picipes*. *Biocontrol Science and Technology* **17**: 313–24.

Connor, E.C., Rott, A.S., Samietz, J. and Dorn, S. (2007) The role of the plant in attracting parasitoids: response to progressive mechanical wounding. *Entomologia Experimentalis et Applicata* **125**: 145–55.

Copolovici, L. and Niinemets, U. (2010) Flooding induced emissions of volatile signalling compounds in three tree species with differing waterlogging tolerance. *Plant Cell and Environment* **33**: 1582–94.

Couty, A., Down, R., Gatehouse, A., Kaiser, L., Pham-Delegue, M. and Poppy, G. (2001) Effects of artificial diet containing GNA and GNA-expressing potatoes on the development of the aphid parasitoid *Aphidius ervi* Haliday (Hymenoptera: Aphidiidae). *Journal of Insect Physiology* **47**: 1357–66.

Cui, L., Francis, F., Heuskin, S., Lognay, G., Liu, Y., Dong, J., Chen, J., Song, X. and Liu, Y. (2012) The functional significance of E-beta-farnesene: does it influence the populations of aphid natural enemies in the fields? *Biological Control* **60**: 108–12.

Dicke, M. and Baldwin, I.T. (2010) The evolutionary context for herbivore-induced plant volatiles: beyond the 'cry for help'. *Trends in Plant Science* **15**: 167–75.

Docherty, M., Salt, D.T. and Holopainen, J.K. (1997) The impacts of climate change and pollution on forest insect pests. In: Watt, A.D., Stork, N.E. and Hunter, M.D. (eds) *Forests and Insects*. Chapman & Hall, London, pp. 229–47.

Dohmen, G., McNeill, S. and Bell, J. (1984) Air-pollution increases *Aphi fabae* pest potential. *Nature* **307**: 52–3.

Elmqvist, T., Folke, C., Nyström, M., Peterson, G., Bengtsson, J., Walker, B. and Norberg, J. (2003) Response diversity, ecosystem change, and resilience. *Frontiers in Ecology and Environment* **1**: 488–94.

Erdelyan, C.N.G., Mahood, T.H., Bader, T.S.Y. and Whyard, S. (2012) Functional validation of the carbon dioxide receptor genes in *Aedes aegypti* mosquitoes using RNA interference. *Insect Molecular Biology* **21**: 119–27.

Ferreira de Almeida, M.A., Pires do Prado, A. and Geden, C.J. (2002) Influence of temperature on development time and longevity of *Tachinaephagus zealandicus* (Hymenoptera: Encyrtidae), and effects of nutrition, and emergence order on longevity. *Environmental Entomology* **31**: 375–80.

Gate, I., McNeill, S. and Ashmore, M. (1995) Effects of air pollution on the searching behaviour of an insect parasitoid. *Water Air and Soil Pollution* **85**: 1425–30.

Gibson, R.H., Pearce, S., Morris, R.J., Symondson, W.O.C. and Memmott, J. (2007) Plant diversity and land use under organic and conventional agriculture: a whole-farm approach. *Journal of Applied Ecology* **44**: 792–803.

Giuliano, G., Tavazza, R., Diretto, G., Beyer, P. and Taylor, M.A. (2008) Metabolic engineering of carotenoid biosynthesis in plants. *Trends in Biotechnology* **26**: 139–45.

Goyret, J., Markwell, P.M. and Raguso, R.A. (2008) Context- and scale-dependent effects of floral CO_2 on nectar foraging by *Manduca sexta*. *Proceedings of the National Academy of Sciences USA* **105**: 4565–70.

Hance, T., van Baaren, J., Vernon, P. and Boivin, G. (2007) Impact of extreme temperatures on parasitoids in a climate change perspective. *Annual Review of Entomology* **52**: 107–26.

Harvey, J., van Nouhuys, S. and Biere, A. (2005) Effects of quantitative variation in allelochemicals in *Plantago lanceolata* on development of a generalist and a specialist herbivore and their endoparasitoids. *Journal of Chemical Ecology* **31**: 287–302.

Hernandez, I., Alegre, L., van Breusegem, F. and Munne-Bosch, S. (2009) How relevant are flavonoids as antioxidants in plants? *Trends in Plant Science* **14**: 125–32.

Herz, A. and Heitland, W. (2005) Species diversity and niche separation of cocoon parasitoids in different forest types with endemic populations of their host, the common pine sawfly *Diprion pini* (Hymenoptera: Diprionidae). *European Journal of Entomology* **102**: 217–24.

Himanen, S.J., Blande, J.D., Klemola, T., Pulkkinen, J., Heijari, J. and Holopainen, J.K. (2010) Birch (*Betula* spp.) leaves adsorb and re-release volatiles specific to neighbouring plants: a mechanism for associational herbivore resistance? *New Phytologist* **186**: 722–32.

Himanen, S.J., Nerg, A., Nissinen, A., Pinto, D.M., Stewart, C.N., Jr, Poppy, G.M. and Holopainen, J.K. (2009a) Effects of elevated carbon dioxide and ozone on volatile terpenoid emissions and multitrophic communication of transgenic insecticidal oilseed rape (*Brassica napus*). *New Phytologist* **181**: 174–86.

Himanen, S.J., Nerg, A., Nissinen, A., Stewart, C.N., Jr, Poppy, G.M. and Holopainen, J.K. (2009b) Elevated atmospheric ozone increases concentration of insecticidal *Bacillus thuringiensis* (Bt) Cry1Ac protein in Bt *Brassica napus* and reduces feeding of a Bt target herbivore on the non-transgenic parent. *Environmental Pollution* **157**: 181–5.

Holopainen, J.K. (2002) Aphid response to elevated ozone and CO_2. *Entomologia Experimentalis et Applicata* **104**: 137–42.

Holopainen, J.K. (2011) Can forest trees compensate for stress-generated growth losses by induced production of volatile compounds? *Tree Physiology* **31**: 1356–77.

Holopainen, J.K. and Gershenzon, J. (2010) Multiple stress factors and the emission of plant VOCs. *Trends in Plant Science* **15**: 176–84.

Holopainen, J.K., Heijari, J., Oksanen, E. and Alessio, G.A. (2010) Leaf volatile emissions of *Betula pendula* during autumn coloration and leaf fall. *Journal of Chemical Ecology* **36**: 1068–75.

Holopainen, J.K., Kainulainen, E., Oksanen, J., Wulff, A. and Karenlampi, L. (1991) Effect of exposure to fluoride, nitrogen-compounds and SO_2 on the numbers of spruce shoot aphids on Norway spruce seedlings. *Oecologia* **86**: 51–6.

Holopainen, J.K., Semiz, G. and Blande, J.D. (2009) Life-history strategies affect aphid preference for yellowing leaves. *Biology Letters* **5**: 603–5.

Holzschuh, A., Steffan-Dewenter, I., Kleijn, D. and Tscharntke, T. (2007) Diversity of flower-visiting bees in cereal fields: effects of farming system, landscape composition and regional context. *Journal of Applied Ecology* **44**: 41–9.

Huberty, A. and Denno, R. (2004) Plant water stress and its consequences for herbivorous insects: a new synthesis. *Ecology* **85**: 1383–98.

Ichiki, R.T., Kainoh, Y., Yamawaki, Y. and Nakamura, S. (2011) The parasitoid fly *Exorista japonica* uses visual and olfactory cues to locate herbivore-infested plants. *Entomologia Experimentalis et Applicata* **138**: 175–83.

IPCC (2007) *Climate Change 2007: The Physical Science Basis*. Contribution of Working Group I to the Fourth Assessment Report of the Intergovernmental Panel on Climate Change. Solomon, S., Qin, D., Manning, M., Chen, Z., Marquis, M., Averyt, K.B., Tignor, M. and Miller, H.L. (eds). Cambridge University Press, Cambridge.

Isidorov, V.A., Smolewska, M., Purzyńska-Pugacewicz, A. and Tyszkiewicz, Z. (2010) Chemical composition of volatile and extractive compounds of pine and spruce leaf litter in the initial stages of decomposition. *Biogeosciences* **7**: 2785–94.

Jacobson, M.Z. and Streets, D.G. (2009) Influence of future anthropogenic emissions on climate, natural emissions, and air quality. *Journal of Geophysical Research–Atmospheres* **114**: D08118.

Johnson, S.N., Staley, J.T., McLeod, F.A.L. and Hartley, S.E. (2011) Plant-mediated effects of soil invertebrates and summer drought on above-ground multitrophic interactions. *Journal of Ecology* **99**: 57–65.

Jones, C.G. (1988) What is chemical ecology? *Journal of Chemical Ecology* **14**: 727–30.

Kaplan, I. (2012) Attracting carnivorous arthropods with plant volatiles: the future of biocontrol or playing with fire? *Biological Control* **60**: 77–89.

Karl, T.R. and Trenberth, K.E. (2003) Modern global climate change. *Science* **302**: 1719–23.

Karnosky, D.F., Werner, H., Holopainen, T., Percy, K., Oksanen, T., Oksanen, E., Heerdt, C., Fabian, P., Nagy, J., Heilman, W., Cox, R., Nelson, N. and Matyssek, R. (2007) Free-air exposure systems to scale up ozone research to mature trees. *Plant Biology* **9**: 181–90.

Köpke, D., Beyaert, I., Gershenzon, J., Hilker, M. and Schmidt, A. (2010) Species-specific responses of pine sesquiterpene synthases to sawfly oviposition. *Phytochemistry* **71**: 909–17.

Lindroth, R.L. (1996) Consequences of elevated atmospheric CO_2 for forest insects. In: Korner, C. and Bazzaz, F.A. (eds) *Carbon Dioxide, Populations and Communities*. Academic Press, San Diego, CA, pp. 347–71.

Macfadyen, S., Craze, P.G., Polaszek, A., van Achterberg, K. and Memmott, J. (2011) Parasitoid diversity reduces the variability in pest control services across time on farms. *Proceedings of the Royal Society B* **278**: 3387–94.

Macfadyen, S., Gibson, R., Polaszek. A., Morris, R.J., Craze, P.G., Planque, R., Symondson, W.O.C. and Memmott, J. (2009) Do differences in food web structure between organic and conventional farms affect the ecosystem service of pest control? *Ecology Letters* **12**: 229–38.

Mackelprang, R., Waldrop, M.P., DeAngelis, K.M., David, M.M., Chavarria, K.L., Blazewicz, S.J., Rubin, E.M. and Jansson, J.K. (2011) Metagenomic analysis of a permafrost microbial community reveals a rapid response to thaw. *Nature* **480**: 368–71.

Martz, F., Jaakola, L., Julkunen-Tiitto, R. and Stark, S. (2010) Phenolic composition and antioxidant capacity of bilberry (*Vaccinium myrtillus*) leaves in northern Europe following foliar development and along environmental gradients. *Journal of Chemical Ecology* **36**: 1017–28.

Mattiacci, L. and Dicke, M. (1995) The parasitoid *Cotesia glomerata* (Hymenoptera: Braconidae) discriminates between first and fifth larval instars of its host *Pieris brassicae*, on the basis of contact cues from frass, silk, and herbivore-damaged leaf tissue. *Journal of Insect Behavior* **8**: 485–98.

McFrederick, Q.S., Fuentes, J.D., Roulston, T., Kathilankal, J.C. and Lerdau, M. (2009) Effects of air pollution on biogenic volatiles and ecological interactions. *Oecologia* **160**: 411–20.

McFrederick, Q.S., Kathilankal, J.C. and Fuentes, J.D. (2008) Air pollution modifies floral scent trails. *Atmospheric Environment* **42**: 2336–48.

McLaughlin, J., Hellmann, J., Boggs, C. and Ehrlich, P. (2002) Climate change hastens population extinctions. *Proceedings of the National Academy of Sciences USA* **99**: 6070–4.

Memmott, J. (2009) Food webs: a ladder for picking strawberries or a practical tool for practical problems? *Philosophical Transactions of the Royal Society B – Biological Sciences* **364**: 1693–9.

Millennium Ecosystem Assessment (2005) *Ecosystems and Human Wellbeing: General Synthesis*. Island Press, Washington, DC [available at http://millenniumassessment.org/en/Synthesis.html, accessed 29 April 2012].

Nieminen, M., Suomi, J., van Nouhuys, S., Sauri, P. and Riekkola, M. (2003) Effect of iridoid glycoside content on oviposition host plant choice and parasitism in a specialist herbivore. *Journal of Chemical Ecology* **29**: 823–44.

Overmyer, K., Brosche, M. and Kangasjarvi, J. (2003) Reactive oxygen species and hormonal control of cell death. *Trends in Plant Science* **8**: 335–42.

Parmesan, C. (2006) Ecological and evolutionary responses to recent climate change. *Annual Review of Ecology Evolution and Systematics* **37**: 637–69.

Parmesan, C., Ryrholm, N., Stefanescu, C., Hill, J., Thomas, C., Descimon, H., Huntley, B., Kaila, L., Kullberg, J., Tammaru, T., Tennent, W., Thomas, J. and Warren, M. (1999) Poleward shifts in geographical ranges of butterfly species associated with regional warming. *Nature* **399**: 579–83.

Peñuelas, J. and Estiarte, M. (1998) Can elevated CO_2 affect secondary metabolism and ecosystem function? *Trends in Ecology & Evolution* **13**: 20–4.

Peñuelas, J. and Staudt, M. (2010) BVOCs and global change. *Trends in Plant Science* **15**: 133–44.

Percy, K., Awmack, C., Lindroth, R., Kubiske, M., Kopper, B., Isebrands, J., Pregitzer, K., Hendrey, G., Dickson, R., Zak, D., Oksanen, E., Sober, J., Harrington, R. and Karnosky, D. (2002) Altered performance of forest pests under atmospheres enriched by CO_2 and O_3. *Nature* **420**: 403–7.

Petschenka, G., Bramer, C., Pankoke, H. and Dobler, S. (2011) Evidence for a deterrent effect of cardenolides on nephila spiders. *Basic and Applied Ecology* **12**: 260–7.

Pierre, P.S., Jansen, J.J., Hordijk, C.A., van Dam, N.M., Cortesero, A. and Dugravot, S. (2011) Differences in volatile profiles of turnip plants subjected to single and dual herbivory above- and belowground. *Journal of Chemical Ecology* **37**: 368–77.

Pinto, D.M., Blande, J.D., Nykänen, R., Dong, W.X., Nerg, A.M. and Holopainen, J.K. (2007a) Ozone degrades common herbivore-induced plant volatiles: does this affect herbivore prey location by predators and parasitoids? *Journal of Chemical Ecology* **33**: 683–94.

Pinto, D.M., Blande, J.D., Souza, S.R., Nerg, A. and Holopainen, J.K. (2010) Plant volatile organic compounds (VOCs) in ozone (O_3) polluted atmospheres: the ecological effects. *Journal of Chemical Ecology* **36**: 22–34.

Pinto, D.M., Himanen, S.J., Nissinen, A., Nerg, A.M. and Holopainen, J.K. (2008) Host location behavior of *Cotesia plutellae* Kurdjumov (Hymenoptera: Braconidae) in ambient and moderately elevated ozone in field conditions. *Environmental Pollution* **156**: 227–31.

Pinto, D.M., Nerg, A. and Holopainen, J.K. (2007b) The role of ozone-reactive compounds, terpenes, and green leaf volatiles (GLVs), in the orientation of *Cotesia plutellae*. *Journal of Chemical Ecology* **33**: 2218–28.

Rasmann, S., Kollner, T., Degenhardt, J., Hiltpold, I., Toepfer, S., Kuhlmann, U., Gershenzon, J. and Turlings, T. (2005) Recruitment of entomopathogenic nematodes by insect-damaged maize roots. *Nature* **434**: 732–7.

Reddy, G., Holopainen, J. and Guerrero, A. (2002) Olfactory responses of *Plutella xylostella* natural enemies to host pheromone, larval frass, and green leaf cabbage volatiles. *Journal of Chemical Ecology* **28**: 131–43.

Reudler, J.H., Biere, A., Harvey, J.A. and van Nouhuys, S. (2011) Differential performance of a specialist and two generalist herbivores and their parasitoids on *Plantago lanceolata. Journal of Chemical Ecology* **37**: 765–78.

Royal Society (2009) *Reaping the Benefits: Science and the Sustainable Intensification of Global Agriculture*. The Royal Society, London.

Rundlöf, M., Bengtsson, J. and Smith, H.G. (2008) Local and landscape effects of organic farming on butterfly species richness and abundance. *Journal of Applied Ecology* **45**: 813–20.

Schaub, A., Blande, J.D., Graus, M., Oksanen, E., Holopainen, J.K. and Hansel, A. (2010) Real-time monitoring of herbivore induced volatile emissions in the field. *Physiologia Plantarum* **138**: 123–33.

Schuler, T., Potting, R., Denholm, I. and Poppy, G. (1999) Parasitoid behaviour and Bt plants. *Nature* **400**: 825–6.

Signoretti, A.G.C., Penaflor, M.F.G.V., Moreira, L.S.D., Noronha, N.C. & Bento, J.M.S. (2012) Diurnal and nocturnal herbivore induction on maize elicit different innate response of the fall armyworm parasitoid, *Campoletis flavicincta. Journal of Pest Science* **85**: 101–7.

Sitch, S., Cox, P.M., Collins, W.J. and Huntingford, C. (2007) Indirect radiative forcing of climate change through ozone effects on the land-carbon sink. *Nature* **448**: 791–4.

Statheropoulos, M., Agapiou, A., Zorba, E., Mikedi, K., Karma, S., Pallis, G.C., Eliopoulos, C. and Spiliopoulou, C. (2011) Combined chemical and optical methods for monitoring the early decay stages of surrogate human models. *Forensic Science International* **210**: 154–63.

Stiling, P. and Cornelissen, T. (2007) How does elevated carbon dioxide (CO_2) affect plant–herbivore interactions? A field experiment and meta-analysis of CO_2-mediated changes on plant chemistry and herbivore performance. *Global Change Biology* **13**: 1823–42.

Stireman, J., Dyer, L., Janzen, D., Singer, M., Lill, J., Marquis, R., Ricklefs, R., Gentry, G., Hallwachs, W., Coley, P., Barone, J., Greeney, H., Connahs, H., Barbosa, P., Morais, H. and Diniz, I. (2005) Climatic unpredictability and parasitism of caterpillars: implications of global warming. *Proceedings of the National Academy of Sciences USA* **102**: 17384–7.

Tamura, Y. (2001) Effects of temperature, shade, and nitrogen application on the growth and accumulation of bioactive compounds in cultivars of *Plantago lanceolata* L. *Japanese Journal of Crop Science* **70**: 548–53.

Thaler, J.S., Humphrey, P.T. and Whiteman, N.K. (2012) Evolution of jasmonate and salicylate signal crosstalk. *Trends in Plant Science* **17**: 260–70.

Thomson, L.J., Macfadyen, S. and Hoffmann, A.A. (2010) Predicting the effects of climate change on natural enemies of agricultural pests. *Biological Control* **52**: 296–306.

Tscharntke, T., Klein, A.M., Kruess, A., Steffan-Dewenter, I., and Thies, C. (2005) Landscape perspectives on agricultural intensification and biodiversity: ecosystem service management. *Ecology Letters* **8**: 857–74.

Turtola, S., Manninen, A., Rikala, R. and Kainulainen, P. (2003) Drought stress alters the concentration of wood terpenoids in Scots pine and Norway spruce seedlings. *Journal of Chemical Ecology* **29**: 1981–95.

Tylianakis, J.M., Didham, R.K., Bascompte, J. and Wardle, D.A. (2008) Global change and species interactions in terrestrial ecosystems. *Ecology Letters* **11**: 1351–63.

United Nations Environmental Programme (2011) *Food and Ecological Security: Identifying Synergy and Trade-offs* [available at http://www.unep.org/ecosystemmanagement/Portals/7/Documents/unep_policy_series/Food and Ecological solutions JS.pdf, accessed 1 July 2012].

van Nouhuys, S. and Lei, G. (2004) Parasitoid–host metapopulation dynamics: the causes and consequences of phenological asynchrony. *Journal of Animal Ecology* **73**: 526–35.

van Nouhuys, S., Reudler, J.H., Biere, A. and Harvey, J.A. (2012) Performance of secondary parasitoids on chemically defended and undefended hosts. *Basic and Applied Ecology* **13**: 241–9.

Verburg, P.H., Koomen, E., Hilferink, M., Perez-Soba, M. and Lesschen, J.P. (2012) An assessment of the impact of climate adaptation measures to reduce flood risk on ecosystem services. *Landscape Ecology* **27**: 473–86.

Vickers, C.E., Gershenzon, J., Lerdau, M.T. and Loreto, F. (2009) A unified mechanism of action for volatile isoprenoids in plant abiotic stress. *Nature Chemical Biology* **5**: 283–91.

von Hoermann, C., Ruther, J., Reibe, S., Madea, B. and Ayasse, M. (2011) The importance of carcass volatiles as attractants for the hide beetle *Dermestes maculatus* (De Geer). *Forensic Science International* **212**: 173–9.

Voss, S.C., Spafford, H. and Dadour, I.R. (2009) Host location and behavioural response patterns of the parasitoid, *Tachinaephagus zealandicus* Ashmead (Hymenoptera: Encyrtidae), to host and host-habitat odours. *Ecological Entomology* **34**: 204–13.

Vuorinen, T., Nerg, A., Ibrahim, M., Reddy, G. and Holopainen, J. (2004) Emission of *Plutella xylostella*-induced compounds from cabbages grown at elevated CO_2 and orientation behavior of the natural enemies. *Plant Physiology* **135**: 1984–92.

Walther, G.-R., Post, E., Convey, P., Menzel, A., Parmesan, C., Beebee, T.J.C., Fromentin, J.-M., Hoegh-Guldberg, O. and Bairlein, F. (2002) Ecological responses to recent climate change. *Nature* **416**: 389–95.

Whittaker, J. (2001) Insects and plants in a changing atmosphere. *Journal of Ecology* **89**: 507–18.

Wilby, A. and Thomas, M.B. (2002) Natural enemy diversity and pest control: patterns of pest emergence with agricultural intensification. *Ecology Letters* **5**: 353–60.

Yuan, J.S., Himanen, S.J., Holopainen, J.K., Chen, F. and Stewart, C.N., Jr (2009) Smelling global climate change: mitigation of function for plant volatile organic compounds. *Trends in Ecology & Evolution* **24**: 323–31.

Zvereva, E.L. and Kozlov, M.V. (2010) Responses of terrestrial arthropods to air pollution: a meta-analysis. *Environmental Science and Pollution Research* **17**: 297–311.

Part 2
Applied concepts

Part 2
Applied concepts

Chemical ecology of insect parasitoids: essential elements for developing effective biological control programmes

Torsten Meiners[1] and Ezio Peri[2]

[1] Department of Applied Zoology/Animal Ecology, Freie Universität Berlin, Germany
[2] Department of Agricultural and Forest Sciences, University of Palermo, Italy

Abstract

Insect parasitoids can find their hosts in complex environments and reproduce through a series of behavioural steps that are regulated mainly by chemical cues, termed semiochemicals. According to functional criteria, stimuli can be classified into four main categories: (i) cues coming from the habitat, the host microhabitat or the food plant; (ii) direct host-related cues; (iii) indirect host-related cues; and (iv) cues coming from the parasitoid itself. In recent years, considerable progress has been made in elucidating the semiochemicals used by parasitoids to locate their hosts. Several studies have provided an interesting perspective for manipulating the foraging behaviour of parasitoids in order to increase their impact on pest populations. However, most of this research has been conducted under laboratory conditions, which differ considerably from field conditions, especially in agroecosystems in which human activities modify the tritrophic interactions between plants, phytophagous insects and their parasitoids. As a consequence, it is often not known how to employ semiochemicals in the field to successfully manipulate parasitoids in order to improve their efficiency in biological control programmes. In order to provide the essential elements for developing effective biological control programmes, we first present the essential elements of parasitoid chemical ecology. We critically review recent research on different strategies to manipulate parasitoid behaviour for the conservation or the recruitment of parasitoids within

Chemical Ecology of Insect Parasitoids, First Edition. Eric Wajnberg and Stefano Colazza.
© 2013 John Wiley & Sons, Ltd. Published 2013 by John Wiley & Sons, Ltd.

agroecosystems. Then we address the essential elements for developing effective biological control programmes using parasitoids and semiochemicals and consider their ecological and methodological implications at different trophic levels. Finally, a cautionary example of the chemical ecology of interspecific competitive interactions in parasitoids should illustrate that chemical ecological interactions between parasitoids and their hosts always involve other organisms. This prohibits the simple application of semiochemicals in the field, which could sometimes even be counterproductive. Clarifying the chemical ecology of insect parasitoids in a multi-species context is therefore important in order to develop effective and successful biological control programmes. When accompanying the introduction and/or conservation of natural enemies within agroecosystems by exploiting tritrophic interactions and manipulation of parasitoid behavioural responses to chemical cues, the complexity of the system needs to be considered.

9.1 Introduction

An alternative to the use of pesticides for controlling insect pests in crops (e.g. van Driesche & Bellows 1996) or wildland (e.g. van Driesche 2012) is the use of parasitoids to control pest organisms. Such a biocontrol approach comprises classical biological control (introducing natural parasitoids of foreign pests), augmentative biological control (increasing the population of introduced or native parasitoids) and conservation biological control (involving manipulation of the environment in such a way that parasitoids benefit from the changes) (Barbosa 1998). A prerequisite for successful introduction and/or environmental manipulation is an understanding of the physiological and behavioural mechanisms of individual organisms within their usual environmental context (Walter 2003).

Successful parasitism of the host by insect parasitoids occurs by means of several steps during host searching, which lead parasitoid females to locate and attack suitable hosts (Vinson 1976, 1998, Godfray 1994). The main steps are: host habitat location, host location, and host selection (Godfray 1994). Upon emergence, a female parasitoid may first search for a suitable environment with adequate physical conditions and food sources (Vinson 1976). Often a parasitoid emerges in an alien habitat and far from a host population if the habitat has been altered (for example by succession, or in response to abiotic or biotic influences) and the host population has declined, or if the hosts have dispersed from the emergence site. Therefore, a female parasitoid must first locate a suitable host habitat. Host habitat location is sometimes defined as a parasitoid finding a patch containing the host plants where its phytophagous host species develops, or sometimes as finding the host plant with the host already present. Because hosts occupy only a fraction of patches containing host plants and only some parts of the host plants, parasitoids would be expected to prefer to locate a 'habitat with hosts' than a 'host habitat' (Hatano *et al.* 2008), as they will need less time to arrive in the close vicinity of the host or to get into contact with it when it is actually present. When the host habitat is reached, parasitoids have to search for the host, and once a host is located, they must decide whether the host can be accepted; in other words, whether it is suitable for supporting the development of their progeny. As a consequence, reproductive success of female wasps, in terms of the number and quality of hosts located and parasitized during their lifetime, is directly influenced by their foraging behaviour.

To efficiently invest their limited time in the location of hosts, parasitoids rely on a series of physical and chemical cues (Vet & Dicke 1992, Godfray 1994, Vinson 1998). Although physical factors such as visual (Battaglia *et al.* 2000, Morehead & Feener 2000), acoustic (Kroder *et al.* 2007) and vibrational (Meyhöfer *et al.* 1994) signals play an important role in mediating parasitoid searching behaviour, chemical cues are instrumental for long- and short-range parasitoid orientation (Vinson 1976, Vet & Dicke 1992, Steidle & van Loon 2003) and we focus only on chemical cues in this chapter.

When determining the essential elements for developing effective biological control programmes, the identification of chemical ecological parameters and their role in a multitrophic context need to be addressed for two reasons: (i) chemical cues can be used directly as means to manipulate parasitoid behaviour or physiology; and (ii) chemicals involved in parasitoid interactions with other organisms in their environment are of major importance in determining parasitoid fitness. Functionally, the chemical signals used by parasitoids can be classified into four main categories:

1 *Cues coming from the habitat, the host microhabitat or the food plant.* Plants in the habitat are not only the setting for parasitoid–host interactions, but also play an important role by conveying information to the parasitoids and their hosts. While plant volatiles attract parasitoids from a distance, non-volatile compounds act via direct contact (taste) as cues for host recognition (Vinson 1976) and host quality evaluation (Conti *et al.* 2004, Fatouros *et al.* 2005a), and may also work as cues for host location at very short distances. Plants might even become active players in multitrophic interactions and specifically offer parasitoids these 'infochemicals' to facilitate host location (Vinson 1976, Vet & Dicke 1992). Specifically induced plant volatiles might be very good indicators of herbivore presence and are of key importance to many parasitoids (De Moraes *et al.* 1998). Since many parasitoids depend on nectar from non-host plants, floral volatiles also have to be considered in this context (Wäckers 2004).

2 *Direct host-related cues.* Chemicals coming from the host growth stage that is directly attacked are the most reliable cues, although they might be hard to detect except in the very close vicinity of the host. These cues can be sex pheromones in the case of adult stages, faeces, aggregation pheromones from larvae, saliva, egg glue, or silk from the pupal cocoon (Quicke 1997).

3 *Indirect host-related cues.* These chemicals, indirectly associated with the presence of the host, are from other host stages than the target and help the parasitoid to get into the vicinity of the attacked stage.

4 *Cues coming from the parasitoid itself.* Parasitoids emit and respond to pheromones in the context of sexual and oviposition behaviour (Anderson 2002; see Ruther, Chapter 6, this volume). These compounds can act to retain parasitoids in certain areas or keep them out of these areas.

Given the importance and the diversity of infochemicals in mediating successful parasitoid foraging for hosts, food or mates, chemical ecological approaches for developing effective biological control programmes are promising. In this chapter, we first identify the essential elements in parasitoid chemical ecology. Then we review studies on the manipulation of natural enemies with semiochemicals at the population level and discuss the limits and perspectives of behavioural manipulation of parasitoids by applying semi-ochemicals, before we address the essential elements for developing effective biological control programmes using parasitoids and semiochemicals. The fact that the chemical ecological interactions between parasitoids and their hosts always involve other organisms prohibits the simple application of semiochemicals without considering the multi-species context. As a caveat, we finally review an example of the chemical ecology of interspecific competitive interactions in parasitoids.

9.2 Essential elements in parasitoid chemical ecology

Semiochemicals or infochemicals are chemical cues that mediate the relationship between two organisms by inducing behavioural and/or physiological responses to one or both of the organisms (Vet & Dicke 1992). According to their ecological roles, semiochemicals are classified as pheromones or allelochemicals (Nordlund & Lewis 1976, Dicke & Sabelis 1988, Vet & Dicke 1992). Pheromones are semiochemicals that mediate the relationships between individuals of the same species. In parasitic wasps, sex pheromones play an important role in allowing long-range orientation towards mates, or in mediating courtship behaviour at close range (see Ruther, Chapter 6, this volume, for a detailed discussion). Allelochemicals are semiochemicals that mediate relationships between individuals belonging to different species. According to the benefits of the signal for the emitter or perceiver, these cues can be classified into different categories: allomones are favourable to the individual that emits the substance; both the emitter and the perceiver benefit from synomones; and kairomones are favourable to the perceiver.

This classification is subject to the specific behavioural interaction. For example, the chemical substance released by females of the tussock moth *Orgyia postica* (Walker) (Lepidoptera: Lymantriidae) to recruit conspecific males is classified as a sex pheromone in an intraspecific context, but is considered a kairomone in interspecific interactions, as it is exploited by the platygastrid parasitoid *Telenomus euproctidis* Wilcox to locate the host eggs (Arakaki *et al.* 2011). Semiochemicals can be volatile substances that are perceived by olfaction at long distances, or contact substances that are mainly perceived by gustatory receptors at short range.

To maximize reproductive success, parasitic wasps must optimize their foraging behaviour; that is, they have to make use of cues that are closely associated with their hosts. During host searching, parasitoids can decide to invest their resources in exploiting cues that are easily detectable but not directly associated with their hosts, or cues that are more reliable in indicating host presence but are less detectable, as explained by Vet & Dicke (1992) in their 'reliability–detectability theory'. Therefore, plant volatiles released in large amounts can be easily detected but are less reliable for locating hosts compared with volatiles from the host that usually occur in low concentrations in the environment. Host insects are naturally much smaller emitters of volatiles than their food plants and, furthermore, it is to their advantage to avoid revealing themselves by the emission of large amounts of compounds.

Synomones

The attraction of parasitoid females to volatiles released by undamaged plants has been shown for several parasitoids (Read *et al.* 1970, Schuster & Starks 1974, Powell & Zhang 1983, Nordlund *et al.* 1985, Kaiser *et al.* 1989, Turlings *et al.* 1991, Bogahawatte & van Emden 1996, Boo & Yang 1998, Powell *et al.* 1998, Reddy *et al.* 2002, Pareja *et al.* 2007, Fontana *et al.* 2011). The importance of 'eavesdropping' on plant volatiles as a tool used by parasitoids to locate their host has been elucidated by Romeis *et al.* (1997), who linked the capacity of a parasitoid to locate a polyphagous host with its response to volatiles emitted by two potential host plants. In the field, *Trichogramma chilonis* Ishii (Hymenoptera: Trichogrammatidae) parasitized about 40% of the eggs laid by *Helicoverpa armigera* (Hubner) (Lepidoptera: Noctuidae) on sorghum (*Sorghum bicolor* L.) plants, and less than 1% of those laid on pigeonpea (*Cajanus cajan* (L.) Millspaugh) plants (Romeis & Shanower 1996). These results are consistent with the laboratory response of the wasp to volatiles emitted by these two crops. In fact, the volatiles from sorghum plants at both the vegetative and reproductive stages induce arrestment behaviour in *T. chilonis*. Conversely, *T. chilonis* does not respond to the volatiles from pigeonpea plants at the vegetative stage, and is repelled by the volatiles from these plants at the reproductive stage (Romeis *et al.* 1997). Moreover, volatile emissions can be modified by the plant as a consequence of phytophagous attacks, and these *de novo*-produced and emitted chemicals, also known as herbivore-induced plant volatiles (HIPVs), attract hymenopteran (Dicke *et al.* 1990, Dicke & Vet 1999, Dicke & van Loon 2000, Turlings *et al.* 2002) and dipteran (see Nakamura, Ichiki & Kainoh, Chapter 7, this volume) parasitoids of the insect herbivores. For example, an attack by the aphid *Acyrthosiphon pisum* (Harris) (Homoptera: Aphididae) induces *Vicia faba* L. to modify its volatile emissions, recruiting the parasitoid *Aphidius ervi* Haliday (Hymenoptera: Braconidae) (Guerrieri *et al.* 1993). Therefore, more *A. ervi* females land on broad beans damaged by *A. pisum* than on undamaged plants (Du *et al.* 1996). Females of the solitary endoparasitoid *Cotesia vestalis* (Haliday) are attracted only by HIPVs released from *Brassica rapa* L. plants infested with larvae of the diamondback moth (Kugimiya *et al.* 2010). However, the composition of HIPVs can also convey information enabling parasitoids to distinguish between cabbage plants (*Brassica oleracea* L.) with different levels of herbivore infestation/damage (Girling *et al.* 2011). To locate host-infested plants, the tachinid fly *Exorista japonica* Townsend, a parasitoid of several lepidopteran larvae, uses a combination of constitutive and host-induced volatiles released by corn plants, *Zea mays* L., infested with larvae of *Mythimna separata* (Walker) (Lepidoptera: Noctuidae) (Ichiki *et al.* 2008). Interestingly, corn plants continue to emit volatiles attracting *E. japonica* for at least 5 hours after the host larvae have stopped feeding. This allows the parasitoids to locate host-infested plants and host larvae in the morning, although *M. separata* larvae are active mainly during the night (Hanyu *et al.* 2009). A similar strategy to bridge the gap between the host's nocturnal activity and the parasitoid's diurnal foraging behaviour has been reported by Signoretti *et al.* (2012) for *Campoletis flavicincta* (Ashmead) (Hymenoptera: Ichneumonidae), a specialist larval parasitoid of the fall armyworm *Spodoptera frugiperda* (JE Smith) (Lepidoptera: Noctuidae). Parasitoid females exploit maize volatiles elicited by the regurgitants of the host larvae, but only up to 5–6 hours after nocturnal volatile induction, thus modifying their response to the HIPVs according to the different volatile profiles that the plants can emit as a consequence of diurnal or nocturnal induction.

The recruitment of parasitoids is also mediated by plant synomones emitted as a consequence of oviposition by phytophagous insects or of a combination of oviposition and feeding activities. In the two tritrophic systems, elm–*Xanthogaleruca luteola* (Muller)– *Oomyzus gallerucae* Fonscolombe, and pine–*Diprion pini* L.–*Closterocerus ruforum* Krausse, the emission of oviposition-induced volatiles allows both eulophid parasitoids to locate trees that carry host eggs (Meiners & Hilker 1997, 2000, Hilker *et al.* 2002, 2005). Also, the platygastrid *Trissolcus basalis* (Wollastone) is recruited by synomones emitted by plants of two different legume species when egg deposition and plant damage from feeding activity of the pentatomid bug *Nezara viridula* L. occur together (Colazza *et al.* 2004).

Moreover, the parasitoid response pattern to HIPVs indicates the existence of host specificity of the parasitoids towards the herbivore–host plant complex, as plants differentially react to herbivore species, and natural enemies are able to distinguish between these different volatile profiles (Turlings *et al.* 1993). For example, *A. ervi* is able to discriminate between plants damaged by the pea aphid *A. pisum* and those damaged by the non-host aphid *Aphis fabae* Scopoli (Powell *et al.* 1998). *Cardiochiles nigriceps* Viereck (Hymenoptera: Braconidae) is attracted to HIPVs emitted by different plant species damaged by larvae of its host *Heliothis virescens* (Fabricius) (Lepidoptera: Noctuidae), but not to volatiles from the same plant species attacked by larvae of *Helicoverpa zea* (Boddie) (Lepidoptera: Noctuidae) (De Moraes *et al.* 1998). For further discussion of the ability of parasitoids to cope with the variability of HIPVs, see Wäschke, Meiners & Rostás (Chapter 3, this volume).

HIPVs can also be perceived by female parasitoids after landing on the host plants. For example, cabbage plants emit oviposition-induced contact synomones as a consequence of oviposition by the large cabbage white butterfly *Pieris brassicae* L. or by the harlequin bug *Murgantia histrionica* (Hahn), which are exploited by the egg parasitoids *Trichogramma brassicae* Bezdenko and *Trissolcus brochymenae* (Ashmead), respectively, after they have landed on the plant (Fatouros *et al.* 2005a, Conti *et al.* 2010).

Kairomones
Several chemical cues that originated from the hosts are exploited by parasitoids to distinguish infested from uninfested areas and to locate the appropriate host stages. Kairomone compounds can be detected by parasitoids over long and/or short distances. In the first case, they are used to find the infested area. In the other case, kairomones can induce a motivated searching behaviour. In addition to the diversity of kairomone-related phenomena, Ruther *et al.* (2002) proposed a further classification of kairomones according to the function for the benefiting organism (i.e. foraging, enemy-avoidance, sexual or aggregation kairomones). Among the volatile chemical cues with kairomonal activity, pheromones have been widely investigated (Powell 1999, Fatouros *et al.* 2008, Colazza *et al.* 2010). Gabrys *et al.* (1997) showed that the sex pheromone of the cabbage aphid *Brevicoryne brassicae* (L.) (Homoptera: Aphididae) is attractive to two of its natural enemies, *Diaeretiella rapae* (McIntosh) and *Praon volucre* (Haliday), specialist and generalist aphidiid parasitoids, respectively. Similarly, *Anagyrus pseudococci* (Girault) in California vineyards is attracted to the pheromone of *Planococcus ficus* (Signoret) (Homoptera: Pseudococcidae) (Millar *et al.* 2002), and a sibling species (*Anagyrus* sp. nov. near *pseudococci*) uses (*S*)-lavandulyl senecioate, the sex pheromone component of *P. ficus*, to locate its hosts (Franco *et al.* 2008).

Parasitoids can also detect volatile host pheromones adsorbed by different substrates: *Trichogramma evanescens* Westwood perceives pheromones from its nocturnal host,

Mamestra brassicae (L.), adsorbed by the epicuticular waxes of Brussels sprout leaves (Noldus *et al.* 1991), while *T. euproctidis* detects the pheromone of its nocturnal host, *Euproctis taiwana* (Shiraki), adsorbed on the scales covering the egg masses (Arakaki & Wakamura 2000). In this way, wasps can extend their host searching time and plug the temporal gap between their diurnal foraging activity and the nocturnal host mating and oviposition activities.

Other interspecific communication can be exploited by parasitic wasps. For example, the aphid alarm pheromone (*E*)-β-farnesene acts as a host-finding kairomone for *Aphidius uzbekistanicus* Luzhetski (Hymenoptera: Aphidiidae) (Micha & Wyss 1996). Fatouros *et al.* (2005b) showed that the phoretic egg parasitoid *T. brassicae* is able to detect an anti-aphrodisiac pheromone, transferred from *P. brassicae* males to females during mating, while another phoretic egg parasitoid, *Telenomus calvus* Johnson (Hymenoptera: Platygastridae), uses the aggregation pheromone emitted by the male spined soldier bug, *Podisus maculiventris* (Say) (Hemiptera: Pentatomidae), to locate the host's mating site and then to move onto mated females (Aldrich 1995). *Leptopilina heterotoma* (Thomson) (Hymenoptera: Eucoilidae), a larval parasitoid of *Drosophila*, is attracted to a volatile aggregation pheromone that is deposited at the oviposition site by recently mated female flies (Wiskerke *et al.* 1993). Finally, *Halticoptera rosae* Burks (Hymenoptera: Pteromalidae) uses the pheromone marking trail of its host, *Rhagoletis basiola* (Osten Sacken) (Diptera: Tephritidae), as a guide to the fly's eggs (Hoffmeister *et al.* 2000). Allomones also possess kairomonal activity. For example, the compounds present in the defensive secretion of *N. viridula* adults are attractive to the females of *T. basalis* (Mattiacci *et al.* 1993). *Lysiphlebus testaceipes* (Cresson) (Hymenoptera: Aphidiidae) increases its attack frequency when in contact with the cornicle wax secretion produced by its host *Rhopalosiphum padi* L. (Homoptera: Aphididae) (Grasswitz & Paine 1992). Contact kairomones from cuticle and cornicle secretions of the pea aphid *A. pisum* appear to be involved in host recognition and acceptance by *A. ervi* females (Powell *et al.* 1998).

In addition, 'host by-products' (for example frass, honeydew, exuviae, mandibular gland secretions, scales) are a good source of host location kairomones (Godfray 1994, Quicke 1997). Mehrnejad & Copland (2006) showed that the honeydew released by the common pistachio psylla, *Agonoscena pistaciae* Burckhardt & Lauterer, acts both as a volatile and a contact kairomone, inducing greater searching activity in the encyrtid parasitoid *Psyllaephagus pistaciae* Ferrière. Similarly, *Encarsia bimaculata* Heraty & Polaszek (Hymenoptera: Aphelinidae) uses honeydew from *Bemisia tabaci* (Gennadius) (Homoptera: Aleyrodidae) as a contact kairomone to locate its host (Mandour *et al.* 2007). *Ephedrus cerasicola* Starý is also attracted to aphid honeydew (Hagvar & Hofsvang 1989). Chemicals from the faeces of both larvae and adults of the elm leaf beetle *X. luteola* attract the egg parasitoid *O. gallerucae* (Meiners & Hilker 1997). The parasitoid *Lariophagus distinguendus* (Förster) (Hymenoptera: Pteromalidae) also recognizes wheat grains infested by its host larvae, the granary weevil, *Sitophilus granarius* (L.) (Coleoptera: Curculionidae) using chemicals from the host's faeces (Steidle & Ruther 2000). Obonyo *et al.* (2010) showed that the frass of the noctuid *Busseola fusca* (Fuller) and the crambid *Chilo partellus* (Swinhoe) are important in short-range host recognition by their natural enemies, the braconid *Cotesia sesamiae* (Cameron) and *Cotesia flavipes* Cameron, respectively, while the regurgitants of these hosts induce ovipositor insertion in *C. flavipes* only.

Upon landing on a plant, wasps exploit chemical cues of lower volatility that are closely associated with the hosts to confirm the presence of suitable hosts. Among these cues, chemical footprints can play an important role. Insect herbivores walking over the leaves may leave chemical traces that are absorbed by the leaf epicuticular wax layer. Rostás *et al.* (2008) found that the braconid *Cotesia marginiventris* (Cresson) can recognize chemical footprints left by walking second-instar caterpillars of *S. frugiperda*, displaying, when in contact, an antennal drumming behaviour on the substrate. Similarly, chemical footprints left on the leaves by *N. viridula* adults induce motivated searching in *T. basalis* that, when encountering a contaminated leaf surface, displays a characteristic arrestment posture, holding the body motionless and antennating the surface, followed by an intense searching behaviour (Colazza *et al.* 2009).

These chemicals most probably originate from the insect cuticle and are lipophilic compounds of low volatility (Colazza *et al.* 2007, Rostás & Wölfling 2009). Footprints left by caterpillars consist mostly of linear alkanes, ranging from heneicosane (nC_{21}) to dotriacontane (nC_{32}) and minor amounts of branched alkanes (Rostás & Wölfling 2009). Footprints of *N. viridula* also contain diglycerides and triglycerides of high molecular weight, and long-chain alcohols and fatty acids (Lo Giudice *et al.* 2011). Since the *n*-alkanes are also compounds of epicuticular waxes, minor constituents may allow parasitoids to distinguish between host-derived chemicals and a background of similar plant hydrocarbons (Rostás & Wölfling 2009). Moreover, some alkanes of insect origin can be also absent in plant waxes. In particular, the absence of nonadecane among the hydrocarbons of epicuticular waxes of several host plants of *N. viridula* is relevant for *T. basalis* host location (Lo Giudice *et al.* 2011), as this parasitoid species is able to discriminate host sex by the presence or absence of nonadecane (nC_{19}), which is present in male host footprints only (Colazza *et al.* 2007). Due to the lipophilic properties of the plant cuticle, chemical footprints can adhere to or be absorbed by plant epicuticular waxes (Müller & Riederer 2005). As a consequence, parasitoids can detect host footprints for a useful length of time (Rostás & Wölfling 2009), although their host recognition behaviour may vary between plants (Lo Giudice *et al.* 2011).

The examples of host searching by exploiting host footprints described above also characterize two different search strategies, systematic or random search, that parasitoids can adopt to use the information they obtained when contacting host chemical residues (Godfray 1994). Parasitoids might use a systematic search when the cues are guiding them directionally to the host. Usually, this search is assumed by parasitoids of an active host stage. In fact, they use cues that reliably indicate the host's presence, as in the system *C. marginiventris–S. frugiperda* (Rostás *et al.* 2008) or as is the case of *Poecilostictus cothurnatus* Gravenhorst, a larval parasitoid of pine looper moth, *Bupalus pinarius* L., which follows the chemical trails left by host larvae to locate its target (Klomp 1981). On the other hand, parasitoids use a random search when the cues are indirectly associated with the hosts and induce them to increase the host searching behaviour in the area where hosts are more likely to be found, as reported for platygastrid egg parasitoids that, in response to chemical footprints left by adults of pentatomid bugs, move around the contaminated area and systematically return to the traces after losing contact with them (Colazza *et al.* 2010).

Using host by-product kairomones, parasitoids can also distinguish between host and non-host or between different hosts. The egg parasitoid *O. gallerucae*, shows attraction towards faecal kairomones from its host elm leaf beetle, *X. luteola*, while it ignores faecal components from the non-host caterpillar *Opisthograptis luteolata* L. (Meiners *et al.* 2000).

The parasitoid *Dibrachys cavus* Walker discriminates between the frass of different potential lepidopteran host species: *Sphinx ligustri* L., *Lobesia botrana* Schiff. and *Eupoecilia ambiguella* Hübner, showing most attraction to the first (Chuche *et al.* 2006). Mattiacci & Dicke (1995) reported that *Cotesia glomerata* L., using host by-products, is able to discriminate between first and fifth larval instars of its host cabbage butterfly, *P. brassicae*. By-products from the first instar elicit significantly longer searching behaviour in the parasitoid than those from the fifth instar. The capacity to discriminate between host and non-host, based on kairomones, is reported also for cues from cocoons (Bekkaoui & Thibout 1993) and exuviae (Battaglia *et al.* 2000). Based on the chemical footprints of adult bugs retained by the epicuticular wax of *Brassica oleracea* leaves, the egg parasitoid *T. basalis* differentiates between hosts of different sexes, following with more intensity the traces left by female hosts so as to obtain reliable information on the presence of host eggs (Lo Giudice *et al.* 2011). Finally, evidence of a sophisticated strategy used by a parasitoid to exploit chemical footprints is reported by Salerno *et al.* (2009), where the egg parasitoid *T. brochymenae* responds to footprints from mated host females that have not yet laid eggs by searching intensely on these chemical traces, which are strongly temporally correlated with the moment of oviposition.

9.3 Manipulation of the population levels of natural enemies by semiochemicals

Many factors in the field can influence the recruitment, colonization and efficiency of natural enemies and, as a consequence, the success of biological control programmes. Recently, numerous studies have combined different approaches, techniques and knowledge to produce exciting and promising research into enhancing the efficiency of parasitoids and predators in agroecosystems (Fig. 9.1). Several studies have concentrated on the

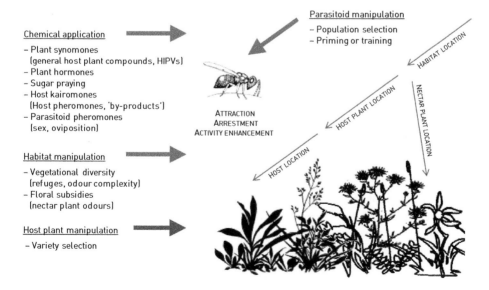

Figure 9.1 Essential elements for developing effective biological control programmes based on the chemical ecology of insect parasitoids.

role played by host plants. Host plants may affect the natural enemies of their phytopha-gous pests directly through their morphology, or indirectly by mediating multitrophic interactions (Romeis *et al.* 2005, Ode 2006, Gols *et al.* 2008a, 2011). Therefore, one of the first possible opportunities to increase the effectiveness of natural enemies is the selection and modification of host plant varieties (Bottrell *et al.* 1998).

More recently, in trying to enhance the impact of biological control in agricultural systems, the concept of habitat manipulation has received increasing attention (Landis *et al.* 2000, Gurr *et al.* 2003, 2004, Rodriguez-Saona *et al.* 2012). This approach aims to create suitable ecological infrastructures within the agroecosystem to recruit natural enemies and to conserve their populations (see Simpson, Read & Gurr, Chapter 12, this volume). The measures suggested for habitat management of arthropod predators and parasitoids include increasing the vegetational diversity to provide alternative food sources such as floral and extrafloral nectar, pollen, alternative prey or hosts, and shelter for the beneficial insects. Manipulation of the vegetation structure at the habitat scale also facili-tates natural enemy action against pests (Thies & Tscharntke 1999, Langellotto & Denno 2004). Recently, these manipulations have also been discussed in the light of changing the chemical features of the vegetation to manipulate the outcome of herbivore–parasitoid interactions (Randlkofer *et al.* 2010), although large spatial scales have usually been ignored in chemical ecological research. A third possible approach is the manipulation of the population levels of natural enemies through the use of semiochemicals (Lewis & Martin 1990, Powell & Pickett 2003). In this approach, the interest has mainly focused on HIPVs (synomones), floral odours, and host-associated cues (kairomones), while the use of parasitoid pheromones in still underexplored and remains at a theoretical level (Suckling *et al.* 2002; see Ruther, Chapter 6, and Colazza, Peri & Cusumano, Chapter 11, this volume).

HIPVs

Over the last decade, the use of HIPV compounds as a tool to improve the control of agricultural pests by manipulating the behaviour of natural enemies has become more common (Khan *et al.* 2008, Gurr & Kvedaras 2010). Numerous studies have explored the potential use of HIPVs to recruit natural enemies in the field (reviewed in Kaplan 2012, Rodriguez-Saona *et al.* 2012). In a pioneering study, Altieri *et al.* (1981) showed that, in fields, *Trichogramma* wasps parasitize more *H. zea* eggs when the crops have been sprayed with a crude extract from corn or *Amaranthus*. Similarly, increases in aphid parasitization rates and/or number of wasps have been observed by Titayavan & Altieri (1990) in the system broccoli plants–*B. brassicae*–*D. rapae*, as a consequence of applying allylisothiocy-anate emulsion to plants. More recently, it has been shown that synthetic HIPVs trigger enhancement of the naturally occurring densities of parasitoids in studies conducted on cotton plants. The mymarid *Anaphes iole* Girault parasitized more eggs of the mirid bug *Lygus lineolaris* (Palisot de Beauvois) when dispensers containing (Z)-3-hexenyl acetate and α-farnesene were placed near the host eggs (Williams *et al.* 2008), and the braconid *Micro-plitis mediator* Haliday parasitized *H. armigera* larvae in field cages treated with 3-7-dimethyl-1,3,6-octatriene (Yu *et al.* 2010). Uefune *et al.* (2012) reported that in *Brassica rapa* crops, the use of dispensers containing a synthetic mixture of (Z)-3-hexenyl acetate, *n*-heptanal, α-pinene and sabinene, four volatiles induced by the feeding activity of diamondback moth, *Plutella xylostella* (L.), larvae, resulted in the attraction of *C. vestalis* to uninfested plants and increased larval parasitism on infested plants.

In addition to increasing the recruitment of natural enemies, treatment with synthetic HIPVs seems to be able to trigger plants to produce endogenous volatiles that attract predator and parasitoid arthropods. Simpson *et al.* (2011a) showed that grapevines and sweet-corn plants sprayed with synthetic HIPVs attracted predators and parasitoids for some days after application, suggesting that HIPV-treated plants may be induced to produce endogenous volatile synomones over an extended period of time. In a greenhouse study, Rodriguez-Saona *et al.* (2011) found that cranberry plants that normally released undetectable quantities of methyl salicylate (MeSA) emitted large quantities of this compound when exposed to synthetic MeSA released by commercial dispensers. Similarly, von Mérey *et al.* (2011) found that maize plants exposed to a mixture of four synthetic green leaf volatiles, (*Z*)-3-hexenal, (*Z*)-3-hexenol, (*Z*)-3-hexenyl acetate and (*E*)-2-hexenal, at a ratio of 80:10:8:2, released from glass vial dispensers, emitted a greater amount of sesquiterpenes than non-exposed plants, although without any influence on overall parasitism rates of *S. frugiperda*.

The production and emission of volatiles similar to HIPVs can be triggered by spraying crops with plant hormones such as jasmonates (Rohwer & Erwin 2008), although the attractiveness of volatiles induced by jasmonates to natural enemies could be lower than that of volatiles induced by herbivores, as evidenced for the carnivorous mite *Phytoseiulus persimilis* Athias-Henriot (Acarina: Phytoseiidae) by Dicke *et al.* (1999). For example, treatments with jasmonic acid (JA) increased the attraction of *Cotesia rubecula* (Marshall) to treated plants of *Arabidopsis thaliana* (L.), whereas treatment with salicylic acid (SA) did not (van Poecke & Dicke 2002). Similarly, treatments with JA increased the attraction of *Cotesia kariyai* Watanabe to treated maize plants for at least 10 days after treatment (Ozawa *et al.* 2004) and, on rice plants, enhanced more than twofold the parasitization of eggs of the brown planthopper, *Nilaparvata lugens* (Stål), by its mymarid parasitoid *Anagrus nilaparvatae* Pang et Wang (Lou *et al.* 2005). Treatment with JA can render plants more attractive to natural enemies of herbivores, as has been demonstrated in the system tomato–*Spodoptera exigua* (Hubner) (Lepidoptera: Noctuidae)–*Hyposoter exiguae* (Viereck) (Hymenoptera: Ichneumonidae). In the field, JA treatment of tomato plants increases parasitism of sentinel host larvae placed near the treated plants 10 days afterwards, as JA induces, during this time interval, the octadecanoid pathway that increases defence systems against herbivores and may regulate the production of volatiles which recruit natural enemies (Thaler 1999).

Floral odours

Most parasitoids are not only dependent on cues from the host or from the host plant for survival and reproduction but also on floral odours from nectar plants. The sugars from these plants provide the parasitoids with energy for maintenance and flight activity (Wäckers 2005). However, although it might be an option to use floral subsidies to enhance biological control in crops (e.g. Gurr *et al.* 2005; see also Simpson, Read & Gurr, Chapter 12, this volume), the volatile bouquets of nectar plants differ in their attractiveness or repellence to parasitoids (Wäckers 2004) and their effects need to be considered among other factors (e.g. accessibility, digestibility of sugars) (Wäckers 2005).

Host-associated cues

Among these cues, interest has mainly been focused on the possible field applications of host pheromones (for detailed discussions see Chapters 10–13, this volume). Unlike HIPVs, these semiochemicals have high specificity, and therefore can be accurate tools to enhance the activities of natural enemies of pests in biological control programmes.

The attractiveness of aphid sex pheromones to female parasitoids in the field was first shown by Hardie *et al.* (1991). Water traps containing $(+)$-(4aS,7S,7aR)-nepetalactone and $(-)$-(1R,4aS,7S,7aR)-nepetalactol at a ratio of 27 : 1, two components of aphid sex pheromones, catch females of three braconid species: *Praon abjectum* (Haliday), *Praon dorsale* (Haliday) and *P. volucre*. In subsequent years, tests in cereal fields showed that the most effective lure was the $(+)$-(4aS,7S,7aR)-nepetalactone and that the attraction was strongest in the autumn (Powell *et al.* 1993, Hardie *et al.* 1994). On the basis of these results, the authors proposed using traps baited with aphid sex pheromones to manipulate natural parasitoid populations in the autumn. It was hoped that this manipulation might attract parasitoids 'into areas seeded with appropriate hosts in order to establish overwintering parasitoid reservoirs in and around cereal fields, thereby promoting synchrony between emerging adult parasitoids and colonizing cereal aphids in the following spring' (Powell *et al.* 1993). This strategy was reported to be tested on commercial farms (Powell & Pickett 2003). Some years later, Glinwood *et al.* (1998) also demonstrated the potential of pheromone components for enhancing the parasitization of aphid populations. After placing glass vial dispensers containing $(+)$-(4aS,7S,7aR)-nepetalactone in a wheat field, they observed that the presence of aphid sex pheromones significantly increased parasitization by two braconid parasitoid species, the generalist *P. volucre* and the specialist parasitoid *Aphidius rhopalosiphi* De Stefani Perez, although the parasitization reached different levels according to the distance from the pheromone-releasing vials.

The role of host sex pheromones in recruiting natural enemies and enhancing their activities has been also evaluated by Franco *et al.* (2008, 2011) in a system including citrus plants, two mealybug species – *Planococcus citri* (Risso) the citrus mealybug, and *P. ficus* the vine mealybug – and the braconid parasitoid *Anagyrus* sp. nov. near *pseudococci*. Host aggregation pheromone can also be used for manipulating parasitoids in agroecosystems. For example, Leal *et al.* (1995) demonstrated that females of *Ooencyrtus nezarae* Ishii (Hymenoptera: Encyrtidae) were caught by traps baited with the synthetic pheromone of their host, *Riptortus clavatus* (Thunberg) (Heteroptera: Alydidae). In particular, wasps can be recruited in crops treated with (E)-2-hexenyl (Z)-3-hexenoate, one of the three components present in the host aggregation pheromone (Mizutani 2006).

A host aggregation pheromone has also been successfully used in monitoring *Gymnosoma rotundatum* (L.) (Diptera: Tachinidae) populations by placing sticky traps baited with methyl (E,E,Z)-2,4,6-decatrienoate, the aggregation pheromone of *Plautia stali* Scott (Hemiptera: Pentatomidae), the brown winged green stink bug, one of the hosts of *G. rotundatum* in the field (Jang & Park 2010, Jang *et al.* 2011). Recently, Teshiba *et al.* (2012) demonstrated the possibility of employing a non-natural chemical compound to control the Japanese mealybug, *Planococcus kraunhiae* (Kuwana) (Homoptera: Pseudococcidae), in persimmon orchards by increasing the parasitism rates of two encyrtid parasitoids, *Anagyrus sawadai* Ishii and *Leptomastix dactylopii* Howard.

9.4 Limits and perspectives of behavioural manipulation of parasitoids by applying semiochemicals

Behavioural manipulation of parasitoids by applying semiochemicals has opened up new opportunities for improving biological control efficiency in agroecosystems, but it is also a subject of controversy with regard to the potential limits and risks of extended use of these compounds in the field (Turlings & Ton 2006, Hilker & McNeil 2008, Khan *et al.* 2008, Kaplan 2012, Rodriguez-Saona *et al.* 2012). Simply applying semiochemicals that have been identified as being attractive to natural enemies of pests on to crops might be ineffective and/or counterproductive. For example, Roland *et al.* (1995) demonstrated that the application of borneol in an apple orchard increased the density of the tachinid fly *Cyzenis albicans* (Fall.), a parasitoid of the winter moth, *Operophtera brumata* L. (Lepidoptera: Geometridae), but not its parasitism rate. Moreover, semiochemical application could disrupt parasitoid population dynamics, since the recruitment of natural enemies in the selected field could result in their removal from the surrounding areas (Vinson 1977, Gross 1981, Jones *et al.* 2011). Semiochemicals might also decrease foraging efficiency, because they could trick parasitoids into following 'false leads' and cause them to waste time and energy searching in patches devoid of hosts (Vinson 1977, Powell & Pickett 2003, Puente *et al.* 2008). Lewis *et al.* (1975a, 1975b) showed that the application of an extract from the scales of *H. zea* or of synthetic kairomones onto soyabean plants increased host egg parasitism by *Trichogramma* species, but only at high host densities. At low to intermediate densities, parasitism rates were high if the moth scales or synthetic kairomones were applied around the host eggs, otherwise parasitism was lower, as parasitoids spent more time searching intensively in areas where the hosts were absent (Lewis *et al.* 1979, Gross 1981). As a consequence, wasps that eavesdrop stimuli without reward can modify their searching behaviour. For example, when *T. basalis* females, which usually detect adult host traces to find host eggs, repeatedly explore a host patch without any oviposition reward, they weakly respond to host footprints and leave the patch, whereas they strongly respond on kairomone patches where oviposition has taken place (Peri *et al.* 2006).

Although success in field trials with surface sprays was inconclusive for *Trichogramma* (Lewis *et al.* 1979), Mills & Wajnberg (2008) suggest that uniform coatings of inexpensive contact kairomones could be effective in suppressing pest density and enhancing crop yields even when the foraging patterns of individual parasitoids are not optimized. By priming insect parasitoids with semiochemicals prior to release (Hare *et al.* 1997), it should be possible to manipulate parasitoid behaviour, for instance to enhance search activity or to retain parasitoids in certain areas (Mills & Wajnberg 2008). In general, it was found that in only 36% of cases did parasitoids become established in the field after having been reared, augmented and introduced against insect pests. This shows that there is potential for improvement by exploring the influence of holding conditions (including priming) and adjusting field/habitat conditions (including spraying of infochemicals) (Mills 2000).

The habituation example mentioned above introduces another factor that cannot be ignored in developing research to manipulate populations of natural enemies: experience. Experience acquired during development and/or during the adult stage is used to modify innate parasitoid behaviours (Vet & Groenewold 1990, Turlings *et al.* 1993, Vet *et al.* 1995). For example, Penaflor *et al.* (2011) showed that the specialist egg parasitoid *Telenomus remus* Nixon (Hymenoptera: Platygastridae) is attracted to volatiles emitted from young maize plants that have been treated with regurgitant from *S. frugiperda* larvae or that have been directly attacked by the caterpillars, only after associating those volatiles with oviposition. On the other hand, the generalist *Trichogramma pretiosum* Westwood (Hymenoptera: Trichogrammatidae) responds innately to HIPVs. Moreover, natural enemies exposed to semiochemicals in fields where hosts are present at a low density, or not at all, could be subject to habituation, resulting in reduced foraging efficiency (Turlings & Ton 2006, Khan *et al.* 2008, Kaplan 2012, Rodriguez-Saona *et al.* 2012). This is likely to be true mainly for specialist natural enemies, since generalists would be less likely to develop a negative association, having more opportunities to find a host in the selected area (Khan *et al.* 2008). On the other hand, experience can have positive effects, as semiochemicals can be used to prime natural enemies in order to enhance their searching behaviours. For example, Hare *et al.* (1997) demonstrated that *Aphytis melinus* DeBach, a parasitoid of the California red scale *Aonidiella aurantii* (Maskell), exposed to the host kairomone O-caffeoyltyrosine prior to being released in the field, improves its parasitoid activity, suggesting a promising use of this kairomone in augmentative release programmes to control the California red scale.

We described above how the direct application of semiochemicals on crops can have negative consequences for natural enemies, such as disrupting population dynamics, decreasing foraging efficiency, or favouring habituation. To limit these negative consequences, different strategies for employing semiochemicals are needed; many of these are under evaluation at the moment and some are outlined below. Synthetic compounds could be applied to prime the plants so that they are induced to emit HIPVs more rapidly in response to pest attack, resulting in a more efficient recruitment of herbivore enemies (Pickett & Poppy 2001, Turlings & Ton 2006, Khan *et al.* 2008). For example, Peng *et al.* (2011) showed that cabbage plants previously exposed to HIPVs and subsequently damaged by *P. brassicae* caterpillars attracted more *C. glomerata* parasitoids than control plants. A much more promising approach would be the production of crop varieties selected by genetic engineering or selective breeding for enhanced HIPV emissions only when attacked by herbivores (Bottrell *et al.* 1998, Powell & Pickett 2003, Turlings & Ton 2006, Dudareva & Pichersky 2008, Khan *et al.* 2008, Rodriguez-Saona *et al.* 2012). In this way, parasitoids would recognize HIPVs as a useful tool to find their hosts, and the energetic costs of emitting HIPVs would remain acceptable to the plants.

A complementary or synergistic approach involving several chemically mediated interactions between trophic levels might be used to enhance the success of measures against pests on crop plants. Attraction of parasitoids by semiochemicals might be accompanied by enhancing the direct defence of plants against their natural enemies. Herbivores on suboptimal diets could suffer higher mortality due to parasitoid attack (Benrey & Denno 1997). By enhancing tissue-specific resistance in plants with allomones, the reduced development of the herbivores would open the opportunity for parasitoids to kill more hosts and build up their own populations (Lewis *et al.* 1997). This might be achieved by rearing plant varieties that attract parasitoids through an enhanced emission of volatile, and reduce herbivore activity directly via toxins (Bottrell *et al.* 1998).

However, plant breeding for increased chemical resistance to herbivores runs the risk of introducing interactions that have negative effects on parasitoids. Most studies examining the effects of plant quality on parasitoid development have shown that the performance of the host and the parasitoid are strongly correlated with plant chemical variation, leading both to resistance against herbivores and to effects on parasitoid fitness (Ode 2006). Parasitoids may be indirectly affected by host plant chemical resistance via altered herbivore size, altered immune defence of the herbivore, and a changed level of plant chemicals in the herbivore (Gols *et al.* 2008b, Bukovinszky *et al.* 2009, Le Guigo *et al.* 2011). The potential effects on the parasitoid include altered development time, survival, size and fertility. Broods of *Copidosoma sosares* (Walker) (Hymenoptera: Encyrtidae) developing in parsnip webworm *Depressaria pastinacella* (Duponchel) (Lepidoptera: Elachistidae) feeding on apiaceous plants containing high levels of furanocoumarins showed smaller clutch sizes and exhibited lower survival rates (due to xanthotoxin) or less successful development (due to isopimpinellin) (Ode *et al.* 2004). Higher concentrations of glucosinolates in the host plant species on which the host *P. brassicae* fed in the field did not influence the cocoon mass of the primary parasitoid *C. glomerata*, while the hyperparasitoid *Lysibia nana* Gravenhorst (Hymenoptera: Ichneumonidae) showed lower survival and a smaller body size when *P. brassicae* fed on the host plant with a lower amount of induced glucosinolates (Harvey *et al.* 2003). Gossypol, a terpenoid produced by cotton, slows down the development of the host *H. virescens*. It has no negative effects on the parasitoid *Campoletis sonorensis* Cameron (Hymenoptera: Ichneumonidae) at natural concentrations (Gunasena *et al.* 1989). The absence of a detrimental effect of toxins to the third trophic level may be due to inherent physiological or biochemical resistance in the parasitoid or to reduced exposure to the allelochemical due to active detoxification by the host. Increased levels of plant defence can either favour or hinder biocontrol agents such as parasitoids. When applying biocontrol measures using plant cultivars showing resistance to herbivores, the effect on parasitoids needs to be taken into account. It is important to know how the resistance mechanism acts on the parasitoid and at least to rear plants in which this mechanism does not affect the parasitoid negatively when parasitoids and herbivore-resistant plant cultivars are combined. If possible, rearing herbivore-resistant plant cultivars with a positive effect on parasitoids should be preferred.

The possibility of using semiochemicals to manipulate parasitoid behaviour is certainly promising and enticing. However, it is still an open question whether their employment improves the effective biological control of crop pests. Further studies are required, taking into account crop-specific aspects and the landscape context of the treatments (Table 9.1). Natural enemy recruitment with semiochemicals should be successful in heterogeneous agricultural landscapes that include natural vegetation areas able to provide 'reservoirs' of beneficial insects. Within this perspective, Simpson *et al.* (2011b, 2011c) defined a strategy called 'attract and reward' that combines the application of HIPVs in crops with habitat manipulation (see Simpson, Read & Gurr, Chapter 12, this volume). Under the ecological approach of this strategy, HIPV application and habitat manipulation should synergistically act on natural enemies' activities. Crops treated with HIPVs recruit natural enemies that may find other plants in the same field that offer them food as a reward. In this way, the negative effects of semiochemical application, such as the absence (or low density) of hosts, should be minimized (for a detailed discussion see Colazza, Peri & Cusumano, Chapter 11, and Simpson, Read & Gurr, Chapter 12, this volume).

As it is important to consider crop- or landscape-specific aspects of semiochemical treatments for biological control using parasitoids, the physiology and behaviour of individuals within their usual environmental context should also be considered (Table 9.1). Following this line of reasoning, the parasitoids must match the environment into which they are to be introduced for biological control (environmental matching). By investigating and defining the relationships of the parasitoid species with the biotic environment, this should ensure that it is functionally equipped for the target situation (Walter 2003). The steps of the host searching behaviour should correspond with the cues associated with the host and its surroundings. Therefore, biological control approaches that relate physiological and behavioural aspects of parasitoid biology to host use and habitat requirements are recommended (Lewis *et al.* 1990, Vinson 1998).

To successfully 'match' a parasitoid to the environment, knowledge of its small-scale complexity is necessary. According to Levin (1992), individual variations in an organism's responses are greater on a small than on a large spatial scale. Therefore, predictions about

Table 9.1 Ecological and methodological implications for manipulation of parasitoids by semiochemicals.

Level	Implications
Spatial scale	
Landscape	• Heterogeneous agricultural landscapes favour natural enemy recruitment by semiochemicals • Natural vegetation areas provide 'reservoirs' for beneficial insects
Habitat	• Habitat manipulation favours recruitment and conservation of parasitoids by ecological infrastructures
Microhabitat	• Environmental matching – selecting parasitoids according to the small-scale infochemical environment

Host plant	• Variety selection influences multitrophic level interactions by:
	  ○ qualitative and quantitative differences in volatile and HIPV emissions
	  ○ additional direct chemical defence against herbivores
	  ○ avoiding negative effects of plant chemicals mediating herbivore resistance

Interspecific interactions	
Multitrophic level interactions	• Total system approach versus simple application of semiochemicals • Considering effects on all trophic levels: attraction of herbivores, hyperparasitoids
Host density	• Semiochemicals decrease parasitoid foraging efficiency at low densities (false leads)
Interspecific competitive interactions	• Extrinsic competition between adult parasitoids ○ superior strategies in locating common hosts, e.g. by exploitation of semiochemicals • Intrinsic competition between parasitoid larvae developing in the same host • Counterbalanced competition (inferior parasitoid at the extrinsic level dominates in intrinsic competition)
Parasitoid traits	

Learning	• Association of environmental cues with rewarding or aversive experiences:
	  ○ habituation/negative association when hosts are at low density (mainly specialized parasitoids)
	  ○ positive association by priming or rewarding during rearing

Host breadth specialization	• Exploitation of more specific cues by specialist parasitoids
Methodology	
Laboratory and field studies	• Measuring infochemical environment in the field • Transferring field situation to laboratory studies • Determining activity window of parasitoids to time application
Multiple approaches	• Combining semiochemical approaches with other biocontrol measures
Multiple success measures	• Parasitoid activity, parasitoid density, parasitism rate, host density, host damage, yield, intimidation effects

the effects of certain factors are more difficult to make on a small scale like the microhabitat. Since small organisms such as parasitoids are limited in their dispersal abilities due to their size, individual variations in their behaviour are important at such small scales (Bommarco & Banks 2003). This is true for small-scale structural complexity (Casas & Djemai 2002, Randlkofer *et al.* 2010), but most probably also for small-scale chemical complexity (see Wäschke, Meiners & Rostás, Chapter 3, this volume). Knowledge of the biology and the host location behaviour of the parasitoids is needed in order to predict how small-scale habitat complexity affects parasitoid–host interactions in the field. Therefore, laboratory studies mimicking small-scale environmental factors are a prerequisite for developing successful biological control measures based on chemical ecology.

Even when semiochemical-based manipulation of parasitoids has been successfully demonstrated in the laboratory, that does not mean that the results can easily be transferred to the field situation. The chemical complexity in the habitat might hamper the efficient use of semiochemicals because of a high 'background noise to signal' ratio (Dicke *et al.* 2003, Hilker & McNeil 2008) or by a large variability in the signal due to biotic or abiotic environmental conditions (Riffell *et al.* 2008, Randlkofer *et al.* 2010; see also Wäschke, Meiners & Rostás, Chapter 3, this volume). Taking chemical ecological investigations from the laboratory to the field should help to clarify the effect of complexity on the parasitoid host location process (Table 9.1). Correlating parasitoid field performance with chemical complexity can be a first step. The collection and identification of volatiles from the field is important in order to elucidate the chemical communication between plants and insects, so that host plant volatiles can be used to attract insect parasitoids of agricultural pests and optimize integrated pest management or novel biological control measures (Metcalf 1994, Reddy & Guerrero 2004, Bruce *et al.* 2005, Norin 2007, Khan *et al.* 2008, Frank 2010).

The ultimate aim of biological control in agricultural environments is to reduce the loss of yield. The successful manipulation of parasitoid behaviour using semiochemicals is a prerequisite for this, but the success of the biocontrol measure needs to be confirmed by actually measuring yield. Moreover, even when semiochemicals do not affect parasitism, they might affect the parasitoids' behaviour and enhance crop yield indirectly when parasitoids negatively influence the herbivores' food consumption, for instance by eliciting time-consuming defensive behaviour in the herbivores (Morrison 1999). Such 'intimidation' effects are thought to act widely on trophic cascades (Schmitz *et al.* 2004).

No specific insect–insect interaction occurs isolated from other interactions in the arthropod community (Price *et al.* 2012). It is known that behavioural differences between parasitoids in the processing of plant-derived infochemicals can affect population and community levels (Vet 2001). Also, the manipulation of a parasitoid by using semiochemicals will most probably affect some of the multiple direct or indirect interactions between species from this community. For successful biological control with the aim of preventing pest outbreaks with parasitoids, it is therefore essential to elucidate the chemical ecology of all trophic partners and to adopt a 'total system approach to sustainable pest management' (Lewis *et al.* 1997) (Table 9.1).

When using semiochemicals directly or via modified plants to manipulate the third trophic level, the cues might affect the second trophic level as well. Plant volatile cues might attract both parasitoids and their hosts, and thus enhanced parasitoid activity might be outweighed by enhanced herbivory. Meiners *et al.* (2005) showed that the elm leaf beetle responds to volatiles induced in elms by the feeding and oviposition of conspecifics. These volatiles are also attractive to the specialized egg parasitoid *O. gallerucae* (Meiners & Hilker

1997, Büchel *et al.* 2011). While the beetles preferred the odours from uninfested elms to that of heavily infested ones, they also preferred mildly infested to uninfested ones.

The fact that parasitoids must simultaneously address multiple decisions in multitrophic interactions is discussed by Ode (see Chapter 2, this volume) who concludes that examining only one-dimensional aspects of parasitoid chemical ecology in relation to lower trophic levels is not enough. In addition, interspecific competitive interactions between parasitoids need to be considered when employing semiochemicals. These are outlined in the following section as a caveat example to raise awareness of the chemical ecological complexity that must be considered when developing effective biological control programmes with insect parasitoids.

9.5 Cautionary example: interspecific competitive interactions in parasitoids

In the field, parasitoids may establish complex ecological interactions with other parasitoids (Table 9.1) that can have a role in sizing and shaping community structures (Godfray 1994) and, as a consequence, they can affect the efficacy of biological control programmes. For example, in the guild of scale parasitoids in California citrus groves, *Aphytis lignanensis* Compere is displaced by *A. melinus*, as the competitor is able to reproduce on smaller hosts (Luck & Podoler 1985, Murdoch *et al.* 1996). On the other hand, different host stage preferences permit the coexistence of *A. melinus* and *Encarsia perniciosi* (Tower) (Yu *et al.* 1990). Bográn *et al.* (2002), studying competitive interactions between *Encarsia formosa* Gahan, *Encarsia pergandiella* Howard and *Eretmocerus mundus* Mercet, three aphelinid parasitoids of the silverleaf whitefly *Bemisia argentifolii* Bellows & Perring (Homoptera: Aleyrodidae), demonstrated that the competition reduces the population growth rates of both *Encarsia* species, but not that of *E. mundus*, in most cases without affecting host population suppression. In general, competitive interactions occur between parasitoid species that have evolved similar ecological strategies (Boivin & Brodeur 2006). These interactions can occur either between adult parasitoids (extrinsic competition) or between parasitoid larvae developing in the same host (intrinsic competition) (Zwolfer 1971, Godfray 1994, De Moraes *et al.* 1999). In particular, extrinsic competition can be evaluated in terms of comparative host location efficiency through exploiting semiochemicals during foraging behaviour. Among species that are searching for the same hosts, the species that displays a high degree of host finding efficiency may have an advantage in finding sufficient hosts, especially if they are present at low density and when the parasitoids are solitary species; that is, when only one individual can develop within a single host (Vinson & Iwantsch 1980, Hagvar 1989). In maize fields in Africa, eggs of the noctuid *Sesamia calamistis* Hampson were parasitized by the platygastrids *Telenomus busseolae* (Gahan), *Telenomus isis* (Polaszek) and the trichogrammatid *Lathromeris ovicida* Risbec, with high levels of parasitism for the platygastrids and very low levels for the trichogrammatid (Schulthess *et al.* 2001). Comparing the attraction of these parasitoids, Fiaboe *et al.* (2003) showed that all three species are attracted to calling *S. calamistis* virgin females, but *T. busseolae* and *T. isis* respond more strongly to moth volatiles than *L. ovicida*. They explained this different strategy by differences in the ecological niches occupied by their host species and by differences in host specificity (*sensu* Vet & Dicke 1992, Steidle & van Loon 2003). Both *Telenomus* species are specialists that attack noctuid stem borers, while *L. ovicida* has mainly been reported from crambids and pyralids. The relationship between host searching efficiency and host breadth specialization is supported by the hypothesis that specialist parasitoid species exploit specific cues more frequently than generalist species do (Sheehan 1986, Vet *et al*. 1990, Vet & Dicke 1992, Steidle & van Loon 2003). Consistent with this hypothesis, specialist parasitoids should be more likely than generalist ones to have developed strategies to optimize their resources in locating a common host, thus gaining an advantage in the extrinsic competition. Moreover, it is necessary to specify what a specific cue for a parasitoid is. By comparing two egg parasitoids of the pine processionary moth, *Thaumetopoea pityocampa* Denis & Schiffermüller, the specialist *Baryscapus servadeii* (Dominichini) and the generalist *Ooencyrtus pityocampae* Mercet, Battisti (1989) showed that only the generalist responds to 'Pityolure', a synthetic host pheromone, although eavesdropping the host sex pheromone is often an adaptive strategy for specialist egg parasitoids. However, in the field, *B. servadeii* was found to achieve higher levels of parasitism than *O. pityocampae*, suggesting that it has evolved a more efficient host location strategy, based on other specific cues that have not yet been elucidated.

Different eavesdropping strategies are also adopted by parasitoids with behavioural differences. *Telenomus calvus* and *Telenomus podisi* (Ashmead) are two egg parasitoids that attack the same pentatomid host *P. maculiventris*. While females of the phoretic egg parasitoid *T. calvus* are able to exploit the synthetic pheromone of the host, the females of the non-phoretic egg parasitoid *T. podisi* are not (Bruni *et al*. 2000).

As previously discussed, innate responses of parasitoids to chemical volatiles may be deeply modified by the experience gained during their life. As a consequence, competitive interactions can also be altered. For example, *T. brassicae* innately exploits anti-aphrodisiac compounds produced during mating by the males of two host species, and oviposition-induced synomones elicited by anti-aphrodisiacs, whereas *T. evanescens* spies on these cues only after an experience associated with oviposition (Fatouros *et al*. 2005b, Huigens *et al*. 2009, 2010, Pashalidou *et al*. 2010). Therefore, rewarding experiences allow *T. evanescens* to plug the gap in host searching efficacy towards *T. brassicae* by learning to exploit host-related cues that are highly reliable for host egg presence.

Understanding how competitor parasitoids use semiochemicals to locate a common host is not enough in order to predict their successful employment in biological control programmes. To obtain a more complete view of parasitoid interactions, intrinsic competition also needs to be thoroughly investigated, as this has great relevance from both an ecological and an applied point of view. An interesting case study that explained how interspecific intrinsic and extrinsic competition has important implications for the community dynamics of a tritrophic plant–herbivore–parasitoid system has been described by De Moraes *et al*. (1999), and by De Moraes & Mescher (2005) for a system including *H. virescens*, the tobacco budworm, and three larval parasitoids: the generalist *C. marginiventris*, and two specialists, *Microplitis croceipes* (Cresson) (Hymenoptera: Braconidae) and *C. nigriceps*. In the guild *C. nigriceps–M. croceipes*, the latter species regularly dominates intrinsic competition by having a shorter hatching time. *C. nigriceps* larvae are dominant when the wasp females oviposited at least 16 hours before the competing female; an interval that corresponds to the difference in hatching time between the species (De Moraes *et al*. 1999). This disadvantage is compensated by *C. nigriceps* at the extrinsic competition level. In fact, it has superior host location efficiency at long range by detecting volatiles induced by host feeding activity, and at short range by locating first-, second- and third-instar larvae of the host. Conversely, the competitor does not use host-specific volatiles from damaged plants and mainly locates the third instars (De Moraes *et al*. 1999).

Microplitis croceipes also exhibits different host searching strategies from *C. marginiventris*. It is attracted to host frass volatiles, while *C. marginiventris* prefers volatiles from damaged plants (Cortesero *et al.* 1997). Moreover, among the competitors, *M. croceipes* appears the weakest at the extrinsic level (Cortesero *et al.* 1997, De Moraes & Lewis 1999). The intrinsic competition is dominated by *M. croceipes* and *C. nigriceps* when they have oviposited prior to *C. marginiventris* (De Moraes & Mescher 2005). The explanation for the outcomes of these interactions has been found by the authors in the concept of counterbalanced competition described by Zwolfer (1971); that is, coexistence among parasitoids attacking the same host(s) is practicable when one species dominates in the intrinsic competition, and the other compensates at the extrinsic level by having higher host finding or dispersal efficacy. This concept was also used to explain other competitive interactions between parasitoids of forest tree pests (Schröder 1974, Hougardy & Grégoire 2003). Recently, an example of counterbalanced competition concerning two sympatric egg parasitoid species, *T. basalis* and *Ooencyrtus telenomicida* (Vassiliev), occurring on a common host, the pentatomid bug *N. viridula*, has been reported. Surveys of *N. viridula* egg masses conducted in the field showed that these species can compete for the same host egg mass, and that *T. basalis* achieves a higher level of parasitization (Peri *et al.* 2011). Laboratory bioassays demonstrated that *T. basalis* is the superior extrinsic competitor in this system. *Trissolcus basalis* not only exploits more cues than *O. telenomicida* to locate *N. viridula* egg masses, but also chooses more reliable ones. It eavesdrops on volatile oviposition-induced synomones, volatile cues from virgin males and preovipositing females, and contact kairomones in the host footprints (Colazza *et al.* 1999, 2004), while *O. telenomicida* exploits volatile kairomones from host virgin males only (Peri *et al.* 2011). On the other hand, when both parasitoid species parasitize the same host egg mass, interspecific larval competition is dominated by *O. telenomicida* regardless of the sequence in which oviposition occurs and whether the parasitoids simultaneously find the host patch (Cusumano *et al.* 2011, 2012). Moreover, *O. telenomicida* can develop as a facultative hyperparasitoid, ovipositing in host eggs already parasitized by *T. basalis* up to 7 days earlier, and developing on the primary parasitoid larva that has totally consumed the host resources (Cusumano *et al.* 2011). These examples provide some insight into the importance of the competitive interactions in determining community structure and dynamics in the plant–herbivore–parasitoid system. Further studies in the field should enhance the knowledge needed to manage these ecological factors so as to avoid alteration of natural communities by natural enemy recruitment and to develop novel strategies for pest management based on parasitoid manipulation by semiochemicals.

9.6 Conclusions

The cautionary example regarding the chemical ecology of interspecific competitive interactions in parasitoids illustrates that chemical ecological interactions between parasitoids and their hosts always involve other organisms. Thus, for sustainable biological control, a 'total systems' approach is needed that prohibits the simple application of semiochemicals in the field, which could sometimes even be counterproductive. Clarifying the chemical ecology of insect parasitoids in a multi-species context is therefore important in order to develop effective and successful biological control programmes. Results from field and laboratory studies on the chemical ecology of agricultural and natural systems and from pioneering biocontrol field trials confirm that the chemical application of constitutive or herbivore-induced plant synomones or of host kairomones to attract, arrest or inform parasitoids can be promising measures for biological control. Also, manipulating plants by hormones or selecting plant varieties that chemically interact with parasitoids in a positive way seem to be viable options for the future. However, more translational research is needed to be able to open the 'chemical ecological toolbox' successfully for biocontrol. Such research requires the competence and cooperation of experts in biological control, plant ecology and plant breeding, insect ecology and behaviour, insect physiology and breeding. It will need to cover different spatial scales and trophic levels of manipulation – from landscape to host plant characteristics to parasitoid traits. Only in that way can the complexity of the system in focus be considered adequately when accompanying the introduction and/or conservation of parasitoids within agroecosystems by exploiting tritrophic interactions and manipulation of parasitoid behavioural or physiological responses to chemical cues.

References

Aldrich, J.R. (1995) Chemical communication in the true bugs and parasitoid exploitation. In: Cardé, R.T. and Bell, W.J. (eds) *Chemical Ecology of Insects II*. Chapman & Hall, New York, pp. 318–63.

Altieri, M.A., Lewis, W.J., Nordlund, D.A., Gueldner, R.C. and Todd, J.W. (1981) Chemical interactions between plants and *Trichogramma* wasps in Georgia soybean fields. *Protection Ecology* **3**: 259–63.

Anderson, P. (2002) Oviposition pheromones in herbivorous and carnivorous insects. In: Hilker, M. and Meiners, T. (eds) *Chemoecology of Insect Eggs and Egg Deposition*. Blackwell Publishing Ltd., Oxford, pp. 235–64.

Arakaki, N. and Wakamura, S. (2000) Bridge in time and space for an egg parasitoid: kairomonal use of trace amount of sex pheromone adsorbed on egg mass scale hair of the tussock moth, *Euproctis taiwana* (Shiraki) (Lepidoptera: Lymantriidae), by egg parasitoid, *Telenomus euproctidis* Wilcox (Hymenoptera: Scelionidae), for host location. *Entomological Science* **3**: 25–31.

Arakaki, N., Yamazawa, H. and Wakamura, S. (2011) The egg parasitoid *Telenomus euproctidis* (Hymenoptera: Scelionidae) uses sex pheromone released by immobile female tussock moth *Orgyia postica* (Lepidoptera: Lymantriidae) as kairomone. *Applied Entomology and Zoology* **46**: 195–200.

Barbosa, P. (1998) *Conservation Biological Control*. Academic Press, San Diego, CA.

Battaglia, D., Poppy, G., Powell, W., Romano, A., Tranfaglia, A. and Pennacchio, F. (2000) Physical and chemical cues influencing the oviposition behaviour of *Aphidius ervi*. *Entomologia Experimentalis et Applicata* **94**: 219–27.

Battisti, A. (1989) Field studies on the behaviour of two egg parasitoids of the pine processionary moth *Thaumetopoea pityocampa*. *Entomophaga* **34**: 29–38.

Bekkaoui, A. and Thibout, E. (1993) Role of the cocoon of *Acrolepiopsis assectella* (Lep., Yponomeutoidae) in host recognition by the parasitoid *Diadromus pulchellus* (Hym., Ichneumonidae). *Entomophaga* **38**: 101–13.

Benrey, B. and Denno, R.F. (1997) The slow-growth–high-mortality hypothesis: a test using the cabbage butterfly. *Ecology* **78**: 987–99.

Bogahawatte, C.N.L. and van Emden, H.F. (1996) The influence of the host plant of diamond-back moth (*Plutella xylostella*) on the plant preferences of its parasitoid *Cotesia plutellae* in Sri Lanka. *Physiological Entomology* **21**: 93–6.

Bográn, C.E., Heinz, K.M. and Ciomperlik, M.A. (2002) Interspecific competition among insect parasitoids: field experiments with whiteflies as hosts in cotton. *Ecology* **83**: 653–68.

Boivin, G. and Brodeur, J. (2006) Intra- and interspecific interactions among parasitoids: mechanisms, outcomes and biological control. In: Brodeur, J. and Boivin, G. (eds) *Trophic and Guild Interactions in Biological Control*. Springer, New York, pp. 123–44.

Bommarco, R. and Banks, J.E. (2003) Scale as modifier in vegetation diversity experiments: effects on herbivores and predators. *Oikos* **102**: 440–8.

Boo, K.S. and Yang, J.P. (1998) Olfactory response of *Trichogramma chilonis* to *Capsicum annum*. *Journal of Asia-Pacific Entomology* **1**: 123–9.

Bottrell, D.G., Barbosa, P. and Gould, F. (1998) Manipulating natural enemies by plant variety selection and modification: a realistic strategy? *Annual Review of Entomology* **43**: 347–67.

Bruce, T.J.A., Wadhams, L.J. and Woodcock, C.M. (2005) Insect host location: a volatile situation. *Trends in Plant Science* **10**: 1360–85.

Bruni, R., Sant'Ana, J., Aldrich, J.R. and Bin, F. (2000) Influence of host pheromone on egg parasitism by scelionid wasps: comparison of phoretic and nonphoretic parasitoids. *Journal of Insect Behavior* **13**: 165–73.

Büchel, K., Malskies, S., Mayer, M., Fenning, T., Gershenzon, J., Hilker, M. and Meiners, T. (2011) How plants give early herbivore alert: volatile terpenoids emitted from elm attract egg parasitoids to plants laden with eggs of the elm leaf beetle. *Basic and Applied Ecology* **12**: 403–12.

Bukovinszky, T., Poelman, E.H., Gols, R., Prekatsakis, G., Vet, L.E.M., Harvey, J.A. and Dicke, M. (2009) Consequences of constitutive and induced variation in plant nutritional quality for immune defence of a herbivore against parasitism. *Oecologia* **160**: 299–308.

Casas, J. and Djemai, I. (2002) Canopy architecture and multitrophic interactions. In: Tscharntke, T. and Hawkins, B.A. (eds) *Multitrophic Level Interactions*. Cambridge University Press, Cambridge, pp. 174–96.

Chuche, J., Xuéreb, A. and Thiéry, D. (2006) Attraction of *Dibrachys cavus* (Hymenoptera: Pteromalidae) to its host frass volatiles. *Journal of Chemical Ecology* **32**: 2721–31.

Colazza, S., Aquila, G., De Pasquale, C., Peri, E. and Millar, J.G. (2007) The egg parasitoid *Trissolcus basalis* uses *n*-nonadecane, a cuticular hydrocarbon from its stink bug host *Nezara viridula*, to discriminate between female and male hosts. *Journal of Chemical Ecology* **33**: 1405–20.

Colazza, S., Fucarino, A., Peri, E., Salerno, G., Conti, E. and Bin, F. (2004) Insect oviposition induces volatile emission in herbaceous plant that attracts egg parasitoids. *Journal of Experimental Biology* **207**: 47–53.

Colazza, S., Lo Bue, M., Lo Giudice, D. and Peri, E. (2009) The response of *Trissolcus basalis* to footprint contact kairomones from *Nezara viridula* females is mediated by leaf epicuticular waxes. *Naturwissenschaften* **96**: 975–81.

Colazza, S., Peri, E., Salerno, G. and Conti, E. (2010) Host searching by egg parasitoids: exploitation of host chemical cues. In: Cônsoli, F.L., Parra, J.R.P. and Zucchi, R.A. (eds) *Egg Parasitoids in Agroecosystems with Emphasis on Trichogramma*. Springer, Dordrecht, pp. 97–147.

Colazza, S., Salerno, G. and Wajnberg, E. (1999) Volatile and contact chemicals released by *Nezara viridula* (Heteroptera: Pentatomidae) have a kairomonal effect on the egg parasitoid *Trissolcus basalis* (Hymenoptera: Scelionidae). *Biological Control* **16**: 310–17.

Conti, E., Salerno, G., Bin, F. and Vinson, S.B. (2004) The role of host semiochemicals in parasitoid specificity: a case study with *Trissolcus brochymenae* and *Trissolcus simoni* on pentatomid bugs. *Biological Control* **29**: 435–44.

Conti, E., Salerno, G., De Santis, F., Leombruni, B., Frati, F. and Bin, F. (2010) Short-range allelochemicals from a plant–herbivore association: a singular case of oviposition-induced synomone for an egg parasitoid. *Journal of Experimental Biology* **213**: 3911–19.

Cortesero, A.M., De Moraes, C.M., Stapel, J.O., Tumlinson, J.H. and Lewis, W.J. (1997) Comparisons and contrasts in host foraging strategies of two larval parasitoids with different degrees of host specificity. *Journal of Chemical Ecology* **23**: 1589–606.

Cusumano, A., Peri, E., Vinson, S.B. and Colazza, S. (2011) Intraguild interactions between two egg parasitoids exploring host patches. *BioControl* **56**: 173–84.

Cusumano, A., Peri, E., Vinson, S.B. and Colazza, S. (2012) The ovipositing female of *Ooencyrtus telenomicida* relies on physiological mechanisms to mediate intrinsic competition with *Trissolcus basalis*. *Entomologia Experimentalis et Applicata* **143**: 155–63.

De Moraes, C.M. and Lewis, W.J. (1999) Analyses of two parasitoids with convergent foraging strategies. *Journal of Insect Behavior* **12**: 571–83.

De Moraes, C.M. and Mescher, M. (2005) Intrinsic competition between larval parasitoids with different degrees of host specificity. *Ecological Entomology* **30**: 564–70.

De Moraes, C.M., Cortesero, A.M., Stapel, J.O. and Lewis, W.J. (1999) Intrinsic and extrinsic competition between two larval parasitoids of *Heliothis virescens*. *Ecological Entomology* **24**: 402–10.

De Moraes, C.M., Lewis, W.J., Paré, P.W., Alborn, H.T. and Tumlinson, J.H. (1998) Herbivore infested plants selectively attract parasitoids. *Nature* **393**: 570–3.

Dicke, M. and Sabelis, M.W. (1988) Infochemical terminology: based on cost–benefit analysis rather than origin of compounds? *Functional Ecology* **2**: 131–9.

Dicke, M. and van Loon, J.J.A. (2000) Multitrophic effects of herbivore-induced plant volatiles in an evolutionary context. *Entomologia Experimentalis et Applicata* **97**: 237–49.

Dicke, M. and Vet, L.E.M. (1999) Plant–carnivore interactions: evolutionary and ecological consequences for plant, herbivore and carnivore. In: Olff, H., Brown, V.K. and Drent, R.H. (eds) *Herbivores: Between Plants and Predators*. Blackwell Science, Oxford, pp. 483–520.

Dicke, M., de Boer, J.G., Höfte, M. and Rocha-Granados, M.C. (2003) Mixed blends of herbivore-induced plant volatiles and foraging success of carnivorous arthropods. *Oikos* **101**: 38–48.

Dicke, M., Gols, R., Ludeking, D. and Posthumus, M.A. (1999) Jasmonic acid and herbivory differentially induce carnivore-attracting plant volatiles in Lima bean plants. *Journal of Chemical Ecology* **25**: 1907–22.

Dicke, M., Sabelis, M.W., Takabayashi, J., Bruin, J. and Posthumus, M.A. (1990) Plant strategies of manipulating predator–prey interactions through allelochemicals: prospects for application in pest control. *Journal of Chemical Ecology* **16**: 3091–118.

Du, Y.J., Poppy, G.M. and Powell, W. (1996) Relative importance of semiochemicals from first and second trophic levels in host foraging behavior of *Aphidius ervi*. *Journal of Chemical Ecology* **22**: 1591–605.

Dudareva, N. and Pichersky, E. (2008) Metabolic engineering of plant volatiles. *Current Opinion in Biotechnology* **19**: 181–9.

Fatouros, N.E., Bukovinszkine'Kiss, G., Kalkers, L.A., Soler Gamborena, R., Dicke, M. and Hilker, M. (2005a) Oviposition-induced plant cues: do they arrest *Trichogramma wasps* during host location? *Entomologia Experimentalis et Applicata* **115**: 207–15.

Fatouros, N.E., Dicke, M., Mumm, R., Meiners, T. and Hilker, M. (2008) Foraging behavior of egg parasitoids exploiting chemical information. *Behavioral Ecology* **19**: 677–89.

Fatouros, N.E., Huigens, M.E., van Loon, J.J.A., Dicke, M. and Hilker, M. (2005b) Chemical communication: butterfly anti-aphrodisiac lures parasitic wasps. *Nature* **433**: 704.

Fiaboe, M.K., Chabi-Olaye, A., Gounou, S., Smith, H., Borgemeister, C. and Schultheiss, F. (2003) *Sesamia calamistis* calling behavior and its role in host finding of egg parasitoids *Telenomus busseolae*, *Telenomus isis*, and *Lathromeris ovicida*. *Journal of Chemical Ecology* **29**: 921–9.

Fontana, A., Held, M.., Fantaye, C.A., Turlings, T.C., Degenhardt, J. and Gershenzon, J. (2011) Attractiveness of constitutive and herbivore-induced sesquiterpene blends of maize to the parasitic wasp *Cotesia marginiventris* (Cresson). *Journal of Chemical Ecology* **37**: 582–91.

Franco, J.C., da Silva, E.B., Fortuna, T., Cortegano, E., Branco, M., Suma, P., La Torre, I., Russo, A., Elyahu, M., Protasov, A., Levi-Zada, A. and Mendel, Z. (2011) Vine mealybug sex pheromone increases citrus mealybug parasitism by *Anagyrus* sp. near *pseudococci* (Girault). *Biological Control* **58**: 230–8.

Franco, J.C., Silva, E.B., Cortegano, E., Campos, L., Branco, M., Zada, A. and Mendel, Z. (2008) Kairomonal response of the parasitoid *Anagyrus* spec. nov. near *pseudococci* to the sex pheromone of the vine mealybug. *Entomologia Experimentalis et Applicata* **126**: 122–30.

Frank, S.D. (2010) Biological control of arthropod pests using banker plant systems: past progress and future directions. *Biological Control* **52**: 8–16.

Gabrys, B.J., Gadomski, H.J., Klukowski, Z., Pickett, J.A., Sobota, G.T., Wadhams, L.J. and Woodcock, C.M. (1997) Sex pheromone of cabbage aphid *Brevicoryne brassicae*: identification and field trapping of male aphids and parasitoids. *Journal of Chemical Ecology* **23**: 1881–90.

Girling, R.D., Stewart-Jones, A., Dherbecourt, J., Staley, J.T., Wright, D.J. and Poppy, G.M. (2011) Parasitoids select plants more heavily infested with their caterpillar hosts: a new approach to aid interpretation of plant headspace volatiles. *Proceedings of the Royal Society B* **278**: 2646–53.

Glinwood, R.T., Powell, W. and Tripathi, C.P.M. (1998) Increased parasitization of aphids on trap plants alongside vials releasing synthetic aphid sex pheromone and effective range of the pheromone. *Biocontrol Science and Technology* **8**: 607–14.

Godfray, H.C.J (1994) *Parasitoids: Behavioral and Evolutionary Ecology*. Monographs in Behavior and Ecology, Princeton University Press, Princeton, NJ.

Gols, R., Bukovinszky, T., van Dam, N.M., Dicke, M., Bullock, J.M. and Harvey, J.A. (2008a) Performance of generalist and specialist herbivores and their endoparasitoids differs on cultivated and wild *Brassica* populations. *Journal of Chemical Ecology* **34**: 132–43.

Gols, R., Bullock, J.M., Dicke, M., Bukovinszky, T. and Harvey, J.A. (2011) Smelling the wood from the trees: non-linear parasitoid responses to volatile attractants produced by wild and cultivated cabbage. *Journal of Chemical Ecology* **37**: 795–807.

Gols, R., Wagenaar, R., Bukovinszky, T., van Dam, N.M., Dicke, M., Bullock, J.M. and Harvey, J.A. (2008b) Genetic variation in defense chemistry in wild cabbages affects herbivores and their endoparasitoids. *Ecology* **89**: 1616–26.

Grasswitz, T.R. and Paine, T.D. (1992) Kairomonal effect of an aphid cornicle secretion on *Lysiphlebus testaceipes* (Cresson) (Hymenoptera, Aphidiidae). *Journal of Insect Behavior* **5**: 447–57.

Gross, H.R. (1981) Employment of kairomones in the management of parasitoids. In: Nordlund, D.A., Jones, R.L. and Lewis, W.J. (eds) *Semiochemicals: Their Role in Pest Control*. John Wiley & Sons, New York, pp. 137–50.

Guerrieri, E., Pennacchio, F. and Tremblay, E. (1993) Flight behavior of the aphid parasitoid *Aphidius ervi* (Hymenoptera, Braconidae) in response to plant and host volatiles. *European Journal of Entomology* **90**: 415–21.

Gunasena, G.H., Vinson, S.B., Williams, H.J. and Stipanovic, R.D. (1989) Development and survival of the endoparasitoid *Campoletis sonorensis* (Hymenoptera: Ichneumonidae) reared from gossypol-exposed *Heliothis virescens* (F.) (Lepidoptera: Noctuidae). *Environmental Entomology* **18**: 886–91.

Gurr, G.M. and Kvedaras, O.L. (2010) Synergizing biological control: scope for sterile insect technique, induced plant defences and cultural techniques to enhance natural enemy impact. *Biological Control* **52**: 198–207.

Gurr, G.M., Wratten, S.D. and Luna, J.M. (2003) Multi-function agricultural biodiversity: pest management and other benefits. *Basic and Applied Ecology* **4**: 107–16.

Gurr, G.M., Wratten, S.D. and Altieri, M.A. (2004) *Ecological Engineering for Pest Management: Advances in Habitat Manipulation for Arthropods*. CABI Press, London.

Gurr, G.M., Wratten, S.D., Tylianakis, J., Kean, J. and Keller, M. (2005) Providing plant foods for natural enemies in farming systems: balancing practicalities and theory. In: Wäckers, F.L., van Rijn, P.C.J. and Bruin, J. (eds) *Plant-Provided Food for Carnivorous Insects: A Protective Mutualism and its Applications*. Cambridge University Press, New York, pp. 341–7.

Hagvar, E.B. (1989) Interspecific competition in parasitoids, with implications for biological control. *Acta Entomologica Bohemoslovaca* **86**: 321–35.

Hagvar, E.B. and Hofsvang, T. (1989) Effect of honeydew and hosts on plant colonization by the aphid parasitoid *Ephedrus cerasicola*. *Entomophaga* **34**: 495–501.

Hanyu, K., Ichiki, R.T., Nakamura, S. and Kainoh, Y. (2009) Duration and location of attraction to herbivore-damaged plants in the tachinid parasitoid *Exorista japonica*. *Applied Entomology and Zoology* **44**: 371–8.

Hardie, J., Hick, A.J., Holler, C., Mann, J., Merritt, L., Nottingham, S.F., Powell, W., Wadhams, L.J., Witthinrich, J. and Wright, A.F. (1994) The responses of *Praon* spp parasitoids to aphid sex-pheromone components in the field. *Entomologia Experimentalis et Applicata* **71**: 95–9.

Hardie, J., Nottingham, S.F., Powell, W. and Wadhams, L.J. (1991) Synthetic aphid sex pheromone lures female parasitoids. *Entomologia Experimentalis et Applicata* **61**: 97–9.

Hare, J.D., Morgan, D.J.W. and Nguyun, T. (1997) Increased parasitization of California red scale in the field after exposing its parasitoid, *Aphytis melinus*, to a synthetic kairomone. *Entomologia Experimentalis et Applicata* **82**: 73–81.

Harvey, J.A., van Dam, N.M. and Gols, R. (2003) Interactions over four trophic levels: foodplant quality affects development of a hyperparasitoid as mediated through a herbivore and its primary parasitoid. *Journal of Animal Ecology* **72**: 520–31.

Hatano, E., Kunert, G., Michaud, J.P. and Weisser, W.W. (2008) Chemical cues mediating aphid location by natural enemies. *European Journal of Entomology* **105**: 797–806.

Hilker, M. and McNeil, J. (2008) Chemical and behavioral ecology in insect parasitoids: how to behave optimally in a complex odorous environment. In: Wajnberg, E., Bernstein, C. and van Alphen, J. (eds) *Behavioral Ecology of Insect Parasitoids: From Theoretical Approaches to Field Applications*. Blackwell Publishing Ltd., New York, pp. 693–705.

Hilker, M., Kobs, C., Varama, M. and Schrank, K. (2002) Insect egg deposition induces *Pinus sylvestris* to attract egg parasitoids. *Journal of Experimental Biology* **205**: 455–61.

Hilker, M., Stein, C., Schroder, R., Varama, M. and Mumm, R. (2005) Insect egg deposition induces defence responses in *Pinus sylvestris*: characterisation of the elicitor. *Journal of Experimental Biology* **208**: 1849–54.

Hoffmeister, T.S., Roitberg, B.D. and Lalonde, R.G. (2000) Catching Ariadne by her thread: how a parasitoid exploits the herbivore's marking trails to locate its host. *Entomologia Experimentalis et Applicata* **95**: 77–85.

Hougardy, E. and Grégoire, J.C. (2003) Cleptoparasitism increases the host finding ability of a poly-phagous parasitoid species, *Rhopalicus tutela* (Hymenoptera: Pteromalidae). *Behavioral Ecology and Sociobiology* **55**: 184–9.

Huigens, M.E., Pashalidou, F.G., Qian, M.H., Bukovinszky, T., Smid, H.M., van Loon, J.J.A., Dicke, M. and Fatouros, N.E. (2009) Hitch-hiking parasitic wasp learns to exploit butterfly anti-aphrodisiac. *Proceedings of the National Academy of Sciences USA* **106**: 820–5.

Huigens, M.E., Woelke, J.B., Pashalidou, F.G., Bukovinszky, T., Smid, H.M. and Fatouros, N.E. (2010) Chemical espionage on species-specific butterfly anti-aphrodisiacs by hitchhiking *Trichogramma* wasps. *Behavioral Ecology* **21**: 470–8.

Ichiki, R.T., Kainoh, Y., Kugimiya, S., Takabayashi, J. and Nakamura, S. (2008) Attraction to herbivore-induced plant volatiles by the host-foraging parasitoid fly *Exorista japonica*. *Journal of Chemical Ecology* **34**: 614–21.

Jang, S.A., Cho, J.H., Park, G.M., Choo, H.Y. and Park, C.G. (2011) Attraction of *Gymnosoma rotundatum* (Diptera: Tachinidae) to different amounts of *Plautia stali* (Hemiptera: Pentatomidae) aggregation pheromone and the effect of different pheromone dispensers. *Journal of Asia-Pacific Entomology* **14**: 119–21.

Jang, S.A. and Park, C.G. (2010) *Gymnosoma rotundatum* (Diptera: Tachinidae) attracted to the aggregation pheromone of *Plautia stali* (Hemiptera: Pentatomidae). *Journal of Asia-Pacific Entomology* **13**: 73–5.

Jones, V.P., Steffan, S.A., Wiman, N.G., Horton, D.R., Miliczky, E., Zhang, Q.H. and Baker, C.C. (2011) Evaluation of herbivore-induced plant volatiles for monitoring green lacewings in Washington apple orchards. *Biological Control* **56**: 98–105.

Kaiser, L., Pham-Delegue, M.H., Bakchine, E. and Masson, C. (1989) Olfactory responses of *Trichogramma maidis* Pint. et Voeg.: effects of chemical cues and behavioral plasticity. *Journal of Insect Behavior* **2**: 701–12.

Kaplan, I. (2012) Attracting carnivorous arthropods with plant volatiles: the future of biocontrol or playing with fire? *Biological Control* **60**: 77–89.

Khan, Z.R., James, D.G., Midega, C.A.O. and Picket, J.A. (2008) Chemical ecology and conservation biological control. *Biological Control* **45**: 210–24.

Klomp, H. (1981) Parasitic wasps as sleuthhounds: response of ichneumonid to the trail of its host. *Netherlands Journal of Zoology* **31**: 762–72.

Kroder, S., Samietz, J., Schneider, D. and Dorn, S. (2007) Adjustment of vibratory signals to ambient temperature in a host-searching parasitoid. *Physiological Entomology* **32**: 105–12.

Kugimiya, S., Shimoda, T., Tabata, J. and Takabayashi, J. (2010) Present or past herbivory: a screening of volatiles released from *Brassica rapa* under caterpillar attacks as attractants for the solitary parasitoid, *Cotesia vestalis*. *Journal of Chemical Ecology* **36**: 620–8.

Landis, D.A., Wratten, S.D. and Gurr, G.M. (2000) Habitat management to conserve natural enemies of arthropod pests in agriculture. *Annual Review of Entomology* **45**: 175–201.

Langellotto, G.A. and Denno, R.F. (2004) Responses of invertebrate natural enemies to complex-structured habitats: a meta-analytical synthesis. *Oecologia* **13**: 1–10.

Leal, W.S., Higushi, H., Mizutani, N., Nakamori, H., Kadosawa, T. and Ono, M. (1995) Multifunctional communication in *Riptortus clavatus* (Heteroptera: Alydidae): conspecific nymphs and egg parasitoid *Ooencyrtus nezarae* use the same adult attractant pheromone as chemical cue. *Journal of Chemical Ecology* **211**: 973–85.

Le Guigo, P., Qu, Y. and Le Corff, J. (2011) Plant-mediated effects on a toxin-sequestering aphid and its endoparasitoids. *Basic and Applied Ecology* **12**: 72–9.

Levin, S.A. (1992) The problem of pattern and scale in ecology. *Ecology* **73**: 1943–67.

Lewis, W.J. and Martin, W.R. (1990) Semiochemicals for use with parasitoids: status and future. *Journal of Chemical Ecology* **16**: 306–9.

Lewis, W.J., Beevers, M., Nordlund, D.A., Gross, H.R., Jr and Hagen, K.S. (1979) Kairomones and their use for management of entomophagous insects. IX. Investigations of various kairomone-treatment patterns for *Trichogramma* spp. *Journal of Chemical Ecology* **5**: 673–80.

Lewis, W.J., Jones, R.L., Nordlund, D.A. and Gross, H.R., Jr (1975a) Kairomones and their use for management of entomophagous insects. II. Mechanisms causing increase in rate of parasitization by *Trichogramma* spp. *Journal of Chemical Ecology* **1**: 349–60.

Lewis, W.J., Jones, R.L., Nordlund, D.A. and Sparks, A.N. (1975b) Kairomones and their use for management of entomophagous insects. I. Evaluation for increasing rates of parasitization by *Trichogramma* spp. in the field. *Journal of Chemical Ecology* **1**: 343–7.

Lewis, W.J., van Lenteren, J.C., Phatak, S.C. and Tumlinson, J.H. (1997) A total systems approach to sustainable pest management. *Proceedings of the National Academy of Sciences USA* **94**: 12243–8.

Lewis, W.L., Vet, L.E.M., Tumlinson, J.H., van Lenteren, J.C. and Papaj, D.R. (1990) Variations in parasitoid foraging behavior: essential element of a sound biological control theory. *Environmental Entomology* **19**: 1183–93.

Lo Giudice, D., Riedel, M., Rostás, M., Peri, E. and Colazza, S. (2011) Host sex discrimination by an egg parasitoid on *Brassica* leaves. *Journal of Chemical Ecology* **37**: 622–8.

Lou, Y.G., Du, M.H., Turlings, T.C., Cheng, J.A. and Shan, W.F. (2005) Exogenous application of jasmonic acid induces volatile emissions in rice and enhances parasitism of *Nilaparvata lugens* eggs by the parasitoid *Anagrus nilaparvatae*. *Journal of Chemical Ecology* **31**: 1985–2002.

Luck, R.F. and Podoler, H. (1985) Competitive-exclusion of *Aphytis lingnanensis* by *Aphytis melinus*: potential role of host size. *Ecology* **66**: 904–13.

Mandour, N.S., Ren, S.X. and Qiu, B.L. (2007) Effect of *Bemisia tabaci* honeydew and its carbohydrates on search time and parasitization of *Encarsia bimaculata*. *Journal of Applied Entomology* **131**: 645–51.

Mattiacci, L. and Dicke, M. (1995) The parasitoid *Cotesia glomerata* (Hymenoptera, Braconidae) discriminates between first and 5th larval instars of its host *Pieris brassicae*, on the basis of contact cues from frass, silk, and herbivore-damaged leaf tissue. *Journal of Insect Behavior* **8**: 485–98.

Mattiacci, L., Vinson, S.B. and Williams, H.J. (1993) A long-range attractant kairomone for egg parasitoid *Trissolcus basalis*, isolated from defensive secretion of its host, *Nezara viridula*. *Journal of Chemical Ecology* **19**: 1167–81.

Mehrnejad, M.R. and Copland, M.J.W. (2006) Behavioral responses of the parasitoid *Psyllaephagus pistaciae* (Hymenoptera: Encyrtidae) to host plant volatiles and honeydew. *Entomological Science* **9**: 31–7.

Meiners, T. and Hilker, M. (1997) Host location in *Oomyzus gallerucae* (Hymenoptera: Eulophidae), an egg parasitoid of the elm leaf beetle *Xanthogaleruca luteola* (Coleoptera: Chrysomelidae). *Oecologia* **112**: 87–93.

Meiners, T. and Hilker, M. (2000) Induction of plant synomones by oviposition of a phytophagous insect. *Journal of Chemical Ecology* **26**: 221–32.

Meiners, T., Hacker, N., Anderson, P. and Hilker, M. (2005) Response of the elm leaf beetle to host plants induced by oviposition and feeding: the infestation rate matters. *Entomologia Experimentalis et Applicata* **115**: 171–7.

Meiners, T., Westerhaus, C. and Hilker, M. (2000) Specificity of chemical cues used by a specialist egg parasitoid during host location. *Entomologia Experimentalis et Applicata* **95**: 151–9.

Metcalf, R.L. (1994) Role of kairomones in integrated pest management. *Phytoparasitica* **22**: 275–9.

Meyhöfer, R., Casas, J. and Dorn, S. (1994) Host location by a parasitoid using leafminer vibrations: characterizing the vibrational signals produced by the leafmining host. *Physiological Entomology* **19**: 349–59.

Micha, S.G. and Wyss, U. (1996) Aphid alarm pheromone (*E*)-β-farnesene: a host finding kairomone for the aphid primary parasitoid *Aphidius uzbekistanicus* (Hymenoptera: Aphidiinae). *Chemoecology* **7**: 132–9.

Millar, J.G., Daane, K.M., McElfresh, J.S., Moreira, J.A., Malakar-Kuenen, R., Guillen, M. and Bentley, W.J. (2002) Development and optimization of methods for using sex pheromone for monitoring the mealybug *Planococcus ficus* (Homoptera: Pseudococcidae) in California vineyards. *Journal of Economic Entomology* **95**: 706–14.

Mills, N.J (2000) Biological control: the need for realistic models and experimental approaches to parasitoid introductions. In: Hochberg, M.E. and Ives, A.R. (eds) *Parasitoid Population Biology*. Princeton University Press, Princeton, NJ, pp. 217–34.

Mills, N.J. and Wajnberg, E. (2008) Optimal foraging behaviour and efficient biological control. In: Wajnberg, E., Bernstein, C. and van Alphen, J.J.M. (eds) *Behavioral Ecology of Insect Parasitoids: From Theoretical Approaches to Field Applications*. Blackwell Science, Oxford, pp. 3–30.

Mizutani, N. (2006) Pheromones of male stink bugs and their attractiveness to their parasitoids. *Japanese Journal of Applied Entomology and Zoology* **50**: 87–99 (in Japanese).

Morehead, S.A. and Feener, D.H., Jr (2000) Visual and chemical cues used in host location and acceptance by a dipteran parasitoid. *Journal of Insect Behavior* **13**: 613–25.

Morrison, L. (1999) Indirect effects of phorid fly parasitoids on the mechanisms of interspecific competition among ants. *Oecologia* **121**: 113–22.

Müller, C. and Riederer, M. (2005) Plant surface properties in chemical ecology. *Journal of Chemical Ecology* **31**: 2621–51.

Murdoch, W.W., Briggs, C.J. and Nisbet, R.M. (1996) Competitive displacement and biological control in parasitoids: a model. *American Naturalist* **148**: 807–26.

Noldus, L.P.J.J., Potting, R.P.J. and Barendregt, H.E. (1991) Moth sex pheromone adsorption to leaf surface: bridge in time for chemical spies. *Physiological Entomology* **16**: 329–44.

Nordlund, D.A. and Lewis, W.J. (1976) Terminology of chemical releasing stimuli in intraspecific and interspecific interactions. *Journal of Chemical Ecology* **2**: 211–20.

Nordlund, D.A., Chalfant, R.B. and Lewis, W.J. (1985) Response of *Trichogramma pretiosum* females to volatile synomones from tomato plants. *Journal of Entomological Science* **20**: 372–6.

Norin, T. (2007) Semiochemicals for insect pest management. *Pure and Applied Chemistry* **79**: 2129–36.

Obonyo, M., Schulthess, F., Le Ru, B., van den Berg, J., Silvain, J.F. and Calatayud, P.A. (2010) Importance of contact chemical cues in host recognition and acceptance by the braconid larval endoparasitoids *Cotesia sesamiae* and *Cotesia flavipes*. *Biological Control* **54**: 270–5.

Ode, P.J. (2006) Plant chemistry and natural enemy fitness: effects on herbivore and natural enemy interactions. *Annual Review of Entomology* **51**: 163–85.

Ode, P.J., Berenbaum, M.R., Zangerl, A.R. and Hardy, I.C.W. (2004) Host plant, host plant chemistry and the polyembryonic parasitoid *Copidosoma sosares*: indirect effects in a tritrophic interaction. *Oikos* **104**: 388–400.

Ozawa, R., Shiojiri, K., Sabelis, M.W., Arimura, G.I., Nishioka, T. and Takabayashi, J. (2004) Corn plants treated with jasmonic acid attract more specialist parasitoids, thereby increasing parasitization of the common armyworm. *Journal of Chemical Ecology* **30**: 1797–808.

Pareja, M., Moraes, M.C.B., Clark, S.J., Birkett, M.A. and Powell, W. (2007) Response of the aphid parasitoid *Aphidius funebris* to volatiles from undamaged and aphid-infested *Centaurea nigra*. *Journal of Chemical Ecology* **33**: 695–710.

Pashalidou, F.G., Huigens, M.E., Dicke, M. and Fatouros, N.E. (2010) The use of oviposition-induced plant cues by *Trichogramma* egg parasitoids. *Ecological Entomology* **35**: 748–53.

Penaflor, M.F.G.V., Erb, M., Miranda, L.A., Werneburg, A.G. and Bento, J.M.S. (2011) Herbivore-induced plant volatiles can serve as host location cues for a generalist and a specialist egg parasitoid. *Journal of Chemical Ecology* **37**: 1304–13.

Peng, J., van Loon, J.J.A., Zheng, S. and Dicke, M. (2011) Herbivore-induced volatiles of cabbage (*Brassica oleracea*) prime defence responses in neighbouring intact plants. *Plant Biology* **13**: 276–84.

Peri, E., Cusumano, A., Agrò, A. and Colazza, S. (2011) Behavioral response of the egg parasitoid *Ooencyrtus telenomicida* to host-related chemical cues in a tritrophic perspective. *BioControl* **56**: 163–71.

Peri, E., Sole, M.A., Wajnberg, E. and Colazza, S. (2006) Effect of host kairomones and oviposition experience on the arrestment behavior of an egg parasitoid. *Journal of Experimental Biology* **209**: 3629–35.

Pickett, J.A. and Poppy, G.M. (2001) Switching on plant genes by external chemical signals. *Trends in Plant Science* **6**: 137–9.

Powell, W. (1999) Parasitoid hosts. In: Hardie, J. and Minks, A.K. (eds) *Pheromones of Non-Lepidopteran Insects Associated with Agricultural Plants*. CABI Publishing, Wallingford, UK, pp. 405–27.

Powell, W. and Pickett, J.A. (2003) Manipulation of parasitoids for aphid pest management: progress and prospects. *Pest Management Science* **59**: 149–55.

Powell, W. and Zhang, Z.L. (1983) The reactions of two cereal aphid parasitoids, *Aphidius uzbekistanicus* and *Aphidius ervi* to host aphids and their food-plants. *Physiological Entomology* **8**: 439–43.

Powell, W., Hardie, J., Hick, A.J., Holler, C., Mann, J., Merritt, L., Nottingham, S.F., Wadhams, L.J., Witthinrich, J. and Wright, A.F. (1993) Responses of the parasitoid *Praon volucre* (Hymenoptera, Braconidae) to aphid sex-pheromone lures in cereal fields in autumn: implications for parasitoid manipulation. *European Journal of Entomology* **90**: 435–8.

Powell, W., Pennacchio, F., Poppy, G.M. and Tremblay, E. (1998) Strategies involved in the location of hosts by the parasitoid *Aphidius ervi* Haliday (Hymenoptera: Braconidae: Aphidiinae). *Biological Control* **11**: 104–12.

Price, P.W., Denno, R.F., Eubanks, M.D., Finke, D.L. and Kaplan, I. (2012) *Insect Ecology: Behaviour, Populations and Communities.* Cambridge University Press, Cambridge.

Puente, M.E., Kennedy, G.G. and Gould, F. (2008) The impact of herbivore-induced plant volatiles on parasitoid foraging success: a general deterministic model. *Journal of Chemical Ecology* **34**: 945–58.

Quicke, D.L.J. (1997) *Parasitic Wasps.* Chapman & Hall, London.

Randlkofer, B., Obermaier, E., Hilker, M. and Meiners, T. (2010) Vegetation complexity: the influence of plant species diversity and plant structures on plant chemical complexity and arthropods. *Basic and Applied Ecology* **11**: 383–95.

Read, D.P., Feeny, P.P. and Root, R.B. (1970) Habitat selection by the aphid parasite *Diaretiella rapae* (Hymenoptera: Braconidae) and hyperparasite *Charips brassicae* (Hymenoptera: Cynipidae). *Canadian Entomologist* **102**: 1567–78.

Reddy, G.V.P. and Guerrero, A. (2004) Interactions of insect pheromones and plant semiochemicals. *Trends in Plant Science* **9**: 1360–85.

Reddy, G.V.P., Holopainen, J.K. and Guerrero, A. (2002) Olfactory responses of *Plutella xylostella* natural enemies to host pheromone, larval frass, and green leaf cabbage volatiles. *Journal of Chemical Ecology* **28**: 131–43.

Riffell, J.A., Abrell, L. and Hildebrand, J.D. (2008) Physical processes and real-time chemical measurement of the insect olfactory environment. *Journal of Chemical Ecology* **34**: 837–52.

Rodriguez-Saona, C., Blaauw, B.R. and Isaacs, R. (2012) Manipulation of natural enemies in agroecosystems: habitat and semiochemicals for sustainable insect pest control. In: Larramendy, M.L. and Soloneski, S. (eds) *Integrated Pest Management and Pest Control: Current and Future Tactics.* InTech, Rijeka, Croatia, pp. 89–126.

Rodriguez-Saona, C., Kaplan, I., Braasch, J., Chinnasamy, D. and Williams, L. (2011) Field responses of predaceous arthropods to methyl salicylate: a meta-analysis and case study in cranberries. *Biological Control* **59**: 294–303.

Rohwer, C.L. and Erwin, J.E. (2008) Horticultural applications of jasmonates: a review. *Journal of Horticultural Science and Biotechnology* **83**: 283–304.

Roland, J., Denford, K.E. and Jiminez, L. (1995) Borneol as an attractant for *Cyzenis albicans*, a tachinid parasitoid of the winter moth, *Operophtera brumata* L (Lepidoptera, Geometridae). *Canadian Entomologist* **127**: 413–21.

Romeis, J. and Shanower, T.G. (1996) Arthropod natural enemies of *Helicoverpa armigera* (Hubner) (Lepidoptera: Noctuidae) in India. *Biocontrol Science and Technology* **6**: 481–508.

Romeis, J., Babendreier, D., Wäckers, F.L. and Shanower, T.G. (2005) Habitat and plant specificity of *Trichogramma* egg parasitoids: underlying mechanisms and implications. *Basic and Applied Ecology* **6**: 215–36.

Romeis, J., Shanower, T.G. and Zebitz, C.P.W. (1997) Volatile plant infochemicals mediate plant preference of *Trichogramma chilonis. Journal of Chemical Ecology* **23**: 2455–65.

Rostás. M. and Wölfling, M. (2009) Caterpillar footprints as host location kairomones for *Cotesia marginiventris*: persistence and chemical nature. *Journal of Chemical Ecology* **35**: 20–7.

Rostás, M., Ruf, D., Zabka, V. and Hildebrandt, U. (2008) Plant surface wax affects parasitoid's response to host footprints. *Naturwissenschaften* **10**: 997–1002.

Ruther, J., Meiners, T. and Steidle, J.L.M. (2002) Rich in phenomena – lacking in terms: a classification of kairomones. *Chemoecology* **12**: 161–7.

Salerno, G., Frati, F., Conti, E., De Pasquale, C., Peri, E. and Colazza, S. (2009) A finely tuned strategy adopted by an egg parasitoid to exploit chemical traces from host adults. *Journal of Experimental Biology* **212**: 1825–31.

Schmitz, O., Krivan, K. and Ovadia, O. (2004) Trophic cascades: the primacy of trait-mediated indirect interactions. *Ecology Letters* **7**: 153–63.

Schröder, D. (1974) A study of the interactions between the internal larval parasites of *Rhyacionia buoliana* [Lepidoptera: Olethreutidae]. *Entomophaga* **19**: 145–71.

Schulthess, F., Chabi-Olaye, A. and Goergen, G. (2001) Seasonal fluctuations of noctuid stemborer egg parasitism in southern Benin with special reference to *Sesamia calamistis* Hampson (Lepidoptera: Noctuidae) and *Telenomus* spp. (Hymenoptera: Scelionidae) on maize. *Biocontrol Science and Technology* **11**: 745–57.

Schuster, D.J. and Starks, K.J. (1974) Response of *Lysiphlebus testaceipes* in an olfactometer to a host and a non-host insect and to plants. *Environmental Entomology* **3**: 1034–5.

Sheehan, W. (1986) Response by specialist and generalist natural enemies to agroecosystem diversification: a selective review. *Environmental Entomology* **15**: 456–61.

Signoretti, A.G.C., Penaflor, M.F.G.V., Moreira, L.S.D., Noronha, N.C. and Bento, J.M.S. (2012) Diurnal and nocturnal herbivore induction on maize elicit different innate response of the fall armyworm parasitoid, *Campoletis flavicincta*. *Journal of Pest Science* **85**: 101–7.

Simpson, M., Gurr, G.M., Simmons, A.T., Wratten, S.D., James, D.G., Leeson, G. and Nicol, H.I. (2011a) Insect attraction to synthetic herbivore-induced plant volatile treated field crops. *Agriculture and Forest Entomology* **13**: 45–57.

Simpson, M., Gurr, G.M., Simmons, A.T., Wratten, S.D., James, D.G., Leeson, G., Nicol, H.I. and Orre-Gordon, G.U.S. (2011b) Field evaluation of the 'attract and reward' biological control approach in vineyards. *Annals of Applied Biology* **159**: 69–78.

Simpson, M., Gurr, G.M., Simmons, A.T., Wratten, S.D., James, D.G., Leeson, G., Nicol, H.I. and Orre-Gordon, G.U.S. (2011c) Attract and reward: combining chemical ecology and habitat manipulation to enhance biological control in field crops. *Journal of Applied Ecology* **48**: 580–90.

Steidle, J.L.M. and Ruther, J. (2000) Chemicals used for host recognition by the granary weevil parasitoid *Lariophagus distinguendus*. *Journal of Chemical Ecology* **26**: 2665–75.

Steidle, J.L.M. and van Loon, J.J.A. (2003) Dietary specialization and infochemical use in carnivorous arthropods: testing a concept. *Entomologia Experimentalis et Applicata* **108**: 133–48.

Suckling, D.M., Gibb, A.R., Burnip, G.M. and Delury, N.C. (2002) Can parasitoid sex pheromones help in insect biocontrol? A case study of codling moth and its parasitoid. *Environmental Entomology* **31**: 947–52.

Teshiba, M., Sugie, H., Tsutsumi, T. and Tabata, J. (2012) A new approach for mealybug management: recruiting an indigenous, but 'non-natural' enemy for biological control using an attractant. *Entomologia Experimentalis et Applicata* **142**: 211–15.

Thaler, J.S. (1999) Jasmonate-inducible plant defenses cause increased parasitism of herbivores. *Nature* **399**: 686–8.

Thies, C. and Tscharntke, T. (1999) Landscape structure and biological control in agroecosystems. *Science* **285**: 893–5.

Titayavan, M. and Altieri, M.A. (1990) Synomone-mediated interactions between the parasitoid *Diaeretiella rapae* and *Brevicoryne brassicae* under field conditions. *Entomophaga* **35**: 499–507.

Turlings, T.C.J. and Ton, J. (2006) Exploiting scents of distress: the prospect of manipulating herbivore-induced plant odours to enhance the control of agricultural pests. *Current Opinion in Plant Biology* **9**: 421–7.

Turlings, T.C.J., Gouinguené, S., Degen, T. and Fritzsche Hoballah, M.E. (2002) The chemical ecology of plant–caterpillar–parasitoid interactions. In: Tscharntke, T. and Hawkins, B.A. (eds) *Multitrophic Level Interactions*. Cambridge University Press, Cambridge, pp. 148–73.

Turlings, T.C.J., Tumlinson, J.H., Eller, F.J. and Lewis, W.J. (1991) Larval-damaged plants: source of volatile synomones that guide the parasitoid *Cotesia marginiventris* to the micro-habitat of its hosts. *Entomologia Experimentalis et Applicata* **58**: 75–82.

Turlings, T.C.J., Wäckers, F.L., Vet, L.E.M., Lewis, W.J. and Tumlinson, J.H. (1993) Learning of host-finding cues by hymenopterous parasitoids. In: Papaj, D.R. and Lewis, A.C. (eds) *Insect Learning: Ecological and Evolutionary Perspectives.* Chapman & Hall, New York, pp. 51–78.

Uefune, M., Choh, Y., Abe, J., Shiojiri, K., Sano, K. and Takabayashi, J. (2012) Application of synthetic herbivore-induced plant volatiles causes increased parasitism of herbivores in the field. *Journal of Applied Entomology* **136**: 561–7.

van Driesche, R.G.V. (2012) The role of biological control in wildlands. *BioControl* **57**: 131–7.

van Driesche, R.G.V. and Bellows, T.S. (1996) Biology of arthropod parasitoids and predators. In: van Driesche, R.G.V. and Bellows, T.S. (eds) *Biological Control.* Chapman & Hall and International Thompson Publishing, New York, pp. 309–35.

van Poecke, R.M.P. and Dicke, M. (2002) Induced parasitoid attraction by *Arabidopsis thaliana*: involvement of the octadecanoid and the salicylic acid pathway. *Journal of Experimental Botany* **53**: 1793–9.

Vet, L.E.M. (2001) Parasitoid searching efficiency links behaviour to population processes. *Applied Entomology and Zoology* **36**: 399–408.

Vet, L.E.M. and Dicke, M. (1992) Ecology of infochemical use by natural enemies in a tritrophic context. *Annual Review of Entomology* **37**: 141–72.

Vet, L.E.M. and Groenewold, A.W. (1990) Semiochemicals and learning in parasitoids. *Journal of Chemical Ecology* **16**: 3119–35.

Vet, L.E.M., Lewis, W.J. and Cardé, R. (1995) Parasitoid foraging and learning. In: Cardé, R. and Bell, W.J. (eds) *Chemical Ecology of Insects.* Chapman & Hall, New York, pp. 65–101.

Vet, L.E.M., Lewis, W.J., Papaj, D.R. and van Lenteren, J.C. (1990) A variable response model for parasitoid foraging behavior. *Journal of Insect Behavior* **3**: 471–90.

Vinson, S.B. (1976) Host selection by insect parasitoids. *Annual Review of Entomology* **21**: 109–33.

Vinson, S.B. (1977) Behavioural chemicals in the augmentation of natural enemies. In: Ridgway, R.L. and Vinson, S.B. (eds) *Biological Control by Augmentation of Natural Enemies.* Plenum, New York, pp. 237–79.

Vinson, S.B. (1998) The general host selection behavior of parasitoid Hymenoptera and a comparison of initial strategies utilized by larvaphagous and oophagous species. *Biological Control* **11**: 79–97.

Vinson, S.B. and Iwantsch, G.F. (1980) Host suitability for insect parasitoids. *Annual Review of Entomology* **25**: 397–419.

von Mérey, G., Veyrat, N., Mahuku, G., Valdez, R.L., Turlings, T.C.J. and D'Alessandro, M. (2011) Dispensing synthetic green leaf volatiles in maize fields increases the release of sesquiterpenes by the plants, but has little effect on the attraction of pest and beneficial insects. *Phytochemistry* **72**: 1838–47.

Wäckers, F.L. (2004) Assessing the suitability of flowering herbs as parasitoid food sources: flower attractiveness and nectar accessibility. *Biological Control* **29**: 307–14.

Wäckers, F.L. (2005) Suitability of (extra-) floral nectar, pollen and honeydew as insect food sources. In: Wäckers, F.L., van Rijn, P.C.J. and Bruin, J. (eds) *Plant-Provided Food for Carnivorous Insects: A Protective Mutualism and its Applications.* Cambridge University Press, Cambridge, pp. 17–74.

Walter, G.H. (2003) *Insect Pest Management and Ecological Research.* Cambridge University Press, Cambridge.

Williams, L., III, Rodriguez-Saona, C., Castle, S.C. and Zhu, S. (2008) EAG-active herbivore induced plant volatiles modify behavioral responses and host attack by an egg parasitoid. *Journal of Chemical Ecology* **34**: 1190–201.

Wiskerke, J.S.C., Dicke, M. and Vet, L.E.M. (1993) Larval parasitoid uses aggregation pheromone of adult hosts in foraging behavior : a solution to the reliability–detectability problem. *Oecologia* **93**: 145–8.

Yu, D.S., Luck, R.F. and Murdoch, W.W. (1990) Competition, resource partitioning and coexistence of an endoparasitoid *Encarsia perniciosi* and an ectoparasitoid *Aphytis melinus* of the California red scale. *Ecological Entomology* **15**: 469–80.

Yu, H., Zhang, Y., Wyckhuys, K.A.G., Wu, K., Gao, X. and Guo, Y. (2010) Electrophysiological and behavioral responses of *Microplitis mediator* (Hymenoptera: Braconidae) to caterpillar-induced volatiles from cotton. *Environmental Entomology* **39**: 600–9.

Zwolfer, H. (1971) The structure and effect of parasite complexes attacking phytophagous host insects. In: den Boer, P.J. and Gradwell, G.R. (eds) *Dynamics of Populations: Proceedings of the Advanced Study Institute on 'Dynamics and Numbers in Populations'*. Centre for Agricultural Publishing and Documentation, Wageningen, The Netherlands, pp. 405–18.

10

The application of chemical cues in arthropod pest management for arable crops

Maria Carolina Blassioli-Moraes, Miguel Borges and Raul Alberto Laumann

Embrapa, Genetic Resources and Biotechnology, Brasília-DF, Brazil

Abstract

Arable crops are the main sources of grains, seeds, oils, fibres and livestock feed. These crops are developed on short temporal and large spatial scales, show low species diversity, and have uniform vegetative and genetic structures. Phenological events, such as germination, flowering and fructification, occur at the same time. Due to these characteristics, pest outbreaks that lower crop yields are frequently recorded. To control pest outbreaks in arable crops, farmers generally use massive quantities of pesticides, which affect all agroecological systems, food safety, the environment (especially the potential biological control agents), and human health. One of the main goals of the agronomic sciences is to minimize or eliminate the use of pesticides and to improve the quality and productivity of the crops. To this end, the application of semiochemicals has been presented as a possible alternative, acting as chemical cues to manipulate the behaviour of parasitoids so as to control pests in biological control programmes. This chapter discusses the use of semiochemicals from plants and insects in arable crops, including cotton, soyabean and maize, mainly as a tool to attract and retain natural enemies. Different viewpoints are presented and discussed, including the characteristics of arable farming systems and trophic interactions mediated by the chemical cues, and also the methodologies for using chemical cues to attract and retain parasitoids in arable crops.

Chemical Ecology of Insect Parasitoids, First Edition. Eric Wajnberg and Stefano Colazza.
© 2013 John Wiley & Sons, Ltd. Published 2013 by John Wiley & Sons, Ltd.

10.1 Arable crops: characteristics of the systems and trophic interactions mediated by chemical cues

In modern times, the substantial development of arable crop farming has mainly been achieved through mechanization and the use of fertilizers, chemical pesticides, highly productive varieties, and improved irrigation. The process usually referred to as the 'green revolution' began in the 1960s and contributed to an increase in food production from 800 million tonnes to more than 2.2 billion tonnes per year worldwide between 1961 and 2000 (FAO 2011). This green revolution aimed at developing an agricultural industry that could quickly provide large amounts of food for human and animal consumption: the cultivation of arable crops expanded all around the world, and food supplies increased thereafter. However, arable crops are artificial ecosystems with specific characteristics: plant uniformity (monocultures), with phenological events (germination, flowering and fruiting) occurring at the same time, the necessity for inputs (fertilizers), and discontinuities over time because most agricultural plants are annual or semi-perennial; in addition, the cultivated areas possess low biodiversity (in terms of plants, animals and other organisms). One of the most critical issues in agroecosystems is the lack of diversity; for instance only eight plant species (sugarcane, maize, wheat, rice, potatoes, sugarbeet, beans and soyabean) are cultivated in 28.7% of the arable land and are responsible for 45.6% of the total agricultural production (FAO 2011).

Due to these characteristics, arable crop systems have proved to be unsustainable and are highly vulnerable, sometimes encouraging the establishment and outbreak of pests and diseases. In recent years, both scientists and members of the public have demanded a more sustainable economic agriculture system. The approach proposed for pest management is one in which pests, diseases and weeds are integrated, using different practices such as the use of the existing biodiversity (in other words, biological control) and selecting low-risk pesticides (FAO 2011).

Within this context, parasitoids are important tools that can be incorporated into pest management using different approaches, such as through their gradual and augmentative release or by manipulating their behaviour using semiochemicals. The foraging behaviour of parasitoids includes several sequential steps: habitat location, host selection, host recognition, host suitability and oviposition (Godfray 1994). During these steps, parasitoids use physical, biochemical and, in particular, semiochemical cues (Vinson 1985, Godfray 1994). Thus, the identification of the cues used by parasitoids in each step could be relevant in selecting and obtaining semiochemicals to be used for manipulating their behaviour. Because the success of natural enemies for controlling pests in arable crops depends on two important factors: (i) attracting the natural enemy in synchronization with the pest population; and (ii) retaining them in the field (Vinson 1984, 1985), studies of potentially useful semiochemicals need to address one or both of these requirements for parasitoid manipulation.

In this chapter, we review current information about the application of chemical cues in arthropod pest management for arable crops, focusing on the cues with a greater potential for use in parasitoid behavioural manipulation. In addition, the main methodologies and techniques developed to use semiochemicals under field conditions, the interaction of semiochemicals with other pest management tactics, and the perspectives and main knowledge gaps with regard to the use of semiochemicals in agricultural systems are discussed.

10.2 Methodologies for using chemical cues to attract and retain parasitoids in arable crops

The potential for natural enemies to manage and control pests is so high and of such importance that researchers have invested great effort into developing methodologies to improve their efficiency. Briefly, a modern biological control method considers two main ways of using natural enemies for pest control:

1 classical biological control, which involves importing exotic enemies in order to control exotic or native pests;
2 the conservation and/or augmentation of the natural enemies that are already present in a particular habitat (De Bach 1964, Barbosa & Benrey 1998).

In the latter strategy, two main approaches could be addressed:

1 conservative biological control, which involves developing cultural practices to enhance the effect of the natural enemies present in cultivated or natural neighbouring areas (Ehler 1998);
2 inundative or augmentative release, which involves increasing the populations of natural enemies by mass rearing and releasing them during critical periods (Barbosa & Benrey 1998).

Since the first discovery of semiochemicals used by natural enemies in the early 1970s, the idea of using these compounds in combination with natural enemies has been investigated. However, after almost 40 years, the application of semiochemicals to manage natural enemies in the field has not been used, in spite of some proposals and research efforts to stimulate their use in pest management (Lewis & Martin 1990, Vet & Dicke 1992, Powell & Pickett 2003). Both the conservative and augmentative biological control techniques offer opportunities to use semiochemicals to improve/aid the efficiency of management practices.

In arable crops, which are usually annual and unstable, the pest populations develop before or at the same time as their natural enemy populations (Ehler & Miller 1978, Borges *et al.* 1998a, 1998b). Occasionally, this situation could even be the cause of failure in biological control programmes. Semiochemicals could be applied in the field to recruit parasitoids, aiming to bring them into synchrony with the pest colonization of the crop. Studies conducted in Brazil to evaluate the population dynamics of the neotropical brown stink bug *Euschistus heros* and its main natural enemies (*Telenomus podisi* and *Trissolcus basalis*) on soyabean crops showed, for example, that these egg parasitoids arrive in the crop fields later than the pests, the population of which has remained high since the beginning of the experiment (Fig. 10.1). In this case, semiochemicals could be used to increase the recruitment of parasitoids just before the pest population arrives in the field.

In addition, semiochemicals related to the location, recognition and acceptance behaviours of the host could be used to achieve goals other than parasitoid attraction. For example, a semiochemical with an arrestment effect on a parasitoid could be applied in the field to retain the parasitoids. Also, semiochemicals involved in the recognition or acceptance of the host could be applied for a more efficient use of alternative hosts in mass

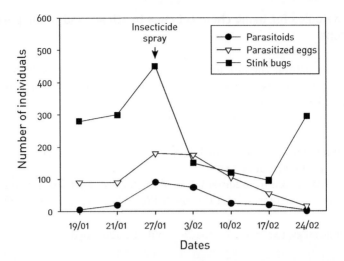

Figure 10.1 Population trends of stink bugs and egg parasitoids in soyabean fields in central Brazil (year 1994), showing that the population of stink bugs developed before the egg parasitoid population (*Telenomus podisi* Ashmead and *Trissolcus basalis* Wollaston (Platygastridae)) (these data corroborate results presented in Borges *et al.* 1998a, 1998b, 1999).

rearing systems or to enable the use of alternative hosts through conditioning or associative learning.

10.2.1 Direct application of semiochemicals

Two strategies are being developed to apply semiochemicals in the field. The first uses synthetic compounds, either individually or in a mixture, and generally uses slow-release dispensers to act directly on parasitoid behaviour, whereas the second uses the chemical application of synthetic compounds or phytohormones as sprays to induce the plant's chemical defence, which also functions in the attraction of natural enemies (Moraes *et al.* 2009, Simpson *et al.* 2011a; see also Ode, Chapter 2, this volume). In addition, the plants not attacked and not treated with chemical compounds may respond to the chemical cues emitted by damaged or treated neighbouring plants that produce and release compounds which also induce a priming response in the plants. Primed plants respond more quickly or strongly when subjected to some type of injury or treatment (Engelberth *et al.* 2004, Arimura *et al.* 2005). Furthermore, the defence induced in neighbouring plants could make them less available for the invasion of new herbivores by acting directly on the herbivores or by emitting volatiles that attract parasitoids (Baldwin *et al.* 2006, Yan & Wang 2006).

A review of current information about the semiochemicals used in field applications for parasitoid behavioural management in arable crops is presented in Table 10.1. Most of the studies using semiochemicals are being conducted with herbivore-induced plant volatiles (HIPVs), and less attention has been paid to the semiochemicals from hosts or host-related compounds. This could be due to some of the characteristics of HIPVs in comparison with host pheromones or host-related compounds. Indeed, HIPVs have a broad-spectrum action due to their low specificity, and some HIPVs have a double action; for instance, directly attracting parasitoids and exerting an effect on neighbouring plants through plant–plant communication.

Much of the work developed in arable crops has concentrated on testing the effects of semiochemicals on parasitoids in general and has usually concentrated on Hymenoptera parasitoids, with some studies specifically centred on a particular group, for example, Platygastridae (Borges & Aldrich 1994a, Peres 2004, Laumann *et al.* 2007, Vieira *et al.* 2010) (Table 10.1). Less attention has been paid to Tachinidae or other dipteran parasitoids, despite several works showing the response of these parasitoids to pheromones of their hosts in pheromone-baited traps but not in arable crops (Aldrich *et al.* 1991, 1995, 2007, Aldrich 1995, Aldrich & Zhang 2002, Higaki & Adachi 2011, Jang *et al.* 2011) and to volatiles from plants damaged by their herbivorous hosts in laboratory bioassays (Mondor & Roland 1997, Stireman 2002, Hanyu *et al.* 2011, Ichiki *et al.* 2011, 2012; see Nakamura, Ichiki & Kainoh, Chapter 7, this volume).

The experimental research has mainly focused on testing the effects of semiochemical application on the attraction of parasitoids. Only recently have studies been oriented towards testing the effects of semiochemicals on the intensity of parasitism and the consequences for herbivore abundance and density and for plant damage and productivity (Glinwood *et al.* 1998, James 2003a, 2003b, 2005, Peres 2004, Lou *et al.* 2005a, 2005b, Laumann *et al.* 2007, Vieira *et al.* 2010, Alim & Lim 2011, Simpson *et al.* 2011b, von Mérey *et al.* 2011, 2012). Moreover, few studies have tested whether the application of plant inductors actually stimulates the release of HIPVs under field conditions (Lou *et al.* 2005b, von Mérey *et al.* 2011, 2012). Some of the inductors used in laboratory and field experiments are methyl jasmonate, methyl salicylate, *cis*-jasmone and (Z)-3-hexenyl acetate (Birkett *et al.* 2000, Lou *et al.* 2005b, Moraes *et al.* 2008a, 2008b, Williams *et al.* 2008, Yu *et al.* 2008, Wang *et al.* 2011).

In Brazil and in the USA, for example, phytophagous stink bugs, mainly the brown stink bug *Euschistus* spp. and the green stink bugs *Nezara viridula*, *Acrosternum* spp., *Dichelops melacanthus* and *Piezodorus guildinii*, are among the main pests of soyabean crops, and egg parasitoids are their main natural enemies, particularly *Telenomus podisi* and *Trissolcus basalis* (Borges *et al.* 1999, 2003, Sujii *et al.* 2002, Laumann *et al.* 2008, Smith *et al.* 2009). To improve the efficiency of the augmentative release of egg parasitoids, as proposed by Corrêa-Ferreira (2002), or to develop conservative biological control programmes, the behavioural manipulation of parasitoids using semiochemicals could be a practical solution to attract and retain them in the field (Borges & Aldrich 1994a).

Field experiments using slow-release dispensers (rubber septa) impregnated with (E)-2-hexenal, a compound present in the metathoracic glands of *Euschitus heros* (Borges & Aldrich 1992, 1994b, Moraes *et al.* 2008c), showed an increased number of parasitoids and increased parasitism of their host eggs laid naturally by caged stink bug females in the treated areas in a week-long experiment (Peres 2004). Similar results were reported by Laumann *et al.* (2007) in soyabean field experiments using rubber septa impregnated with 5.0 mg of (E)-2-hexenal, as shown in Figure 10.2. However, a full crop season in a more extensive experimental area (soyabean plots of 400 m²), with applications of four septa per plot impregnated with (E)-2-hexenal at two different doses (4.0 mg and 10.0 mg), showed that the abundance of parasitoids did not differ from the control plots without the compound. In the (E)-2-hexenal-treated plots, however, the recruitment of parasitoids started at an early phenological stage of the crop (Vieira *et al.* 2010).

Table 10.1 The semiochemicals tested in arable crops for parasitoid management, the application method, and the main effects on the plants, herbivore population and parasitoids. Only the significant effects reported by the authors are considered. Some of the studies also include reported effects on predators, which are not included in this table. The table includes only those studies intended to test the practical use of semiochemicals under field conditions. Parasitoid identification is at the taxonomic level, as reported in the original works. HIPVs = herbivore-induced plant volatiles, GLV = green leaf volatiles.

Crop	Semiochemical		Application	Effect	Reference
	Chemicals	Origin/ function			
Cereals (wheat and barley)	4aS,7S,7aR Nepetalactone	Aphid sex pheromone	Lures with synthetic compounds	Attraction of *Praon volucre* [Haliday], *P. dorsale* [Haliday] and *P. abjectum* [Haliday] [Braconidae]	Powell *et al.* [1993]
Cotton	3,7, Dichlorometil,1,3,6 octadiene, (Z)-3-hexenyl acetate, nonanal, octanal, decanal, (Z)-3-hexen-1-ol, methyl salicylate, nonanal + (Z)-3-hexen-1-ol	HIPVs	Glass vial dispensers	No effects on the parasitoid *Campoletis chloridae* [Uchida] [Ichneumonidae]. *Macrocentrus lineares* [Nees] [Braconidae] was attracted to (Z)-3-hexen-1-ol and 3,7-dimethyl-1,3,6, octatriene	Yu *et al.* [2008]
	(Z)-3-Hexenyl acetate, methyl salicylate, α-farnesene	HIPVs	Glass vial dispensers	Parasitism was greater in the areas treated with α-farnesene and (Z)-3-hexenyl acetate	Williams *et al.* [2008]

Crop	Category	Compounds	Application	Effect	Reference
Maize	HIPVs	Not identified	Induction: scratching leaf + S. littoralis larvae regurgitant	Increased abundance of Hymenoptera parasitoids. Increased parasitism	Bernasconi et al. (2001)
	HIPVS/inductor Pathogen resistance/inductor	Methyl jasmonate, benzo-[1,2,3]-thiadiazole-7-carbothioic acid S-methyl ester (BTH)	Plant spraying	Methyl jasmonate increased the production of HIPVs (sesquiterpenes) and had a marginal positive effect on the Spodoptera frugiperda parasitoids (Ichneumnidae, Braconidae, Diptera: Tachinidae). No effect on herbivore infestation	von Mérey et al. (2012)
	HIPVs (GLV)	(Z)-3-Hexenal, (Z)-3-hexenol, (Z)-3-hexenyl acetate, (E)-2-hexenal	Glass vial dispensers	Increased HIPV (sesquiterpenes) emission. No effect on the parasitoids of S. frugiperda. Herbivores: higher frequency and damage on treated plants	von Mérey et al. (2011)
	HIPVs/inductors	Methyl salicylate, methyl anthranilate methyl jasmonate, benzaldehyde,(Z)-3-hexenyl acetate, (Z)-3-hexen-1-ol	Plant spraying	Increased abundance of total parasitoids using all of the compounds; Encyrtidae (methyl anthranilate) and Braconidae ((Z)-3-hexenyl acetate)	Simpson et al. (2011a)
	HIPVs/inductors	Methyl salicylate, methyl jasmonate, methyl anthranilate. Mixture (3 compounds)	Plant spraying + Fagopyrum esculentum as plant reward	Increased abundance of total parasitoids and Eulophidae (methyl salicylate, methyl jasmonate and mixture). Increased abundance of Encyrtidae (methyl jasmonate). Reduction of Helicoverpa spp. abundance and damage to cobs	Simpson et al. (2011b)

(cont'd)

Table 10.1 (cont'd)

Crop	Semiochemical			Effect	Reference
	Chemicals	**Origin/ function**	**Application**		
Rice	Jasmonic acid	HIPVs/ inductor	Plant spraying	Increased parasitism of the eggs of *Nilaparvata lugens* by *Anagrus nilaparvatae* Pang et Wang (Mymaridae) under greenhouse and field conditions	Lou *et al.* (2005b)
Soyabean	Methyl-2,6,10-trimethyltridecanoate	*E. heros* sex pheromone	Baited septa	Attraction of *Telenomus podisi, Trissolcus teretis, Trissolcus urichi* and *Trissolcus brachymenae* in field experiments	Borges *et al.* (1998a)
	(*E*)-2-Hexenal	GLV and defensive secretions of stink bugs	Slow-release dispensers (rubber septa)	Increased attraction of *Trissolcus* sp. and *Telenomus* sp. and increased parasitism of the natural and sentinel eggs of stink bugs	Peres (2004), Laumann *et al.* (2007), Vieira *et al.* (2010)
	(*E*)-2-Hexenyl-(*Z*)-3-hexanoate	*Riptortus clavatus* aggregation pheromones	Plastic pellets impregnated with pheromone	Early attraction of *Ooencyrtus nezarae* Ishii (Encyrtidae) and increased parasitism of sentinel eggs	Mizutani (2006)

Crop	Compound	Category	Application	Effect	Reference
	cis-Jasmone	inductor	Plant spraying	Increased abundance of Platygastridae. No effects on the parasitism intensity or the stink bug populations	Vieira et al. (2010)
	(E)-2-Hexenyl-(E)-2-hexenaoate, (E)-2-hexenyl-(Z)-3-hexenoate, tetradecyl isobutyrate	Aggregation pheromone of Riptortus pedestris	Pheromone lure + (frozen) host (eggs)	Attraction of Ooencyrtus nezarae (Encyritidae) and increased egg parasitism. Reduction of stink bug damage on pods	Alim & Lim (2011)
Wheat	cis-Jasmone	HIPVs/inductor	Plant spraying	Reduction of Sitobium avenae infestations. No reported effect on parasitoids	Bruce et al. (2003)
	4aS,7S,7aR Nepetalactone	Aphid pheromone	Glass vial dispensers	Increased parasitism by Praon volucre (Braconidae)	Glinwood et al. (1998)
	Mixture of sucatol:sulcatone and plant extract	HIPVs	Plant spraying	Increased parasitism of Sitobium avenae by Aphidius avenae (Haliday) (Braconidae)	Yong et al. (2009)
	Intercropped with oilseed rape and/or slow release of methyl salicylate	HIPVs	Slow-release dispensers	Plots intercropped with oilseed rape and methyl salicylate induced a higher parasitism rate of Sitobium avenae	Wang et al. (2011)

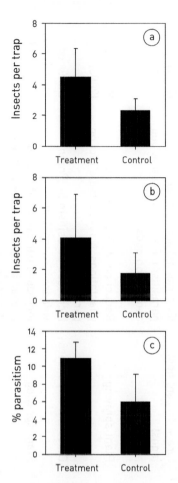

Figure 10.2 (a) Mean number (+SD) of Platygastridae parasitoids found in yellow sticky traps baited with (*Treatment*) or without (*Control*) (*E*)-2-hexenal in an experiment conducted in 2006; (b) similar data for an experiment conducted in 2007; (c) mean percentage (+SD) of *Euschistus heros* eggs, glued onto cardboard strips, parasitized by Platygastridae wasps when placed near rubber septa with 5 mg of (*E*)-2-hexenal (*Treatment*) or with *n*-hexane (*Control*) in experiments performed in 2006 in soyabean fields in central Brazil. (ANOVA, $p < 0.05$, for the number of insects on sticky traps; Mann–Whitney test, $p < 0.005$, for the percentage of parasitism).

The use of *cis*-jasmone, a phytohormone that induces indirect defences in wheat, cotton and soyabean (Birkett *et al.* 2000, Bruce *et al.* 2003, 2008, Moraes *et al.* 2008b, 2009, Hegde *et al.* 2012), has been tested using different systems (Bruce *et al.* 2003, Vieira *et al.* 2010). In field experiments in which soyabean plants in small plots (2.0 m²) were sprayed with a solution of *cis*-jasmone (6 ml of a solution of 0.25 g *cis*-jasmone + 0.1 g of Tween 20 in 1.0 litre of water), the abundance of stink bug egg parasitoids (Platygastridae) was higher than in control untreated plots during the first 3 weeks of the experiments (Fig. 10.3) (Vieira *et al.* 2010).

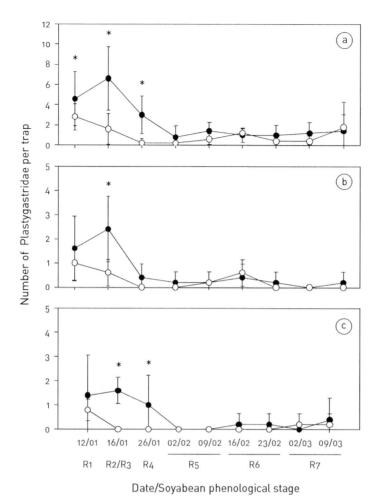

Figure 10.3 Mean (±SD) number of egg parasitoids captured in sticky traps in soya-bean plots in which the plants were treated with *cis*-jasmone (*black circles*) or not (control, *white circles*). *Stars* indicate significant differences in the mean number of egg parasitoids between the *cis*-jasmone and control plots for each survey date (GLM; $p < 0.05$). (a) Total Plastygastridae; (b) *Telenomus* spp.; (c) *Trissolcus* spp.

Similar observations were obtained in experiments using semiochemicals on maize. In Australia, Simpson *et al.* (2011a) applied synthetic HIPVs (methyl anthranilate, benzalde-hyde, (Z)-3-hexenyl acetate, (Z)-3-hexen-1-ol, methyl salicylate and methyl jasmonate) to plants in a 10 ha experimental field. Considering the total abundance of parasitoids, all of the compounds tested were efficient in attracting Braconidae species, whereas Encyrtidae were attracted only to the area treated with methyl anthranilate. Working in maize fields in Mexico, von Mérey *et al.* (2011) used a specific slow-release dispenser with green leaf volatiles (GLVs) ((Z)-3-hexenal, (Z)-3-hexenol, (Z)-3-hexenyl acetate and (E)-2-hexenal) inside an amber glass vial filled with fibreglass material. In three fields, both of the dispens-ers containing the GLVs (treated plants, $n = 12$) or not (untreated plants, control, $n = 12$)

were placed near the plants (<10 cm). The maize plants exposed to the GLV mixture showed an increased release of sesquiterpenes, but the parasitism rates on *S. frugiperda* larvae were not affected by the treatments when measured 10–13 days after the treatments. In cotton fields (1.5 ha) in China, Yu *et al.* (2008) tested a series of compounds (3,7-dimethyl-1,3,6-octatriene, (*Z*)-3-hexenyl acetate, nonanal, octanal, decanal, (*Z*)-3-hexen-1-ol, methyl salicylate, and a mixture of nonanal and (*Z*)-3-hexen-1-ol) for the attraction of beneficial insects. The compounds were applied using slow-release dispensers, which consisted of glass vials containing 1 ml of a solution of 10 mg ml⁻¹ of each compound in lanolin, and the insects were monitored using sticky traps. The vials were placed at various distances (5, 10, 20 and 30 m) from the plants, and were replaced weekly. Eleven different species of beneficial insects were trapped on the sticky cards, and two parasitoids were identified: a Braconidae, *Macrocentrus lineares* Nees (Hymenoptera Braconidae), which was significantly attracted to the areas baited with (*Z*)-3-hexen-1-ol and 3,7-dimethyl-1,3,6-octatriene; and an Ichneumonidae, *Campoletis chloridae* Uchida, which was not affected by the treatments. Another experiment in maize fields tested the effect of spraying methyl jasmonate and benzo-(1,2,3)-thiadiazole-7-carbothioic acid *S*-methyl ester (BTH) on experimental plots consisting of four 4-m-long rows in experimental areas of approximately 160 m² (von Mérey *et al.* 2012). Seven days after spraying, surveys of the herbivorous insects (*S. frugiperda* larvae), their parasitism rates and the plant damage were conducted. The elicitors did not affect the infestation of the maize seedlings by the pest, other herbivores or pathogens. Furthermore, the treatments did not show consistent effects on the parasitism rates. Considering all of the fields tested, there was a marginal effect of BTH on parasitism rates and, in one replicate, the larval parasitism by *Chelonus insularis* Cresson (Hymenoptera: Braconidae) was higher for the BTH-treated plants and showed a marginal increase in the methyl jasmonate-treated plants (von Mérey *et al.* 2012).

10.2.2 Environmental manipulation

The manipulation of the environment as a method for managing natural enemies has been extensively examined and discussed (Landis *et al.* 2000, Gurr *et al.* 2004, Jonsson *et al.* 2010; see also Simpson, Read & Gurr, Chapter 12, this volume). Here we discuss the interaction and integration of semiochemicals with plants in order to attract and retain natural enemies or to attract them using an additional food supply.

One strategy proposes using plants as a food reward for natural enemies (Khan *et al.* 2008, Simpson *et al.* 2011b). This 'attract and reward' strategy is based on the idea that most of the semiochemicals that attract natural enemies are inadequate for retaining them in the treated areas because these compounds might not be the only cue used by the natural enemies for host/prey location. Furthermore, the response to a semiochemical without a satisfactory oviposition site or attack on a host/prey could result in a weakened response. Therefore, the integration of semiochemical application with food resources could help to improve the response of the natural enemies. The presence of a food supply could indirectly improve the action of the natural enemies, positively influencing their longevity, fecundity, locomotion and host-searching abilities (Barbosa & Benrey 1998, Winkler *et al.* 2009). Simpson *et al.* (2011b) showed the potential of this strategy in maize using several HIPV formulations as the attractants and buckwheat (*Fagopyrum esculentum* (Moench) (Caryophyllales: Polygonaceae)) as the reward. Eulophidae were more numerous in the plots treated with the attractants methyl anthranilate, methyl jasmonate, methyl salicylate and an HIPV mix, whereas Encyrtidae were more abundant near the methyl jasmonate-treated plants. In addition, the presence of plants acting as a source of nectar increased the abundance of both parasitoids and predators in maize plots. The recruitment and retention of parasitoids in the 'attract and reward' plots significantly reduced the density of the herbivore *Helicoverpa* spp., and plant damage was lower in the plots with rewards.

Another strategy proposes the use of plant semiochemicals in 'push–pull' systems. This strategy has a wide spectrum of action and is not centred exclusively on the behavioural manipulation of natural enemies. Push–pull strategies involve the behavioural manipulation of pests and natural enemies using different semiochemical stimuli. Several push–pull systems have been proposed (Cook *et al.* 2007, Khan *et al.* 2011) but the only one currently in practice by farmers was developed in Africa for maize intercropped with a plant that is repellent to stemborers, *Desmodium uncinatum* (Jacq.) (Leguminosa: Fabaceae), and an attractive plant, Napier grass, *Pennisetum purpureum* Schumach (Poales: Poaceae). The latter is cultivated as a trap crop around the maize field, while *Desmodium* is cultivated between the maize plants. The volatile semiochemicals emitted by the *Desmodium* plants act by repelling the gravid female moths, which are attracted by the volatiles emitted by the Napier grass. There was a higher predation rate of eggs and larvae by natural enemies compared with monoculture maize plots (Khan *et al.* 2001, 2011, Midega *et al.* 2006). The studies did not identify whether the natural enemies were attracted by the plant volatiles or pest cues, but the natural enemies were more abundant and effective in the polyculture systems than in monocultures. All push–pull systems are based on the use of different plants (traps and repellents) in the same area.

10.3 Final considerations

The information presented in this chapter shows that the use of chemical cues for parasitoid behaviour manipulation may be an excellent tool for improving parasitoid efficiency in the biological control of arable crops. Various aspects should be considered to improve the use of semiochemicals in arable crops. Among them, the selection of the compounds (individual or mixtures) and the dosages (Kaplan 2012), time and form of application (dispenser, sprays, plants emitting semiochemicals) and other technical aspects need to be better established. All of these considerations probably require specific technological development. For field applications, new dispensers for slow-release chemicals and methodologies for the formulations for sprays could help to improve the efficiency of applications.

The large-scale production of semiochemicals demanded by the extensive use of land by arable crops needs specific solutions from the chemical industry. An interesting prospect could be the use of natural products from plants, for example, as a source of semiochemicals or as a starter material for chemical synthesis (Birkett & Pickett 2003). In addition, new approaches, such as classical breeding or the genetic engineering of plants to emit volatiles that could affect both herbivores and parasitoids, could be used to manipulate the emission of semiochemicals (Poppy & Sutherland 2004, Beale *et al.* 2006, Schnee *et al.* 2006, Kunert *et al.* 2010, Michereff *et al.* 2011). Although there is no example in the literature of transgenically transformed arable crops for such a purpose, *Arabidopsis thaliana* (L.) Heynh. (Brasssicales: Brassicaceae) plants that overexpress a terpene synthase, TPS10, identified from maize plants, released a profile of volatile terpenes that was similar to that of maize plants. Using olfactometer assays, these transformed plants attracted the parasitoid *Cotesia marginiventris* (Schnee *et al.* 2006). One of the major questions that arises when modified plants continuously release volatiles that, in a natural context, are released only when the plant is disturbed is whether the effects of the volatiles on the natural enemies will be the same (Kunert *et al.* 2010).

Another point to establish is whether the parasitoid response to semiochemicals is innate or acquired by associative learning, and therefore whether such a response could change over time (Kaplan 2012). If semiochemicals are applied indiscriminately and not coupled with the presence of the host, the parasitoids could rapidly adapt to this situation, for example by avoiding the chemical cue. This effect could be reduced by controlling the dosage of the semiochemical application, the extent of the area, and the association of the semiochemical application with the presence of the host in the field (coupling signals with rewards) (Kaplan 2012). However, the existence of associative learning could be useful for augmentative biological control because, in such a case, the natural enemies should be trained to respond to specific semiochemicals before being released in the field. Chemical cues are relevant for the searching behaviour of parasitoids, but the importance of other cues needs to be considered as well. Some studies have discussed the integration of chemical with visual, mechanical or physical cues (Godfray 1994, Wäckers & Lewis 1994, Ichiki *et al.* 2011, Laumann *et al.* 2011). The use of multimodal cues for the behavioural manipulation of parasitoids could improve the recruitment and retention of the natural enemies in the crops and needs to be considered in the future.

Finally, for the long-term use of semiochemicals with parasitoids on arable crops, some ecological aspects need to be considered. First, the effect of the semiochemicals on the communities needs to be addressed to determine any negative or positive effects on the other components of the system (Dicke 2009). For example, the HIPVs from plants could benefit herbivores (Bolter *et al.* 1997, Kalberer *et al.* 2001, Pope *et al.* 2007, von Mérey *et al.* 2011, 2012) or parasitic plants (Ruyon *et al.* 2006). The HIPVs can also have a repellent or deterrent action or no effect on herbivores (De Moraes *et al.* 2001, Kessler & Baldwin 2001, Bruce *et al.* 2003, Vieira *et al.* 2010, Simpson *et al.* 2011b, von Mérey *et al.* 2012). HIPVs could also affect mutualism (e.g. pollinators) either positively or negatively (Kessler & Halitschke 2007). These different effects of semiochemicals can have various impacts on the food chain and community structure (Kessler & Halitschke 2007, Dicke 2009), and this needs to be evaluated before going ahead with the broad use of semiochemicals for pest management.

The landscape's influence on the natural enemies must also be taken into account. The landscape characteristics of both cultivated and natural areas could affect the composition of the ecosystems and the habitat diversity (Turner 1989) and may influence trophic interactions (Gardiner *et al.* 2009). In general, studies suggest that polyculture systems provide a higher biodiversity of plants, natural enemies and microorganisms to an ecosystem, thus enhancing the natural control of pests (Andow 1991, Gurr *et al.* 2004, Hole *et al.* 2005, Jonsson *et al.* 2010). However, one question that needs to be addressed is whether the natural enemies attracted within a natural context are present in the correct numbers or density to provide efficient pest control in arable crops under monoculture and polyculture systems. Another question examines how the size of the cultivated area could affect the use of semiochemicals and both the immigration and emigration rates (i.e. the turnover of species) of the parasitoids (Kaplan 2012). Most of the main arable crops are established in large areas, thus reducing the perimeter-to-area ratios, with obvious effects on edge processes and the reduction of individual exchange.

The above considerations show that the use of semiochemicals in arable crops needs to be linked with the reformulation of some agronomic practices, including a reorganization of the landscape with the restoration/maintenance of native areas, and the incorporation of new production technologies. Such efforts could help to create a new paradigm based on an ecosystem approach to agriculture (FAO 2011). An increase in the human population to 9 billion people is expected by 2050, and the new paradigm could be decisive for the sustainable production of food while conserving natural resources and reducing negative impacts on the environment in order to ensure the well-being of future generations.

Acknowledgements

Eric Wajnberg is thanked for drawing the figures.

References

Aldrich, J.R. (1995) Testing the 'new associations' biological control concept with a tachinid parasitoid (*Euclytia flava*). *Journal of Chemical Ecology* **21**: 1031–42.

Aldrich, J.R. and Zhang, A. (2002) Kairomone strains of *Euclytia flava* (Townsend), a parasitoid of stink bugs. *Journal of Chemical Ecology* **28**: 1565–82.

Aldrich, J.R., Hoffmann, M.P., Kochansky, J.P., Lusby, W.R., Eger, J.E. and Payne, J.A. (1991) Identification and attractiveness of a major pheromone component for Nearctic *Euschistus* spp. stink bugs (Heteroptera: Pentatomidae). *Environmental. Entomology* **20**: 477–83.

Aldrich, J.R., Khrimian, A. and Camp, M.J. (2007) Methyl 2,4,6-decatrienoates attract stink bugs (Hemiptera: Heteroptera: Pentatomidae) and tachinid parasitoids. *Journal of Chemical Ecology* **33**: 801–15.

Aldrich, J.R., Rosi, M.C. and Bin, F. (1995) Behavioral correlates for minor volatile compounds from stink bugs (Heteroptera: Pentatomidae). *Journal of Chemical Ecology* **21**: 1907–20.

Alim, M.A. and Lim, U.T. (2011) Refrigerated eggs of *Riptortus pedestris* (Hemiptera: Alydidae) added to aggregation pheromone traps increase field parasitism in soybean. *Journal of Economic Entomology* **104**: 1833–9.

Andow, D.A. (1991) Vegetational diversity and arthropod population response. *Annual Review of Entomology* **36**: 561–86.

Arimura, G., Kost, C. and Boland, W. (2005) Herbivore-induced, indirect plant defences. *Biochimica et Biophysica Acta* **1734**: 91–111.

Baldwin, I.T., Halitschke, R., Paschold, A., von Dahl, C.C. and Preston, C.A. (2006) Volatile signaling in plant–plant interactions: 'talking trees' in the genomics era. *Science* **311**: 812–15.

Barbosa, P. and Benrey, B. (1998) The influence of plants on insect parasitoids: implications for conservation biological control. In: Barbosa, P. (ed). *Conservation Biological Control*. Academic Press, San Diego, CA, pp. 55–82.

Beale, M.H., Birkett, M.A., Brice, T.J.A., Chamberlain, K., Field, L.M., Huttly, A.K., Martin, J.L., Parker, R., Phillips, A.L., Pickett, J.A., Prosser, I.A., Shewry, P.R., Smart, L.E., Wadhams, L.J., Woodcock, C.M. and Zhang, Y. (2006) Aphid alarm pheromone produced by transgenic plants affects aphid and parasitoid behavior. *Proceedings of the National Academy of Sciences USA* **103**: 10509–13.

Bernasconi, M.L.O., Turlings, T.C.J., Edwards, P.J., Fritzsche-Hoballah, M.E., Ambrosetti, L., Bassetti, P. and Dorn, S. (2001) Response of natural populations of predators and parasitoids to artificially induced volatile emissions in maize plants (*Zea mays* L.). *Agricultural and Forest Entomology* **3**: 1–10.

Birkett, M.A. and Pickett, J.A. (2003) Aphid sex pheromones: from discovery to commercial production. *Phytochemistry* **62**: 651–6.

Birkett, M.A., Campbell, C.A.M., Chamberlain, K., Guerrieri, E., Hick, A.J., Martin, J.L., Matthes, M., Napier, J., Pettersson, J., Pickett, J.A., Poppy, G.M., Pow, E.M., Pye, B.J., Smart, L.E., Wadhams, G.E., Wadhams, L.J. and Woodcock, C.M. (2000) New roles for *cis*-jasmone as an insect semiochemical and in plant defence against insects. *Proceedings of the National Academy of Sciences USA* **97**: 9329–34.

Bolter, C.J., Dicke, M., van Loon, J.J.A., Visser, J.H. and Posthumus, M.A. (1997) Attraction of Colorado potato beetle to herbivore-damaged plants during herbivory and after its termination. *Journal of Chemical Ecology* **23**: 1003–20.

Borges, M. and Aldrich, J.R. (1992) Instar-specific defensive secretions of stink bugs (Heteroptera: Pentatomidae). *Experientia* **48**: 893–6.

Borges, M. and Aldrich, J.R. (1994a) Estudos de semioquímicos para o manejo de Telenominae. *Anais da Sociedade Entomológica do Brasil* **23**: 575–7.

Borges, M. and Aldrich, J.R. (1994b) Attractant pheromone for Nearctic stink bug, *Euschistus obscurus* (Heteroptera: Pentatomidae): insight into a Neotropical relative. *Journal of Chemical Ecology* **20**: 1095–102.

Borges, M., Colazza, S., Ramirez-Lucas, P., Chauhan, K.R., Aldrich, J.R. and Moraes, M.C.B. (2003) Kairomonal effect of walking traces from *Euschistus heros* (Heteroptera: Pentatomidae) on two strains of *Telenomus podisi* (Hymenoptera: Scelionidae). *Physiological Entomology* **28**: 349–55.

Borges, M., Costa, M.L.M., Sujii, E.R., Cavalcanti, M. das G., Redígolo, G.F., Resck, I.S. and Vilela, E.F. (1999) Semiochemical and physical stimuli involved in host recognition by *Telenomus podisi* (Hymenoptera: Scelionidae) toward *Euschistus heros* (Heteroptera: Pentatomidae). *Physiological Entomology* **24**: 227–33.

Borges, M., Mori, K., Costa, M.L.M. and Sujii, E.R. (1998b) Behavioural evidence of methyl-2,6,10-trimethyltridecanoate as a sex pheromone of *Euschistus heros* (Het., Pentatomidae). *Journal of Applied Entomology* **122**: 335–8.

Borges, M., Schmidt, F.G.V., Sujii, E.R., Medeiros, M.A., Mori, K., Zarbin, P.H.G. and Ferreira, J.T.B. (1998a) Field responses of stink bugs to the natural and synthetic pheromone of the Neotropical brown stink bug, *Euschistus heros* (Heteroptera: Pentatomidae). *Physiological Entomology* **23**: 202–7.

Bruce, T.J.A., Martin, L.J., Pickett, J.A., Pye, B.J., Smart, L.E. and Wadhams, L.J. (2003) *cis*-Jasmone treatment induces resistance in wheat plants against the grain aphid, *Sitobion avenae* (Fabricius) (Homoptera: Aphididae). *Pest Management Science* **59**: 1031–6.

Bruce, T.J.A., Matthes, M.C., Chamberlain, K., Woodcock, C.M., Mohib, A., Webster, B., Smart, L.E., Birkett, M.A., Pickett, J.A. and Napier, J.A. (2008) cis-Jasmone induces *Arabidopsis* genes that affect the chemical ecology of multitrophic interactions with aphids and their parasitoids. *Proceedings of the National Academy of Sciences USA* **105**: 4553–8.

Cook, S.M., Khan, Z.R. and Pickett, J.A. (2007) The use of push–pull in integrated pest management. *Annual Review of Entomology* **52**: 375–400.

Corrêa-Ferreira, B.S. (2002) *Trissolcus basalis* para o controle de percevejos da soja. In: Parra, J.R.P., Botelho, P.S., Corrêa-Ferreira, B. and Bento, J.M.S. (eds) *Controle Biológico no Brasil, Parasitóides e Preadores*. Manole Ltda., São Paulo, pp. 449–76.

De Bach, P. (1964) The scope of biological control. In: De Bach, P. (ed) *Biological Control of Insect Pests and Weeds*. Chapman & Hall, London, pp. 3–20.

De Moraes, C.M., Mescher, M.C. and Tumlinson, J.H. (2001) Caterpillar induced nocturnal plant volatiles repel conspecific females. *Nature* **410**: 577–80.

Dicke, M. (2009) Behavioural and community ecology of plants that cry for help. *Plant, Cell and Environment* **32**: 654–65.

Ehler, L.E. (1998) Conservative biological control: past, present and future. In: Barbosa, P. (ed) *Conservation Biological Control*. Academic Press, London, pp. 1–8.

Ehler, L.E. and Miller, J.C. (1978) Biological control in temporary agroecosystems. *Entomophaga* **3**: 207–12.

Engelberth, J., Alborn, H.T., Schmelz, E.A. and Tumlinson, J.H. (2004) Airborne signals prime plants against insect herbivore attack. *Proceedings of the National Academy of Sciences USA* **101**: 1781–5.

FAO (2011) *Save and Grow: A Policymaker's Guide to the Sustainable Intensification of Smallholder Crop Production*. Food and Agriculture Organization of the United Nations, Rome.

Gardiner, M.M., Landis, D.A., Gratton. C., DiFonzo, C.D., O'Neal. M., Chacon, J.M., Wayo, M.T., Schmidt, N.P., Mueller, E.E. and Heimpel, G.E. (2009) Landscape diversity enhances the biological control of an introduced crop pest in the north-central USA. *Ecological Applications* **19**: 143–54.

Glinwood, R.T., Powell, W. and Tripathi, C.P.M. (1998) Increased parasitization of aphids on trap plants alongside vials releasing synthetic aphid sex pheromone and effective range of the pheromone. *Biocontrol Science and Technology* **8**: 607–14.

Godfray, H.C.J. (1994) *Parasitoids: Behavioural and Evolutionary Ecology*. Princeton University Press, Princeton, NJ.

Gurr, G.M., Wratten, S.D. and Altieri, M.A. (2004) Ecological engineering: a new direction for agricultural pest management. *AFBM Journal* **1**: 28–35.

Hanyu, K., Ichiki, R.T., Nakamura, S. and Kainoh, Y. (2011) Behavior of the tachinid parasitoid *Exorista japonica* (Diptera: Tachinidae) on herbivore-infested plants. *Applied Entomology and Zoology* **46**: 565–71.

Hegde, M., Oliveira, J.N., da Costa, J.G., Loza-Reyes, E., Bleicher, E., Santana, A.E.G., Caulfield, J.C., Mayon, P., Dewhirst, S.Y., Bruce, T.J.A., Pickett, J.A. and Birkett, M.A. (2012) Aphid antixenosis in cotton is activated by the natural plant defence elicitor *cis*-jasmone. *Phytochemistry* **78**: 81–8.

Higaki, M., and Adachi, I. (2011) Response of a parasitoid fly, *Gymnosoma rotundatum* (Linnaeus) (Diptera: Tachinidae) to the aggregation pheromone of *Plautia stali* Scott (Hemiptera: Pentatomidae) and its parasitism of hosts under field conditions. *Biological Control* **58**: 215–21.

Hole, D.G., Perkins, A.J., Wilson, J.D., Alexander, I.H., Grice, P.V. and Evans, A.D. (2005) Does organic farming benefit biodiversity? *Biological Conservation* **122**: 113–30.

Ichiki, R.T., Ho, G.T.T., Wajnberg, E., Kainoh, Y., Tabata, J. and Nakamura, S. (2012) Different uses of plant semiochemicals in host location strategies of the two tachinid parasitoids. *Naturwissenschaften* **99**: 687–94.

Ichiki, R.T., Kainoh, Y., Yamawaki, Y. and Nakamura, S. (2011) The parasitoid fly *Exorista japonica* uses visual and olfactory cues to locate herbivore-infested plants. *Entomologia Experimentalis et Applicata* **138**: 175–83.

James, D.G. (2003a) Field evaluation of herbivore-induced plant volatiles as attractants for beneficial insects: methyl salicylate and the green lacewing, *Chrysopa nigricornis. Journal of Chemical Ecology* **29**: 1601–9.

James, D.G. (2003b) Synthetic herbivore-induced plant volatiles as field attractants for beneficial insects. *Environmental Entomology* **32**: 977–82.

James, D.G. (2005) Further field evaluation of synthetic herbivore-induced plant volatiles as attractants for beneficial insects. *Journal of Chemical Ecology* **31**: 481–95.

Jang, S.A., Cho, J.H., Park, G.M., Choo, H.Y. and Park, C.G. (2011) Attraction of *Gymnosoma rotundatum* (Diptera: Tachinidae) to different amounts of *Plautia stali* (Hemiptera: Pentatomidae) aggregation pheromone and the effect of different pheromone dispensers. *Journal of Asia-Pacific Entomology* **14**: 119–21.

Jonsson, M., Wratten, S.D., Landis, D.A., Tompkins, J.M.L. and Cullen, R. (2010) Habitat manipulation to mitigate the impacts of invasive arthropod pests. *Biological Invasions* **12**: 2933–45.

Kalberer, N.M., Turlings, T.C.J. and Rahier, M. (2001) Attraction of a lead beetle (*Oreina cacaliae*) to damaged host plants. *Journal of Chemical Ecology* **27**: 647–61.

Kaplan, I., (2012) Attracting carnivorous arthropods with plant volatiles: the future of biocontrol or playing with fire? *Biological Control* **60**: 77–89.

Kessler, A. and Baldwin, I.T. (2001) Defensive function of herbivore-induced plant volatile emissions in nature. *Science* **291**: 2141–4.

Kessler, A. and Halitschke, R. (2007) Specificity and complexity: the impact of herbivore-induced plant responses on arthropod community structure. *Current Opinion in Plant Biology* **10**: 409–14.

Khan, Z.R., Midega, C.A.O., Amudavi, D.M.J., Hassanali, A. and Pickett, J.A. (2008) On-farm evaluation of the 'push–pull' technology for the control of stemborers and striga weed on maize in western Kenya. *Field Crops Research* **106**: 224–33.

Khan, Z., Midega, C., Pittchar, J., Pickett, J. and Bruce, T. (2011) Push–pull technology: a conservation agriculture approach for integrated management of insect pests, weeds and soil health in Africa. UK Government's Foresight Food and Farming Futures project. *International Journal of Agricultural Sustainability* **9**: 162–70.

Khan, Z.R., Pickett, J.A., Wadhams, L.J. and Muyekho, F. (2001) Habitat management strategies for the control of cereal stemborers and striga in maize in Kenya. *Insect Science and its Application* **21**: 375–80.

Kunert, G., Reinhold, C. and Gershenzon, J. (2010) Constitutive emission of the aphid alarm pheromone, (E)-β-farnesene, from plants does not serve as a direct defense against aphids. *BMC Ecology* **10**: 23.

Landis, D.A., Wratten, S.D. and Gurr, G.M. (2000) Habitat management to conserve natural enemies of arthropod pests in agriculture. *Annual Review of Entomology* **45**: 175–201.

Laumann, R.A., Čokl, A., Lopes, A.P.S., Moraes, M.C.B. and Borges, M. (2011) Silent singers are not safe: selective response of a parasitoid to substrate-borne vibratory signals of stink bugs. *Animal Behavior* **82**: 1175–83.

Laumann, R.A., Moraes, M.C.B., Corrêa Ferreira, M.B., Caetano, L.D., Vieira, A.R.A., Pires, C.S.S. and Borges, M. (2007) Field applications of (E)-2-hexenal: a kairomone for egg parasitoids (Hymenoptera: Scelionidae). In: *23rd Annual Meeting of the International Society of Chemical Ecology*, Jena, Book of Abstracts, p. 177.

Laumann, R.A., Moraes, M.C.B, Pareja, M, Alarcão, G.C., Botelho, A.C., Maia, A.H.N., Leonardez, E. and Borges, M. (2008) Comparative biology and functional response of *Trissolcus* spp. (Hymenoptera: Scelionidae) and implications for stink bugs (Hemiptera: Pentatomidae) biological control. *Biological Control* **44**: 32–41.

Lewis, W.J. and Martin, W.R. (1990) Semiochemicals for use with parasitoids: status and future. *Journal of Chemical Ecology* **16**: 306–9.

Lou, Y.G., Du, M.H., Turlings, T.C.J., Cheng, J.A. and Shan, W.F. (2005b) Exogenous application of jasmonic acid induces volatile emissions in rice and enhances parasitism of *Nilaparavata lugnes* eggs by the parasitoid *Anagrus nilaparvatae*. *Journal of Chemical Ecology* **31**: 1985–2002.

Lou, Y.G, Ma, B. and Cheng, J.A. (2005a) Attraction of the parasitoid *Anagrus nilaparvatae* to rice volatiles induced by the rice brown planthopper *Nilparavata lugens*. *Journal of Chemical Ecology* **31**: 2357–72.

Michereff, M.F.F., Laumann, R.A., Borges, M., Michereff-Filho, N., Diniz, I.R., Farias Neto, A.L. and Moraes, M.C.B. (2011) Volatiles mediating a plant–herbivore–natural enemy interaction in resistant and susceptible soybean cultivars. *Journal of Chemical Ecology* **37**: 273–285.

Midega, C.A.O., Khan, Z.R., van Den Berg J., Ogol, C.K.P.O., Pickett, J.A. and Wadhams, L.J. (2006) Maize stemborer predator activity 'push–pull' system and Bt maize: a potential component in managing Bt resistance. *International Journal of Pest Management* **52**: 1–10.

Mizutani, N. (2006) Pheromones of male stink bugs and their attractiveness to their parasitoids. *Japanese Journal of Applied Entomology and Zoology* **50**: 87–99.

Mondor, E.B. and Roland, J. (1997) Host locating behaviour of *Leschenaultia exul* and *Patelloa pachypyga*: two tachinid parasitoids of the forest tent caterpillar, *Malacosoma disstria*. *Entomologia Experimentalis et Applicata* **85**: 161–8.

Moraes, M.C.B., Birkett, M.A., Gordon-Weeks, R., Smart, L.E., Bromilow, R. and Pickett, J.A. (2008b) *cis*-Jasmone induces accumulation of defence compounds in wheat, *Triticum aestivum*. *Phytochemistry* **69**: 9–17.

Moraes, M C.B., Laumann, R.A., Pareja, M., Sereno, F.P.S., Michereff, M.F.F., Birkett, M.A., Pickett, J.A. and Borges, M. (2009) Attraction of the stink bug egg parasitoid *Telenomus podisi* to defence signals from soybean activated by treatment with *cis*-jasmone. *Entomologia Experimentalis et Applicata* **131**: 178–88.

Moraes, M.C.B., Pareja, M., Laumann, R.A. and Borges, M. (2008c) The chemical volatiles (semiochemicals) produced by neotropical stink bugs (Hemiptera: Pentatomidae). *Neotropical Entomology* **37**: 489–505.

Moraes, M.C.B., Pareja, M., Laumann, R.A., Hoffmann-Campo, C.B. and Borges, M. (2008a) Response of the parasitoid *Telenomus podisi* to induced volatiles from soybean damaged by stink bug herbivory and oviposition. *Journal of Plant Interaction* **3**: 1742–56.

Peres, W.A.A. (2004) *Aspectos bioecológicos e táticas de manejo dos percevejos* Nezara viridula *(Linnaeus)*, Euschistus heros *(Fabricius) e* Piezodorus guildinii *(Westwood) (Hemiptera: Pentatomidae) em cultivo orgânico de soja.* PhD thesis, Universidade Federal do Paraná, Brazil.

Pope, T.W., Campbell, C.A.M., Hardie, J. and Wadhams, L.J. (2007) Treating hop plants with (*Z*)-jasmone increases colonization by *Phorodon humuli* (Hemiptera: Aphididae) spring migrants. *Bulletin of Entomological Research* **7**: 317–19.

Poppy, G.M. and Sutherland, J.P. (2004) Can biological control benefit from genetically modified crops: tritrophic interactions on insect-resistant transgenic plants. *Physiological Entomology* **29**: 257–68.

Powell, W. and Pickett, J.A. (2003) Manipulation of parasitoids for aphid pest management: progress and prospects. *Pest Management Science* **59**: 149–55.

Powell, W., Hardie, J., Hick, A.J., Holler, C., Mann, J., Merritt, L., Nottingham, S.F., Wadhams, L.J., Witthinrich, J. and Wright, A.F. (1993) Responses of the parasitoid *Praon volucre* (Hymenoptera, Braconidae) to aphid sex-pheromone lures in cereal fields in autumn: implications for parasitoid manipulation. *European Journal of Entomology* **90**: 435–8.

Ruyon, J.B., Mescher, M.C. and De Moraes, C.M. (2006) Volatile chemical cues guide host location and host selection by parasitic plants. *Science* **313**: 1964–7.

Schnee, C., Kollner, T.G., Held, M., Turlings, T.C.J., Gershenzon, J. and Degenhardt, J. (2006) The products of a single maize sesquiterpene synthase form a volatile defense signal that attracts natural enemies of maize herbivores. *Proceedings of the National Academy of Sciences USA* **103**: 1129–34.

Simpson, M., Gurr, G.M., Simmons, A.T., Wratten, S.D., James, D G., Leeson, G., Nicol, H.I. and Orre-Gordon, G.U.S. (2011a) Insect attraction to synthetic herbivore-induced plant volatile-treated field crops. *Agricultural and Forest Entomology* **13**: 45–57.

Simpson, M., Gurr, G.M., Simmons, A.T., Wratten, S.D., James, D.G., Leeson, G., Nicol, H.I. and Orre-Gordon, G.U.S. (2011b) Attract and reward: combining chemical ecology and habitat manipulation to enhance biological control in field crops. *Journal of Applied Ecology* **48**: 580–90.

Smith, L.E., Luttrell, R.G. and Greene, L.K. (2009) Seasonal abundance, species composition, and population dynamics of stink bugs in production fields of early and late soybean in South Arkansas. *Journal of Economic Entomology* **102**: 229–36.

Stireman, J.O. (2002) Learning in the generalist tachinid parasitoid *Exorista mella* Walker (Diptera: Tachinidae). *Journal of Insect Behavior* **15**: 689–706.

Sujii, E.R., Costa, M.L.M., Pires, C.S.S., Colazza, S. and Borges, M. (2002) Inter- and intra-guild interactions in egg parasitoid species of the soybean–stink bug complex. *Pesquisa Agropecuária Brasileira* **37**: 1541–9.

Turner, M.G. (1989) Landscape ecology: the effect of pattern on process. *Annual Review of Ecology, Evolution, and Systematics* **20**: 171–97.

Vet, L.E.M. and Dicke, M. (1992) Ecology of infochemical use by natural enemies in a tritrophic context. *Annual Review of Entomology* **37**: 141–72.

Vieira, C.R., Moraes, M.C.B., Borges, M., Ferreira, J.C., Sujii, E.R. and Laumann, R. (2010) Behavioral manipulation of stink bug egg parasitoids (Hymenoptera: Scelionidae) by field applications of semiochemicals. In: *26th Annual Meeting of the International Society of Chemical Ecology*, Tours, Book of Abstracts, p. 323.

Vinson, S.B. (1984) Parasitoid–host relationship. In: Bell, W.J. and Cardé, R.T. (eds) *Chemical Ecology of Insects*. Chapman & Hall, New York, pp. 205–33.

Vinson, S.B. (1985) The behaviour of parasitoids. In: Kerkut, G.A. and Gilbert, L.I. (eds) *Comprehensive Insect Physiology, Biochemistry and Pharmacology*. Pergamon Press, New York, pp. 417–69.

von Mérey G.E., Veyrat, N., Mahuku, G., López Valdez, R., Turlings, T.C.J. and D'Alessandro, M. (2011) Dispensing synthetic green leaf volatiles in maize fields increases the release of sesquiterpenes by the plants, but has little effect on the attraction of pest and beneficial insects. *Phytochemistry* 72: 1838–47.

von Mérey, G.E., Veyrat, N., de Lange, E., Degen, T., Mahuku, G., López Valdez, R., Turlings, T.C.J. and D'Alessandro, M. (2012) Minor effects of two elicitors of insect and pathogen resistance on volatile emissions and parasitism of *Spodoptera frugiperda* in Mexican maize fields. *Biological Control* 60: 7–15.

Wäckers, F.L. and Lewis, W.J. (1994) Olfactory and visual learning and their interactive influence on host site location by the parasitoid, *Microplitis croceipes*. *Biological Control* 4: 105–12.

Wang, G., Cui, L.L., Dong, J., Francis, F., Liu, Y. and Tooker, J. (2011) Combining intercropping with semiochemical releases: optimization of alternative control of *Sitobion avenae* in wheat crops in China. *Entomologia Experimentalis et Applicata* 140: 189–95.

Williams, L., Rodriguez-Saona, C., Castle, S.C. and Zhu, S. (2008) EAG-active herbivore-induced plant volatiles modify behavioral responses and host attack by an egg parasitoid. *Journal of Chemical Ecology* 34: 1190–201.

Winkler, K., Wäckers, F.L., Kaufman, L.V., Larraz, V. and van Lenteren, J.C. (2009) Nectar exploitation by herbivores and their parasitoids is a function of flower species and relative humidity. *Biological Control* 50: 299–306.

Yan, Z.G. and Wang, C.Z. (2006) Wound-induced green leaf volatiles cause the release of acetylated derivatives and a terpenoid in maize. *Phytochemistry* 67: 34–42.

Yong, L., Wang, W.-L., Guo, G.-X. and Ji, X.-L. (2009) Volatile emission in wheat and parasitism by *Aphidius avenae* after exogenous application of salivary enzymes of *Sitobion avenae*. *Entomologia Experimentalis et Applicata* 130: 215–21.

Yu, H., Zhang, Y., Wu, K., Gao, X.W. and Guo, Y.Y. (2008) Field-testing of synthetic herbivore-induced plant volatiles as attractants for beneficial insects. *Environmental Entomology* 37: 1410–15.

11

Application of chemical cues in arthropod pest management for orchards and vineyards

Stefano Colazza, Ezio Peri and Antonino Cusumano

Department of Agricultural and Forest Sciences, University of Palermo, Italy

Abstract

Semiochemicals are naturally occurring chemical signals that are widely used in orchards and vineyards to monitor insect pest populations and/or to interfere with their behaviour to reduce agricultural damage. Some semiochemicals also are exploited by insect parasitoids to locate and recognize their hosts. Therefore, an indirect benefit of semiochemical-based management strategies is the conservation and enhancement of the efficacy of natural enemies against insect pests. In this chapter we review the literature on semiochemical–parasitoid systems in perennial crops. In the first part, we analyse the possible effects of pheromone-based tactics on parasitoid efficacy. In both orchards and vineyards, the most successful application of synthetic sex pheromones has been their use as mating disruptants. We then examine how the application of mating disruption can indirectly affect the efficacy of natural and augmented parasitoid populations. We also discuss the possibility of using parasitoid pheromones to directly affect parasitoid behaviour. In the second part, we summarize research in orchards and vineyards that has tested whether synthetic chemical cues such as herbivore-induced plant volatiles (HIPVs) and host-associated volatiles (HAVs) can be used to manipulate parasitoid behaviour, and specifically to enhance their searching efficacy. Finally, we consider the interaction of semiochemical–parasitoid systems with habitat management as novel combined strategies that have recently been developed to improve the success of semiochemical-based integrated pest management (IPM).

Chemical Ecology of Insect Parasitoids, First Edition. Eric Wajnberg and Stefano Colazza.
© 2013 John Wiley & Sons, Ltd. Published 2013 by John Wiley & Sons, Ltd.

11.1 Introduction

Chemical ecology is an interdisciplinary research area that investigates chemical interactions and signalling processes between organisms and their biotic and abiotic environment (Dicke & Takken 2006, Wortman-Wunder & Vivanco 2011). The best-known and most-intensive studies of chemical ecology have examined semiochemicals that mediate changes in the behaviour and development of a wide range of insects for their application in controlling pest populations (Pickett *et al.* 1997, Rodriguez-Saona & Stelinski 2009, Rodriguez-Saona *et al.* 2012). Pheromones are widely used by growers in integrated pest management (IPM) programmes to monitor pest populations, to interfere with pest behaviour by disruption of mating, for mass-trapping, or in 'attract and kill' strategies with insecticide-treated bait (Thomson *et al.* 1999, Suckling 2000). Synomones and kairomones are the main chemical cues used by insect parasitoids to locate their potential hosts (Vet & Dicke 1992, Vinson 1998; see Meiners & Peri, Chapter 9, this volume). Parasitoid females locate their hosts by exploiting herbivore-induced plant volatiles (HIPVs), the synomones produced by the plant in response to herbivore attack, and/or host-associated volatiles (HAVs), the kairomones emitted by the hosts (Vet & Dicke 1992, Vinson 1998, Colazza *et al.* 2010). It is generally accepted that HIPVs are more detectable to parasitoids over longer ranges than HAVs (Vet & Dicke 1992). However, the former are not considered as reliable indicators of host presence, whereas the latter are more definite indicators of host presence (Vet & Dicke 1992). Semiochemical-based IPM programmes should evaluate the indirect effects of the use of semiochemicals for pest control in terms of conserving and/or enhancing the efficacy of a pest's natural enemies (DeBach 1964, Bellows & Fisher 1999). However, the effective application of synomones and kairomones for direct manipulation of insect parasitoids requires a detailed understanding of agroecosystems rather than 'trial and error'-based approaches applied in the hope that something will work.

In the last decade, there has been increasing interest in the possible application of semiochemicals to improve the efficacy of insect parasitoids for the control of agricultural pests (James 2003, James & Price 2004, Simpson *et al.* 2011a). In this chapter we review recent advances in semiochemical–parasitoid systems in orchards and vineyards. In the first part, we analyse the possible effects of pheromone-based tactics on parasitoid efficacy. In orchards and vineyards, pheromone-based mating disruption has been the most successful application of synthetic sex pheromones for the control of pests. We then examine how mating disruption can indirectly affect the efficacy of natural and augmented parasitoid populations. We also discuss the possibility of using synthetic parasitoid pheromones to directly affect parasitoid behaviour. In the second part, we discuss research in orchards and vineyards using other types of semiochemicals, specifically herbivore-induced plant volatiles (HIPVs) and host-associated volatiles (HAVs), to directly manipulate the behaviour of parasitoids to enhance their efficacy. Although successful host location by parasitoids can be achieved through chemical cues acting over different ranges, we focus on volatile chemicals that act as medium- and long-range cues. Short-range volatile and contact chemical cues associated with host recognition are not discussed here. Several authors have pointed out the possibility of using HIPVs and HAVs for pest suppression within sustainable crop management programmes (Thaler 1999, Dicke & Hilker 2003, Halitschke *et al.* 2008, Khan *et al.* 2008). However, the effect of these volatile cues in the attraction of natural enemies has mainly been studied in laboratory experiments using simple linear food chains with plants infested by a single herbivore species, rather than in the much more complex natural environment. Moreover, in some cases the volatile cues used in experiments were crude samples obtained from airborne collections and their constituents were not fully identified. We therefore focus on examples in which the relevant chemical compounds have been properly identified and tested for insect management in the field. Finally, we consider the interaction of semiochemical–parasitoid systems with habitat management as a novel combined strategy aimed at enhancing the success of semiochemical-based IPM efforts. Overall, our intention is to synthesize what is known about semiochemical–parasitoid systems and to facilitate the development of more-sustainable management techniques in perennial crops.

11.2 Pheromone-based tactics in orchards and vineyards

Insect pheromones are the basis of most strategies that aim to control pest insects through manipulation of their behaviour. Among these compounds, sex pheromones, particularly those of Lepidoptera, have been used most successfully in IPM in perennial cropping systems (Ridgway *et al.* 1990, Witzgall *et al.* 2010). However, because the behaviour of several species of parasitoids is known to be affected by the sex pheromones of their hosts, the possible indirect effects of pheromone-based control of pest insects on their natural enemies should also be considered when using pheromones for pest control. Additionally, in recent years, the identification of pheromones for a number of parasitoid species has presented an opportunity to apply pheromone-based tactics to the direct manipulation of parasitoid behaviour (Suckling 2000, Hardy & Goubault 2007).

11.2.1 Host sex pheromones

The identification of 1000 or more volatile pheromones of insect pests has resulted in widespread and reliable use of these attractants in IPM (Cardé 2007). In orchards and vineyards, sex pheromones are mainly employed for disruption of mating of the target pest (Cardé 2007). Although mating disruption has proved to be a valuable alternative to insecticidal control of pests, it raises the question as to whether the behaviour of natural enemies might inadvertently be disrupted as well (Walton *et al.* 2006, Vacas *et al.* 2012).

Answering this question is not an easy task, considering the complexity of multitrophic interactions in perennial cropping systems, and few studies have attempted to address these effects with specifically designed experiments. In the USA, Walton *et al.* (2006) conducted experiments in vineyards to test the effectiveness of mating disruption on the vine mealybug, *Planococcus ficus* Signoret, and possible effects on its main parasitoid *Anagyrus pseudococci* (Girault), which is attracted by (*S*)-(+)-lavandulyl senecioate (LS), the sex pheromone of the vine mealybug (Millar *et al.* 2002). The results showed that parasitism levels were not significantly different between mating-disruption and control plots. To explain these unexpected results, the authors suggested that the negative effects of wide distribution of host sex pheromone on host location by parasitoids might have been counterbalanced by the attraction of parasitoids from nearby fields (Walton *et al.* 2006). In apple orchards, Biddinger *et al.* (1994) investigated the effects of mating disruption on the tufted apple bud moth, *Platynota idaeusalis* (Walker), and on its parasitoids. The study showed that in mating-disruption orchards, the total parasitism was similar to the control orchards except during the first moth generation, when the host larvae suffered from a higher parasitization rate. In vineyards, Williamson & Johnson (2005) showed the effects of mating disruption to control the grape berry moth, *Endopiza vitana* Clemens, on parasitoids. The peculiarity of this experiment was that mating disruption was based on 'auto-confusion' dispensers (where males responding to the dispensers become contaminated with pheromone, and thus become secondary dispensers). The results showed a negative effect, since the parasitism level of the target pest larvae in plots with the mating-disruption treatment was reduced by half. In Spanish citrus orchards, Vacas *et al.* (2012) investigated the effects of mating disruption of the California red scale, *Aonidiella aurantii* Maskell, in combination with augmentative releases of the parasitoid *Aphytis melinus* (DeBach). The results showed that parasitoid efficacy was higher in mating-disruption plots combined with parasitoid augmentative release than on plots treated with augmentative release of parasitoids only. However, this positive effect cannot be unequivocally explained in terms of parasitoid response to host sex pheromone because the kairomonal function of California red scale sex pheromone is uncertain (Sternlicht 1973, Morgan & Hare 1998). Alternatively, the authors suggested that the higher parasitism rates might have been a result of the prolonged vulnerability of the scale hosts, due to the delayed development of the immature stages of *A. aurantii* induced by the large amounts of sex pheromone (Vacas *et al.* 2012). Similar positive results have been reported in American peach orchards, where the egg parasitoids *Trichogramma minutum* Riley and *Trichogramma exiguum* Pinto & Platner parasitized more sentinel egg masses of the oriental fruit moth, *Grapholita molesta* (Busck), in mating-disruption plots than in control plots (Atanassov *et al.* 2003). Again, the lack of information on the effect of the oriental fruit moth sex pheromone on the host location behaviour of *Trichogramma* spp. limits the interpretation of these results.

In conclusion, although the experimental evidence is still limited, the use of sex pheromones for mating disruption of perennial crop pests does not seem to affect parasitoid efficacy, with the exception of the experiments in vineyards by Williamson & Johnson (2005). However, in this case study the possible influences on insect parasitoid efficacy of the pheromone itself and the methodology used for its application (auto-confusion) cannot be disentangled. With the limited number of systems tested and experiments run, the efficacy of strategies that integrate semiochemicals and parasitoids remains inconclusive.

11.2.2 Parasitoid pheromones

The field application of parasitoid pheromones in semiochemical-based tactics is still limited. To a large extent, parasitoid sex pheromones can be deployed for monitoring parasitoid populations, in a similar manner to the use of the sex pheromones of pests. An example was provided by Suckling *et al.* (2002), who used the sex pheromone of female *Ascogaster quadridentata* Wesmael in apple orchards in New Zealand. Sticky traps baited with 100 µg of the sex pheromone (*Z,Z*)-9,12-octadecadienal were used to assess parasitoid establishment, abundance and phenology synchronization with the host, the codling moth *Cydia pomonella* (L.). However, data based on male catches should be used with caution, because the sex ratio of parasitoids is subject to field conditions.

In biological control programmes with bethylid wasps, a new control tactic may be provided by direct application of pheromones released by females (Hardy & Goubault 2007). In Mexico, *Cephalonomia stephanoderis* Betrem is mass-reared for augmentative release against the coffee berry borer *Hypothenemus hampei* (Ferrari). Wasps released as adults that were reared in groups tended to disperse, whereas wasps released as immatures inside berries did not (Damon & Valle 2002). Such differences could be explained by a pheromone released during the contest behaviour of bethylid wasps. It has been shown that female wasps stressed during transportation release 3-methylindole (skatole), a compound associated with contest behaviour that could induce parasitoids to disperse from the target field (Gómez *et al.* 2005, Hardy & Goubault 2007). Further studies should focus on the possibility of manipulating such pheromones in order to improve the efficiency of bethylids as biological control agents.

11.3 Allelochemical-based manipulation in orchards and vineyards

Allelochemical-based manipulations employ volatile synomones and kairomones to directly improve parasitoid efficacy. The main synomones applied in IPM are herbivore-induced plant volatiles (HIPVs), the composition of which depends on several factors, such as the species of herbivore, the plant genotype and abiotic conditions (Schoonhoven *et al.* 1998, Dicke *et al.* 2009). HIPVs are a blend of various volatile organic chemicals (VOCs) that can include alkanes, alkenes, aldehydes, alcohols, ketones, ethers, esters and carboxylic acids. The main kairomones applied in IPM are host-associated volatiles (HAVs), which consist mainly of host sex or aggregation pheromones (Hardie *et al.* 1991, Krupke & Brunner 2003, Franco *et al.* 2008, 2011, Jang & Park 2010, Bayoumy *et al.* 2011, Jang *et al.* 2011).

11.3.1 Herbivore-induced plant volatiles (HIPVs)

To date, field demonstrations on the use of synthetic HIPVs to manipulate the behaviour of parasitoids in perennial cropping systems are limited to vineyards and hops (Table 11.1). In vineyards, the first experiments conducted in the USA showed that braconid species were significantly attracted to methyl salicylate (James & Price 2004). Subsequently, James & Grasswitz (2005) showed that three HIPVs, methyl salicylate, methyl jasmonate and (Z)-3-hexenyl acetate, attracted *Methaphycus* spp. and *Anagrus* spp., which are natural enemies of scale insects and grape leafhoppers, respectively. Parasitoid attraction to HIPVs has also been demonstrated in vineyards in Australia with methyl salicylate, methyl jasmonate, methyl anthranilate, benzaldehyde, (Z)-3-hexenyl acetate and (Z)-3-hexen-1-ol (Simpson *et al.* 2011a). Treatments were applied twice in the season, early and late, and at three different concentrations, 0.5%, 1% and 2% by volume sprayed on the leaves until runoff. Significant effects were found only in the 'early' treatment, where trichogrammatid wasps were attracted by benzaldehyde and methyl anthranilate; encyrtids and bethylids were attracted by methyl anthranilate; and platygastrids were attracted by all HIPVs. The authors explained the results by suggesting that a seasonal influence and plant phenology effect played a role in parasitoid attraction. In hop gardens, the application of HIPVs had specific effects in different groups of parasitoids (James 2003, 2005). The mymarid wasp *Anagrus daanei* Triapitsyn responded to farnesene, octyl aldehyde, and (Z)-3-hexen-1-ol; braconid wasps responded to (Z)-3-hexenyl acetate, geraniol, methyl anthranilate, methyl jasmonate, *cis*-jasmone and (Z)-3-hexen-1-ol; sarcophagid flies responded to benzaldehyde, nonanal, geraniol and *cis*-jasmone; and tachinid flies responded to benzaldehyde.

Table 11.1 Studies on allelochemical-based manipulations in perennial cropping systems to improve parasitoid efficacy. 'ND' indicates that the experimental design did not focus on any specific host species.

Crop	Chemical compound[a]	Allelochemicals	Host species	Parasitoid species	Aim	Comments[a]	Reference
Apple	Methyl [2E,4Z]-decadienoate	Host sex pheromone component	Euschistus conspersus	Gymnosoma filiola Gymnosoma occidentalis	Monitoring adult parasitoid populations	Bucket traps baited with the pheromone captured significantly more G. occidentalis than controls	Krupke & Brunner [2003]
				Trissolcus utahensis Trissolcus euschisti Telenomus podisi	Test for increased egg parasitism rates	No differences in parasitism rates were observed between baited and non-baited plants with host egg masses	
Citrus	PcA LS	Host sex pheromone Planococcus ficus sex pheromone	Planococcus citri	Anagyrus sp. nov. near pseudococci	Monitoring parasitoid populations	Anagyrus sp. nov. near pseudococci was not attracted to PcA but was attracted to LS	Franco et al. [2008]
	PcA LS	Host sex pheromone P. ficus sex pheromone	Planococcus citri	Anagyrus sp. nov. near pseudococci	Test for increased parasitism rates	Parasitism by Anagyrus sp. nov. near pseudococci of sentinel mealybugs placed in 'potato traps' was significantly increased by LS in Portugal, Italy, and Israel, whereas PcA had no significant effect on wasp parasitism rates except for the Italian population	Franco et al. [2011]
Deciduous trees	Mix of (E)-2-hexenal, α-terpineol, benzyl alcohol	Host sex pheromone	Podisus maculiventris	Telenomus podisi Telenomus calvus	Test for increased discovery efficiency	Traps containing P. maculiventris egg masses and baited with the pheromone increased discovery efficiency only for T. calvus	Bruni et al. [2000]

Crop	Compound	Type	Target species	Parasitoid	Purpose	Results	Reference
Fig	LS LI	Host sex pheromone Compound produced under mass-rearing conditions	*Planococcus ficus*	*Anagrus* sp. nov. near *pseudococci*	Monitoring parasitoid populations	Pheromone traps baited with LS attracted significantly more wasps than traps baited with LI	Franco *et al.* (2008)
Grapevine	MeSA	HIPVs	ND	Braconidae Ichneumonidae *Anagrus* spp.	Test for increased population density of natural enemies	Sticky traps placed in blocks of grape plants baited with MeSA captured significantly more braconid parasitoids than traps in unbaited blocks. No significant effect for Ichneumonidae and *Anagrus* spp.	James & Price (2004)
	HA MeJA MeSA	HIPVs	ND	*Metaphycus* spp. *Anagrus* spp.	Test for increased population density of natural enemies	Sticky traps placed in blocks of grape plants baited with MeSA, MeJA captured significantly more *Metaphycus* spp. and *Anagrus* spp. than traps in unbaited blocks. HA attracted only *Anagrus* spp.	James & Grasswitz (2005)
	LS LI	Host sex pheromone Compound produced under mass-rearing condition	*Planococcus ficus*	*Anagrus* sp. nov. near *pseudococci*	Monitoring parasitoid populations	Pheromone traps baited with LS attracted significantly more wasps than traps baited with LI	Franco *et al.* (2008)

(cont'd)

Table 11.1 (cont'd)

Crop	Chemical compound[a]	Allelochemicals	Host species	Parasitoid species	Aim	Comments[a]	Reference
	LS	*Planococcus ficus* sex pheromone	*Planococcus citri* (sentinel host)	*Anagyrus* sp. nov. near *pseudococci*	Test for increased parasitism rates	The number of *Anagyrus* sp. nov. near *pseudococci* females captured inside traps consisting of potato sprouts infested with sentinel *P. citri* (=potato traps) was significantly higher in LS baited traps	Mansour *et al.* (2011)
					Test for combined effect of parasitoid inundative release + LS application on parasitism rates	The number of *Anagyrus* sp. nov. near *pseudococci* females captured in potato traps baited with LS was significantly higher in plots with parasitoid inundative releases compared to traps baited with LS placed in plots without parasitoid releases	
	Be HA He MeA MeJA MeSA	HIPVs	ND	Bethylidae Braconidae Encyrtidae Platygastridae Trichogrammatidae	Test for increased population density of natural enemies	Sticky traps placed in blocks of grape plants and baited with HIPVs captured significantly more Trichogrammatidae (Be and MeA), Encyrtidae (MeA), Bethylidae (MeA) and Platygastridae (all compounds) than controls	Simpson *et al.* (2011a)

Volatile	Treatment		Natural enemy taxa	Objective	Results	Reference
HA MeA MeSA	HIPVs + reward plant	ND	Bethylidae Braconidae Ceraphronidae Encyrtidae Eulophidae Platygastridae	Test for increased population density of natural enemies	Analysis of parasitoid response was based on number of insect captured in sticky traps from plots treated with HIPV+reward, HIPV only and controls. No significant effect of the attractant treatments (HIPV) was apparent for any taxa. However, numbers of Eulophidae, Braconidae, Encyrtidae, and Ceraphronidae were significantly increased in treatments with rewards	Simpson *et al.* (2011b)
MeSA MeA	HIPVs + reward plants	ND	Bethylidae Braconidae Ceraphronidae Encyrtidae Eulophidae Platygastridae	Test for increased population density of natural enemies	Analysis of parasitoid response was based on number of insects captured in sticky traps from plots treated with HIPVs+reward, HIPVs only, and controls. Trap catches of platygastrids were increased by MeSA and MeA whereas buckwheat reward increased catches of eulophids	Simpson *et al.* (2011c)
(*E*)-4,8-dimethyl-1,3,7-nonatriene HA MeSA Hop	HIPVs	ND	*Anagrus* spp. parasitic Hymenoptera	Test for increased population density of natural enemies	Sticky traps baited with HIPVs did not capture significantly more *Anagrus* spp. or other wasps (pooled together as Hymenoptera) than controls	James (2003)

(cont'd)

Table 11.1 [cont'd]

Crop	Chemical compound[a]	Allelochemicals	Host species	Parasitoid species	Aim	Comments[a]	Reference
	3-octanone Be (E)-2-hexenal farnesene geraniol HA He indole J linalool MeA MeJA MeSA nonanal Oa	HIPVs	ND	Anagrus spp. Braconidae Sarcophagidae Tachinidae Platygastridae Encyrtidae Mymaridae (excluding Anagrus spp.)	Test for increased population density of natural enemies	Sticky traps baited with HIPVs attracted: Anagrus daanei (3-octanone; He; farnesene; Oa); Braconidae (HA; He; geraniol; MeA; J; MeJA); Sarcophagidae (Be; geraniol; J; nonanal);Tachinidae (Be)	James (2005)
Mixed species orchard	(Z)-3,9-dimethyl-6-isopropenyl-3,9-decadien-1-yl propanoate	Host sex pheromone component	Pseudaulacaspis pentagona	Thomsonisca amathus Encarsia berlesei Aphytis proclia Coccophagous sp.	Monitoring parasitoid populations	Only T. amathus was significantly attracted to pheromone compared with unbaited controls	Bayoumy et al. (2011)

Crop	Compound	Type	Pest	Parasitoid species	Aim	Result	Reference
	Mix of: [Z]-3,7-dimethyl-2,7-octadien-1-yl propanoate; 7-methyl-3-methylene-7-octen-1-yl propanoate	Host sex pheromone	*Diaspidiotus perniciosus*	*Encarsia perniciosi* *Aphytis proclia* *Coccophagous* sp.	Monitoring parasitoid populations	Only *E. perniciosi* was significantly attracted to pheromone compared with unbaited controls	Bayoumy *et al.* [2011]
Persimmon	CLB	Non-natural parasitoid attractant	*Planococcus kraunhiae*	*Anagyrus sawadai* *Anagyrus fujikona* *Anagyrus subnigricornis* *Leptomastix dactylopii*	Monitoring parasitoid populations Test for increased acceptance and parasitism rates	Female wasps of *A. sawadai* and of *L. dactylopii* were significantly more attracted to CBL baited traps than to controls Sentinel mealybugs placed on trees with CLB-impregnated rubber septa were parasitized significantly more often than those on non-treatment trees. Under natural conditions *A. sawadai* usually does not attack *P. kraunhiae*. However, this wasp parasitized *P. kraunhiae* in the presence of CLB in field experiments	Teshiba *et al.* [2012]
	Methyl [E,E,Z-2,4,6-decatrienoate	Host aggregation pheromone	*Plautia stali*	*Gymnosoma rotundatum*	Monitoring parasitoid populations	Traps baited with the synthetic host pheromone attracted significantly more flies than controls	Jang & Park (2010), Jang *et al.* (2011)

(cont'd)

Table 11.1 (cont'd)

Crop	Chemical compound[a]	Allelochemicals	Host species	Parasitoid species	Aim	Comments[a]	Reference
Spindle tree	Mix of (+)-(4aS, 7S, 7aR)-nepetalactone and (−)-(1R, 4aS, 7S, 7aR)-nepetalactol	Host sex pheromone	*Aphis fabae*	*Praon abjectum* *Praon dorsale* *Praon volucre*	Monitoring parasitoid populations	Traps with the host sex pheromone captured significantly more parasitoid species than controls	Hardie *et al.* [1991]

[a]Be = benzaldehyde; CLB = (2,4,4-trimethyl-2-cyclohexenyl)-methyl butyrate (= cyclolavandulyl butyrate); HA = (Z)-3-hexenyl acetate; He = (Z)-3-hexen-1-ol; J = *cis*-jasmone; Ll = (S)-(+)-lavandulyl isovalerate; LS = (S)-(+)-lavandulyl senecioate; MeA = methyl anthranilate; MeJA = methyl jasmonate; MeSA = methyl salicylate; Oa=octyl aldehyde; PcA = (+)-[1R,3R]-*cis*-2,2-dimethyl-3-isopropenyl-cyclobutanemethanol acetate (= planococcyl acetate).

The encouraging results obtained in terms of parasitoid attraction to HIPVs need to be explored further in terms of the impact observed on pests. A major constraint is the difficulty in highlighting a specific effect of parasitoids on a target pest, as many HIPVs, such as the widely used methyl salicylate, often attract a wide range of generalist predators and parasitoids that could parasitize alternative hosts (Rodriguez-Saona *et al.* 2012). Another factor that may hamper research in this area is the lack of knowledge about parasitoid response to HIPVs in the absence of their hosts. In fact, considering that many natural enemies attracted by HIPVs are likely to be generalists, they might actually learn to associate HIPVs with the absence of their hosts, which is exactly the opposite of the intended result. Furthermore, the technologies for distributing HIPVs for semiochemical-based manipulation of parasitoids have only recently become available and time is required to better develop and improve this approach. Nonetheless, the positive findings to date have attracted commercial interest, and methyl salicylate is now available to growers in the USA as a specific product for the attraction of natural enemies of insect pests. Whether or not the possible benefit from attracting parasitoids and the resulting effects on crop yields outweigh the cost of the semiochemical treatment has not yet been fully addressed.

11.3.2 Host-associated volatiles (HAVs)

The most common host-associated volatiles (HAVs) used to attract parasitoids in orchards are host sex or aggregation pheromones. In general, HAVs have been applied as tools for monitoring parasitoid populations, while applications to enhance parasitism levels have not yet been developed (Table 11.1). The first experiment that monitored parasitoid population density with pheromones in perennial crops was conducted in the UK in spindle trees, a primary host plant of *Aphis fabae* Scopoli (Hardie *et al.* 1991). The authors documented the attraction of three parasitoid species – *Praon abjectum* (Haliday), *Praon dorsale* (Haliday) and *Praon volucre* (Haliday) – using water traps baited with the sex pheromone of *A. fabae*, a blend of (+)-(4a*S*, 7*S*, 7a*R*)-nepetalactone and (−)-(1*R*, 4a*S*, 7*S*, 7a*R*)-nepetalactol. Another example of monitoring parasitoid population densities with host sex pheromones was provided by Bayoumy *et al.* (2011) in a mixed species orchard (*Prunus* sp., *Malus pumila* Mill. and *Syringa vulgaris* L.) in Hungary. In the orchard, sticky traps baited with different host sex pheromones caught different parasitoid species. For example, traps baited with the sex pheromone of the white peach scale, *Pseudaulacaspis pentagona* (Targioni-Tozzetti), (*Z*)-3,9-dimethy-6-isopropenyl-3,9-decadien-1-yl propanoate, preferentially captured adults of *Thomsonisca amathus* (Walker), whereas traps baited with the sex pheromone of the San José scale, *Diaspidiotus perniciosus* (Comstock), a blend of (*Z*)-3,7-dimethy1-2,7-octadien-1-yl propanoate and 7-methyl-3-methylene-7-octen-1-yl propanoate, preferentially captured adults of *Encarsia perniciosi* (Tower).

In citrus orchards located in the Mediterranean basin, semiochemical investigations have been carried out to verify that the synthetic sex pheromones of two mealybug species, the citrus mealybug *Planococcus citri* (Risso) and the vine mealybug *Planococcus ficus* (Signoret), attract the parasitoid *Anagyrus* sp. nov. near *pseudococci* (Girault) (Franco *et al.* 2008, 2011). In experiments conducted in Portugal, sticky traps baited with 200 μg of the sex pheromone of *P. citri*, (+)-(1*R*,3*R*)-*cis*-2,2-dimethyl-3-isopropenyl-cyclobutanemethanol acetate (also called planococcyl acetate, PcA), or with 200 μg of the sex pheromone of *P. ficus*, (*S*)-(+)-lavandulyl senecioate (LS) were used. The results showed that parasitoids were attracted only to the sex pheromone of the vine mealybug. This differential response of parasitoid females to host sex pheromones was explained in terms of the evolutionary host–parasitoid relationship (Franco *et al.* 2008). That is, both *A.* sp. nov. near *pseudococci* and *P. ficus* probably originated from the Mediterranean basin, whereas *P. citri* may be native to other regions. Moreover, *P. citri* appears to be a less suitable host because it more readily encapsulates parasitoid eggs than the other two mealybug species (Franco *et al.* 2008). The recruitment of *Anagyrus* sp. nov. near *pseudococci* was not affected by the cropping systems because similar results in terms of caught parasitoids were obtained in both vineyards and fig orchards (Franco *et al.* 2008, 2011).

Recently, the sex pheromone of vine mealybug was also tested to evaluate whether it could be used in citrus orchards to enhance parasitism levels of the citrus mealybug (Franco *et al.* 2011). In experiments replicated in Portugal, Italy and Israel, the authors placed traps containing a sprouted potato infested with sentinel citrus mealybugs and a rubber septum baited with PcA or LS. The levels of mealybug parasitism by *Anagyrus* sp. nov. near *pseudococci* were significantly increased by LS in all three parasitoid populations, whereas PcA had significant effects only on the Italian parasitoid population, suggesting differences in the genetic backgrounds of the parasitoid populations (Triapitsyn *et al.* 2007). The attraction of parasitoids to mealybug sex pheromones suggests that new HAV-based strategies to control the citrus mealybug might be possible in orchards in Portugal and Israel. It would be interesting to test the combined effects of parasitoid recruitment using LS and mating disruption against the citrus mealybug using PcA (Franco *et al.* 2008, 2011).

In the USA, Krupke & Brunner (2003) tested the attractiveness of stink bug pheromones to adult and egg parasitoids of the consperse stink bug, *Euschistus conspersus* Uhler, in apple orchards. In this study, the main component of the consperse stink bug aggregation pheromone, methyl (2*E*,4*Z*)-decadienoate, deployed in bucket traps, caught significant numbers of the tachinid parasitoid *Gymnosoma occidentalis* Townsend. However, when this compound was sprayed on vegetation, it did not increase parasitism of sentinel egg masses by egg parasitoids. In another study, the role played by the attractant pheromone of *Podisus maculiventris* (Say) on two egg parasitoids, *Telenomus podisi* Ashmead and *Telenomus calvus* Johnson, was investigated. Here the host pheromone, a blend of (*E*)-2-hexanal, α-terpineol and benzyl alcohol, significantly improved parasitism levels of the phoretic egg parasitoid, *T. calvus*, but not of *T. podisi* (Bruni *et al.* 2000).

In Korea, methyl (*E,E,Z*)-2,4,6-decatrienoate, the aggregation pheromone of the brown winged green stink bug (*Plautia stali* Scott) was tested as an attractant for the tachinid parasitoid *Gymnosoma rotundatum* (L.) in persimmon orchards (Jang & Park 2010, Jang *et al.* 2011). Yellow sticky traps baited with the synthetic host pheromone caught significantly more parasitoids than the controls did. The pheromone dispenser design influenced parasitoid attraction, with rubber septa and polytubes attracting significantly higher numbers of parasitoids than polyethylene vials. Considering that the brown winged green stink bug aggregation pheromone attracts both sexes of *P. stali* and *G. rotundatum*, and that all types of releaser were equally attractive to the host (Park *et al.* 2010), the authors suggested that appropriate traps capable of reducing pest population densities while simultaneously increasing parasitism rates could be developed (Jang & Park 2010, Jang *et al.* 2011).

It might also be possible to develop allelochemical-based tactics with 'non-natural' chemical compounds. For example, the synthetic compound (2,4,4-trimethyl-2-cyclohexenyl)-methyl butyrate, or cyclolavandulyl butyrate (CLB), has been demonstrated to attract the parasitoids *Anagyrus sawadai* Ishii and *Leptomastix dactylopii* Howard (Tabata *et al.* 2011, Teshiba *et al.* 2012). This compound was tested in persimmon orchards to control the Japanese mealybug *Planococcus kraunhiae* (Kuwana) by recruiting parasitoids (Teshiba *et al.* 2012). Sticky traps baited with 0.8 and 0.16 mg of CLB attracted both parasitoids, while sticky traps baited with 0.032 mg of CLB and unbaited controls did not. Moreover, by hanging 'fruit infested traps' (i.e. the fruit of kabosu (*Citrus sphaerocarpa* hort. ex Tanaka) artificially infested with *P. kraunhiae*) on plants, the authors showed that the parasitism rate significantly increased on 'fruit infested traps' placed near a wire with 10 rubber septa loaded with 0.16 mg of CLB. Also, the total parasitism level increased significantly when 'fruit infested traps' were placed near 10 wires, making a total of 100 dispensers (Fig. 11.1). Because *A. sawadai* is not likely to parasitize *P. kraunhiae* under natural conditions, CLB could be a promising tool to manipulate parasitoid behaviour in IPM programmes for the Japanese mealybug (Teshiba *et al.* 2012).

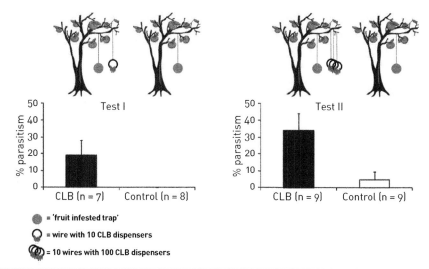

= 'fruit infested trap'

= wire with 10 CLB dispensers

= 10 wires with 100 CLB dispensers

	CLB-treated plots			Control plots		
	Test I	Test II	Total no. (%)	Test I	Test II	Total no. (%)
Anagyrus sawadai	8	8	16 (20)	0	0	0
Anagyrus fujikona	0	1	1 (1)	0	8	8 (73)
Anagyrus subnigricornis	0	7	7 (9)	0	0	0
Leptomastix dactylopii	11	34	45 (57)	0	0	0
Unemerged mummy	0	10	10 (13)	0	3	3 (27)

Figure 11.1 (Top) Mean (+SE) parasitism rate of mealybugs in the presence or absence (Control) of cyclolavandulyl butyrate (CLB); 10 or 100 rubber septa per tree were impregnated with 0.16 mg of CLB and used as attractants in Test I and Test II, respectively. A significant difference was observed between treatments and controls in both tests (GLMMs: $P < 0.005$). (Bottom) Number of wasps that emerged from parasitized mealybugs in field experiments, in the presence of 10 (Test I) or 100 (Test II) rubber septa impregnated with 0.16 mg of cyclolavandulyl butyrate (CLB) or in the absence of CLB (Control) (from Teshiba, M., Sugie, H., Tsutsumi, T. and Tabata, J. (2012) A new approach for mealybug management: recruiting an indigenous, but 'non-natural' enemy for biological control using an attractant. *Entomologia Experimentalis et Applicata* **142**: 211–215, © 2012 The Netherlands Entomological Society) .

11.4 Conclusions

Manipulation of pest behaviour is one of the prime means of protecting crops, their products and the environment from pesticide pollution, which is a global problem. Furthermore, the available data suggest that it should be possible to improve the efficacy of natural enemies by combining more than one strategy to manipulate their behaviour in crops. For example, semiochemical-based manipulations combined with habitat manipulations show considerable promise. The ecological basis of this approach relies on the possibility that adult parasitoids attracted to the crop treated with allelochemicals might find food sources such as nectar and pollen. In this way, the association of high concentrations of allelochemicals with the absence of hosts should not occur, and the possible negative effects of semiochemical manipulation should be minimized. A good example of combining semiochemicals and habitat manipulation is offered by the strategy called 'attract and reward' (A&R), which aims to synergistically attract natural enemies into the crop and maintain them within the crop (Simpson *et al.* 2011b, 2011c). In the experiments carried out so far, natural enemies were 'attracted' by synthetically produced HIPVs and were then 'rewarded' by providing them with floral resources. To date, the application of this combined approach in perennial cropping systems has been tested only in vineyards, using HIPVs and buckwheat plants (Simpson *et al.* 2011b, 2011c). Buckwheat is an annual plant, the nectar and pollen resources of which are able to increase the abundance, fitness and parasitism rates of natural enemies (Lavandero *et al.* 2005, Lee & Heimpel 2005, Irvin *et al.* 2006). However, contrary to expectations, A&R in this agroecosystem did not significantly improve parasitoid recruitment (Simpson *et al.* 2011b, 2011c). These results highlighted the complexity of A&R experiments and stressed the need for further research in vineyards to better understand the potential of such combined strategies in perennial cropping systems.

Semiochemical manipulation to increase natural enemy efficiency can also be achieved by use of transgenic plants (Degenhardt *et al.* 2003, Turlings & Ton 2006, Kos *et al.* 2009). One approach would be to create transgenic plants with enhanced production of HIPVs after herbivore attack (Turlings & Ton 2006). To date, research in orchard crops has lagged behind comparable research with arable crops due to a lack of model study systems and technical limitations such as long ontogenetic cycles. In fact, most research on transgenic plants has been conducted with annual plants such as maize and cotton, or model organisms such as *Arabidopsis thaliana* (L.) (Degenhardt *et al.* 2003, Kappers *et al.* 2005, Schnee *et al.* 2006; see Blassioli-Moraes, Borges & Laumann, Chapter 10, this volume). Even in annual plants, research has mainly focused on basic aspects of plant signalling pathways from a tritropic perspective, and extension of these laboratory studies to practical field applications has been limited (Thaler *et al.* 2002, van Poecke & Dicke 2003, Ament *et al.* 2004, Shiojiri *et al.* 2006). One possible complicating factor with transgenic perennial crops is whether the risks associated with implementation outweigh the potential benefits (Gurr *et al.* 2004).

In conclusion, the expansion of monocultural cropping systems at the expense of natural vegetation has caused instability within agroecosystems because decreasing local habitat diversity results in increases in insect pest problems (Altieri & Letourneau 1982). Within this viewpoint, perennial crops are considered more stable ecosystems than annual crops and are more abundant in arthropod fauna (Altieri 1999, Simon *et al.* 2010). In general, orchards and vineyards contain a high diversity of vegetation and multi-strata habitats under the plants, such as cover crops, intercropping or, on their boundaries, ornamental or windbreak plants. The coexistence of these different habitats in and around perennial crops is favourable to beneficial insects, providing refuges and areas where they can reproduce and find alternate prey, hosts, and nectar and pollen (Altieri 1999, Landis *et al.* 2000, Simon *et al.* 2010). Therefore, orchards and vineyards, with fewer disturbances and having greater structural diversity, are favourable to the establishment and effective use of biological control agents. As a consequence, in these agroecosystems the manipulation of parasitoids by semiochemicals offers a good opportunity to maximize ecosystem services for pest control.

Acknowledgement

We thank Jocelyn Millar for providing comments on the manuscript.

References

Altieri, M.A. (1999) The ecological role of biodiversity in agroecosystems. *Agriculture, Ecosystems and Environment* **74**: 19–31.

Altieri, M.A. and Letourneau, D.K. (1982) Vegetation management and biological control in agroecosystems. *Crop Protection* **1**: 405–30.

Ament, K., Kant, M.R., Sabelis, M.W., Haring, M.A. and Schuurink, R.C. (2004) Jasmonic acid is a key regulator of spider mite-induced volatile terpenoid and methyl salicylate emission in tomato. *Plant Physiology* **135**: 2025–37.

Atanassov, A., Shearer, P.W. and Hamilton, G.C. (2003) Peach pest management programs impact beneficial fauna abundance and *Grapholita molesta* (Lepidoptera: Tortricidae) egg parasitism and predation. *Environmental Entomology* **32**: 780–8.

Bayoumy, M.H., Kaydan, B.A. and Kozár, F. (2011) Are synthetic pheromone captures predictive of parasitoid densities as a kairomonal attracted tool? *Journal of Entomological and Acarological Research* **43**: 23–31.

Bellows, T.S. and Fisher, T.W. (1999) *Handbook of Biological Control*. Academic Press, San Diego, CA.

Biddinger, D.J., Fellan, C.M. and Hill, L.A. (1994) Parasitism of the tufted bud apple moth (Lepidoptera: Tortricidae), in conventional insecticide and pheromone treated Pennsylvania apple orchards. *Environmental Entomology* **23**: 1568–79.

Bruni, R., Sant'Ana, J., Aldrich, J.R. and Bin, F. (2000) Influence of host pheromone on egg parasitism by scelionid wasps: comparison of phoretic and nonphoretic parasitoids. *Journal of Insect Behavior* **13**:165–73.

Cardé, R.T. (2007) Using pheromones to disrupt mating of moth pests. In: Kogan, M. and Jepson, P. (eds) *Perspectives in Ecological Theory and Integrated Pest Management*. Cambridge University Press, Cambridge, pp. 122–69.

Colazza, S., Peri, E., Salerno, G. and Conti, E. (2010) Host searching by egg parasitoids: exploitation of host chemical cues. In: Parra, J.R.P., Cônsoli, F.L. and Zucchi, R.A. (eds) *Egg Parasitoids in Agroecosystems with Emphasis on Trichogramma*. Springer, Dordrecht, pp. 97–147.

Damon, A. and Valle, J. (2002) Comparison of two release techniques for the use of *Cephalonomia stephanoderis* (Hymenoptera: Bethylidae), to control the coffee berry borer *Hypothenemus hampei* (Coleoptera: Scolytidae) in Soconusco, southeastern México. *Biological Control* **24**: 117–27.

DeBach, P. (1964) *Biological Control by Natural Enemies*. Cambridge University Press, London.

Degenhardt, J., Gershenzon, J., Baldwin, I.T. and Kessler, A. (2003) Attracting friends to feast on foes: engineering terpene emission to make crop plants more attractive to herbivore enemies. *Current Opinion in Biotechnology* **14**: 169–76.

Dicke, M. and Hilker, M. (2003) Induced plant defences: from molecular biology to evolutionary ecology. *Basic and Applied Ecology* **4**: 3–14.

Dicke, M. and Takken, W. (2006) *Chemical Ecology: From Gene to Ecosystem.* Wageningen UR Frontis Series 16, Wageningen, The Netherlands.

Dicke, M., van Loon, J.J.A. and Soler, R. (2009) Chemical complexity of volatiles from plants induced by multiple attack. *Nature: Chemical Biology* **5**: 317–24.

Franco, J.C., Silva, E.B., Cortegano, E., Campos, L., Branco, M., Zada, A. and Mendel, Z. (2008) Kairomonal response of the parasitoid *Anagyrus* spec. nov. near *pseudococci* to the sex pheromone of the vine mealybug. *Entomologia Experimentalis et Applicata* **126**: 122–30.

Franco, J.C., Silva, E.B., Fortuna, T., Cortegano, E., Branco, M., Suma, P., La Torre, I., Russo, A., Elyahu, M., Protasov, A., Levi-Zada, A. and Mendel, Z. (2011) Vine mealybug sex pheromone increases citrus mealybug parasitism by *Anagyrus* sp. near *pseudococci* (Girault). *Biological Control* **58**: 230–38.

Gómez, J., Barrera, J.F., Rojas, J.C., Macias-Samano, J., Liedo, J.P., Badii, M.H. (2005) Volatile compounds released by disturbed females of *Cephalonomia stephanoderis* (Hymenoptera: Bethylidae): a parasitoid of the coffee berry borer *Hypothenemus hampei* (Coleoptera: Scolytidae). *Florida Entomologist* **88**: 180–87.

Gurr, G.M., Wratten, S.D. and Altieri, M.A. (2004) Ecological engineering for enhanced pest management: toward a rigorous science. In: Gurr, G.M., Wratten, S.D. and Altieri, M.A. (eds) *Ecological Engineering for Pest Management Advances in Habitat Manipulation for Arthropods.* CSIRO Publishing, Australia, pp. 219–25.

Halitschke, R., Stenberg, J.A., Kessler, D., Kessler, A. and Baldwin, T. (2008) Shared signals: 'alarm calls' from plants increase apparency to herbivores and their enemies in nature. *Ecology Letters* **11**: 24–34.

Hardie, J., Nottingham, S.F., Powell, W. and Wadhams, L.J. (1991) Synthetic aphid pheromone lures female parasitoids. *Entomologia Experimentalis et Applicata* **61**: 97–9.

Hardy, I.C.W. and Goubault, M. (2007) Wasp fights: understanding and utilizing agonistic bethylid behaviour. *Biocontrol News and Information* **28**: 11–15.

Irvin, N.A., Scarratt, S.L., Wratten, S.D., Frampton, C.M., Chapman, R.B. and Tylianakis, J.M. (2006) The effects of floral understoreys on parasitism of leafrollers (Lepidoptera: Tortricidae) on apples in New Zealand. *Agricultural and Forest Entomology* **8**: 25–34.

James, D.G. (2003) Synthetic herbivore-induced plant volatiles as attractants for beneficial insects. *Environmental Entomology* **32**: 977–82.

James, D.G. (2005) Further field evaluation of synthetic herbivore-induced plant volatiles as attractants for beneficial insects. *Journal of Chemical Ecology* **31**: 481–95.

James, D.G. and Grasswitz, T.R. (2005) Synthetic herbivore-induced plant volatiles increase field captures of parasitic wasps. *BioControl* **50**: 871–80.

James, D.G. and Price, T.S. (2004) Field-testing of methyl salicylate for recruitment and retention of beneficial insects in grapes and hops. *Journal of Chemical Ecology* **30**: 1613–28.

Jang, S.A. and Park, C.G. (2010) *Gymnosoma rotundatum* (Diptera: Tachinidae) attracted to the aggregation pheromone of *Plautia stali* (Hemiptera: Pentatomidae). *Journal of Asia-Pacific Entomology* **13**: 73–5.

Jang, S.A., Cho, J.H., Park, G.M., Choo, H.Y. and Park, C.G. (2011) Attraction of *Gymnosoma rotundatum* (Diptera: Tachinidae) to different amounts of *Plautia stali* (Hemiptera: Pentatomidae) aggregation pheromone and the effect of different pheromone dispensers. *Journal of Asia-Pacific Entomology* **14**: 119–21.

Kappers, I.F., Aharoni, A., van Herpen, T.W.J.M., Luckerhoff, L.L.P., Dicke, M. and Bouwmeester, H.J. (2005) Genetic engineering of terpenoid metabolism attracts bodyguards to *Arabidopsis*. *Science* **309**: 2070–2.

Khan, Z.R., James, D.G., Midega, C.A.O. and Pickett, C.H. (2008) Chemical ecology and conservation biological control. *Biological Control* **45**: 210–24.

Kos, M., van Loon, J.J.A., Dicke, M. and Vet, L.E.M. (2009) Transgenic plants as vital components of integrated pest management. *Trends in Biotechnology* **27**: 621–7.

Krupke, C.H. and Brunner, J.F. (2003) Parasitoids of the consperse stink bug (Hemiptera: Pentatomidae) in North Central Washington and attractiveness of a host-produced pheromone component. *Journal of Entomological Science* **38**: 84–92.

Landis, D.A., Wratten, S.D. and Gurr, G.M. (2000) Habitat management to conserve natural enemies of arthropod pests in agriculture. *Annual Reviews of Entomology* **45**: 175–201.

Lavandero, B., Wratten, S., Shishehbor, P. and Worner, S. (2005) Enhancing the effectiveness of the parasitoid *Diadegma semiclausum* (Helen): movement after use of nectar in the field. *Biological Control* **34**: 152–8.

Lee, J.C. and Heimpel, G.E. (2005) Impact of flowering buckwheat on lepidopteran cabbage pests and their parasitoids at two spatial scales. *Biological Control* **34**: 290–301.

Mansour, R., Suma, P., Mazzeo, G., Buonocore, E., Lebdi, G.K. and Russo, A. (2011) Using a kairomone-based attracting system to enhance biological control of mealybugs (Hemiptera: Pseudococcidae) by *Anagyrus* sp. near *pseudococci* (Hymenoptera: Encyrtidae) in Sicilian vineyards. *Journal of Entomological and Acarological Research* **42**: 161–70.

Millar, J.G., Daane, K.M., McElfresh, J.S., Moreira, J., Malakar-Kuenen, R., Guillen, M. and Bentley, W.J. (2002) Development and optimization of methods for using sex pheromone for monitoring the mealybug *Planococcus ficus* (Homoptera: Pseudococcidae) in California vineyards. *Journal of Economic Entomology* **95**: 706–14.

Morgan, D.J.W. and Hare, J.D. (1998) Volatile cues used by the parasitoid, *Aphytis melinus*, for host location: California red scale revisited. *Entomologia Experimentalis et Applicata* **88**: 235–45.

Park, G.M., Jang, S.A., Choi, S.H. and Park, C.G. (2010) Attraction of *Plautia stali* (Hemiptera: Pentatomidae) to different amounts and dispensers of its aggregation pheromone. *Korean Journal of Applied Entomology* **49**: 123–7.

Pickett, J.A., Wadhams, L.J. and Woodcock, C.M. (1997) Developing sustainable pest control from chemical ecology. *Agriculture, Ecosystems and Environment* **64**: 149–56.

Ridgway, R.L., Silverstein, R.M. and Inscoe, M.N. (1990) *Behavior-Modifying Chemicals for Insect Management: Applications of Pheromones and Other Attractants*. Marcel Dekker, New York.

Rodriguez-Saona, C.R. and Stelinski, L.L. (2009) Behavior-modifying strategies in IPM: theory and practice. In: Peshin, R. and Dhawan, A.K. (eds) *Integrated Pest Management: Innovation-Development Process, Vol.1*. Springer, Dordrecht, pp. 263–315.

Rodriguez-Saona, C., Blaauw, B.R. and Isaacs, R. (2012) Manipulation of natural enemies in agroecosystems: habitat and semiochemicals for sustainable insect pest control. In: Larramendy, M.L. and Soloneski, S. (eds) *Integrated Pest Management and Pest Control: Current and Future Tactics*. InTech, Rijeka, Croatia, pp. 89–126.

Schnee, C., Köllner, T.G., Held, M., Turlings, T.C.J., Gershenzon, J. and Degenhardt, J. (2006) The products of a single maize sesquiterpene synthase form a volatile defense signal that attracts natural enemies of maize herbivores. *Proceedings of the National Academy of Sciences USA* **103**: 1129–34.

Schoonhoven, L.M., Jermy, T. and van Loon, J.J.A. (1998) *Insect–Plant Biology: From Physiology to Evolution*. Chapman & Hall, London.

Shiojiri, K., Kishimoto, K., Ozawa, R., Kugimiya, S., Urashimo, S., Arimura, G., Horiuchi, J., Nishiokan, T., Matsuin, K. and Takabayashi, J. (2006) Changing green leaf volatile biosynthesis in plants: an approach for improving plant resistance against both herbivores and pathogens. *Proceedings of the National Academy of Sciences USA* **103**: 16672–6.

Simon, S., Bouvier, J.C., Debras, J.F. and Sauphanor, B. (2010) Biodiversity and pest management in orchard systems: a review. *Agronomy for Sustainable Development* **30**: 139–52.

Simpson, M., Gurr, G.M., Simmons, A.T., Wratten, S.D., James, D.G., Leeson, G., Nicol, H.I. and Orre-Gordon, G.U.S. (2011a) Insect attraction to synthetic herbivore-induced plant volatile treated field crops. *Agriculture and Forest Entomology* **13**: 45–57.

Simpson, M., Gurr, G.M., Simmons, A.T., Wratten, S.D., James, D.G., Leeson, G., Nicol, H.I. and Orre-Gordon, G.U.S. (2011b) Attract and reward: combining chemical ecology and habitat manipulation to enhance biological control in field crops. *Journal of Applied Ecology* **48**: 580–90.

Simpson, M., Gurr, G.M., Simmons, A.T., Wratten, S.D., James, D.G., Leeson, G., Nicol, H.I. and Orre-Gordon, G.U.S. (2011c) Field evaluation of the 'attract and reward' biological control approach in vineyards. *Annals of Applied Biology* **159**: 69–78.

Sternlicht, M. (1973) Parasitic wasps attracted by the sex pheromone of their coccid host. *Entomophaga* **18**: 339–42.

Suckling, D.M. (2000) Issues affecting the use of pheromones and other semiochemicals in orchards. *Crop Protection* **19**: 677–83.

Suckling, D.M., Gibb, A.R., Burnip, G.M. and Delury, N.C. (2002) Can parasitoid sex pheromones help in insect biocontrol? A case study of codling moth and its parasitoid. *Environmental Entomology* **31**: 947–52.

Tabata, J., Teshiba, M., Hiradate, S., Tsutsumi, T., Shimizu, N. and Sugie, H. (2011) Cyclolavandulyl butyrate: an attractant for a mealybug parasitoid, *Anagyrus sawadai* (Hymenoptera: Encyrtidae). *Applied Entomology and Zoology* **46**: 117–23.

Teshiba, M., Sugie, H., Tsutsumi, T. and Tabata, J. (2012) A new approach for mealybug management: recruiting an indigenous, but 'non-natural' enemy for biological control using an attractant. *Entomologia Experimentalis et Applicata* **142**: 211–15.

Thaler, J. (1999) Jasmonate-inducible plant defences cause increased parasitism of herbivores. *Nature* **399**: 686–8.

Thaler, J.S., Farag, M.A., Paré, P.W. and Dicke, M. (2002) Jasmonate-deficient plants have reduced direct and indirect defences against herbivores. *Ecology Letters* **5**: 764–74.

Thomson, D.R., Gut, L.J. and Jenkins, J.W. (1999) Pheromones for insect control: strategies and successes. In: Hall, F.R. and Menn, J.J. (eds) *Methods in Biotechnology 5. Biopesticides: Use and Delivery.* Humana Press, NJ, pp. 385–412.

Triapitsyn, S.V., González, D., Danel, B., Vickerman, D.B., Noyes, J.S. and Ernest, B.W. (2007) Morphological, biological, and molecular comparisons among the different geographical populations of *Anagyrus pseudococci* (Hymenoptera: Encyrtidae), parasitoids of *Planococcus* spp (Hemiptera: Pseudococcidae), with notes on *Anagyrus dactylopii. Biological Control* **41**: 14–24.

Turlings, T.C.J. and Ton, J. (2006) Exploiting scents of distress: the prospect of manipulating herbivore-induced plant odours to enhance the control of agricultural pests. *Current Opinion in Plant Biology* **9**: 421–7.

Vacas, S., Vanaclocha, P., Alfaro, C., Primo, J., Verdú, M.J., Urbaneja, A. and Navarro-Llopis, V. (2012) Mating disruption for the control of *Aonidiella aurantii* Maskell (Hemiptera: Diaspididae) may contribute to increased effectiveness of natural enemies. *Pest Management Science* **68**: 142–8.

van Poecke, R.M.P. and Dicke, M. (2003) Signal transduction downstream of salicylic and jasmonic acid in herbivory-induced parasitoid attraction by *Arabidopsis* is independent of JAR1 and NPR1. *Plant, Cell and Environment* **26**: 1541–8.

Vet, L.E.M. and Dicke, M. (1992) Ecology of infochemical use by natural enemies in a tritrophic context. *Annual Review of Entomology* **37**: 141–72.

Vinson, S.B. (1998) The general host selection behavior of parasitoid Hymenoptera and a comparison of initial strategies utilized by larvaphagous and oophagous species. *Biological Control* **11**: 79–97.

Walton, V.M., Daane, K.M., Bentley, W.J., Millar, J.G., Larsen, T.E. and Malakar-Kuenen, R. (2006) Pheromone-based mating disruption of *Planococcus ficus* (Hemiptera: Pseudococcidae) in California vineyards. *Journal of Economic Entomology* **99**: 1280–90.

Williamson, J.R. and Johnson, D.T. (2005) Effects of grape berry moth management practices on arthropod diversity in grape vineyards in the southern United States. *HortTechnology* **15**: 232–8.

Witzgall, P., Kirsch, P. and Cork, A. (2010) Sex pheromones and their impact on pest management. *Journal of Chemical Ecology* **36**: 80–100.

Wortman-Wunder, E. and Vivanco, J.M. (2011) Chemical ecology: definition and famous examples. In: Vivanco, J.M. and Weir, T. (eds) *Chemical Biology of the Tropics: An Interdisciplinary Approach.* Springer-Verlag, Berlin Heidelberg, pp. 15–26.

12

Application of chemical cues in arthropod pest management for organic crops

Marja Simpson[1], Donna M.Y. Read[2] and Geoff M. Gurr[1]

[1]E.H. Graham Centre for Agricultural Innovation, NSW Department of Primary Industries and Charles Sturt University, Orange, New South Wales, Australia
[2]School of Agriculture and Wine Science, Charles Sturt University, Orange, New South Wales, Australia

Abstract

Organic agriculture is becoming progressively more popular as a result of widespread concerns about the safety of foods treated with synthetic pesticides, and wider issues surrounding the sustainability of agricultural systems based on synthetic inputs. The use of chemical cues in pest management for organic crops is constrained by the fact that all definitions of organic agriculture prohibit the use of synthetic compounds. The majority of studies of herbivore-induced plant volatiles (HIPVs) and other chemical cues have used synthetic compounds and have thus taken place within conventional agricultural systems. There is, however, good scope for chemical cues to become an important strategy for organic pest management. Oil of wintergreen is a natural product that would be allowable in organic agriculture, and a major constituent of this is methyl salicylate, one of the most well researched and useful HIPVs. Silicon compounds are also allowable in organic agriculture and some research suggests that these can boost HIPV production by plants when they are attacked by pests. Plant breeding and a closer examination of heirloom varieties and landraces is also likely to yield crop cultivars with a greater ability to mount induced defences that are based on the activity of third trophic level components of agricultural biodiversity. This chapter also looks more broadly at chemical cue-based plant protection methods, and two of these, 'attract and reward' and 'push–pull', offer good scope to better exploit parasitoids and other natural enemies in organic agriculture.

Chemical Ecology of Insect Parasitoids, First Edition. Eric Wajnberg and Stefano Colazza.
© 2013 John Wiley & Sons, Ltd. Published 2013 by John Wiley & Sons, Ltd.

12.1 Introduction: organic farming and compatibility of chemical cues

Worldwide demand for organic food products continues to increase, especially in developed countries. This is largely due to consumer fears over food safety from pesticides and genetically modified crops, although wider issues related to the environmental impact and even animal welfare of conventional farming systems are also significant. In 2010, organic agriculture was practiced in approximately 141 countries of the world, with 37 million hectares managed organically worldwide by 1.6 million producers (FiBL 2012). A major part of this land is in Australia (12 million ha), followed by Argentina (4.2 million ha) and the USA (1.9 million ha) (FiBL 2012). In Australia and Argentina, most of the organically managed land is used for extensive grazing. The world's largest organically managed farming enterprise, almost 1 million ha in area, is located in Australia. In Asia, the total organic area is approximately 600,000 ha. Organic farming in Africa is also increasing, partly due to the demand for imports of organic products into developed countries (Yussefi & Willer 2003).

In organic farming, the cycling of materials on the farm is an important principle and is achieved without the use of synthetic fertilizers, pesticides and genetically modified organisms. Organic farming aims to minimize pollution of air, soil and water, and to optimize the health and productivity of interdependent communities of plants, animals and people (CAC 2001) (Table 12.1). The International Federation of Organic Agriculture Movements (IFOAM), which promotes and establishes international standards for organic agriculture, says:

> Organic farming is a production system that sustains the health of soils, ecosystems and people. It relies on ecological processes, biodiversity and cycles adapted to local conditions, rather than the use of inputs with adverse effects.

Table 12.1 Principles of organic agriculture (based on data from IFOAM 2009b).

- Improvement and maintenance of natural resources and agroecosystems
- Avoiding overexploitation and pollution of natural resources
- Minimizing consumption of non-renewable energy and resources
- Production of sufficient quantities of nutritious, wholesome and high-quality food
- Providing adequate returns within a safe, secure and healthy working environment
- Acknowledgement of indigenous knowledge and traditional farming systems
- Maintenance and improvement of long-term soil fertility
- Enhancement of biological cycles within the farm (e.g. nutrient cycles)
- Providing nitrogen supply through the use of nitrogen-fixing plants
- Biological plant protection based on prevention rather than eradication
- Crop diversity and animal diversity appropriate to the local conditions
- Animal husbandry appropriate to the needs of the animals
- Prohibition of synthetic chemical fertilizers, plant protection products, hormones and growth regulators
- Prohibition of genetic engineering and its products
- Prohibition of synthetic or harmful methods, processing aids, and ingredients in food processing

> Organic agriculture combines tradition, innovation and science to benefit the shared environment and promote fair relationships and a good quality of life for all involved (IFOAM 2009a).

Similarly, the Council of the European Union (2007) defines organic agriculture as an:

> overall system of farm management and food production that combines best environmental practices, a high level of biodiversity, the preservation of natural resources, the application of high animal welfare standards and a production method in line with the preference of certain consumers for products produced using natural substances and process.

To meet the necessary production conditions, organic farmers have to implement practices that optimize nutrient and energy flows and minimize risks. Such practices include crop rotations and enhanced crop ecosystem diversity through habitat manipulation, different combinations of livestock and plants, symbiotic nitrogen fixation with legumes, application of organic manure, use of pest-resistant or pest-tolerant cultivars, and biological pest control (Zehnder *et al.* 2007, El-Hage Scialabba & Müller-Lindenlauf 2010). The two main principles that distinguish arthropod pest management practices in organic farming from those of other farming systems are that synthetic pesticides are rejected in favour of natural pesticides and biodiversity is promoted.

In an in-depth review, Shrivastava *et al.* (2010) discussed a variety of strategies in the context of plant volatiles for arthropod pest management compatible with organic farming principles, including resistant cultivars, polyculture, beneficial microorganisms such as mycorrhizal fungi and endophytes, and plant-derived pesticides. Herbivore-induced plant volatiles (HIPVs) have the potential to attract natural enemies into a crop and thereby increase their abundance and diversity. However, more research is necessary into the use of HIPV-based strategies for pest management (Shrivastava *et al.* 2010, Kaplan 2012), particularly to convert the ever expanding body of biological information into practicable techniques.

A more fundamental issue is that the majority of the studies into the effectiveness of HIPVs have been carried out using synthetic compounds (Kaplan 2012). To fulfil the requirements of the IFOAM Basic Standards for Organic Production and Processing, any HIPVs used in a certified organic system would have to be naturally derived, and this may present challenges for the development of inputs that are effective and allowable while also being economically viable. Accordingly, this chapter first reviews the current knowledge about HIPVs for field use and then looks at arthropod pest management strategies currently used in organic systems, and finally at the opportunities for the extended application of HIPVs and related technologies involving chemical cues in organic agriculture.

12.2 Overview of plant defences involving plant volatiles

Plants have evolved various direct and indirect defence mechanisms against attacking organisms (Pieterse & Dicke 2007, Dicke 2009; see Ode, Chapter 2, this volume). As part of constitutive and induced defence mechanisms, plants release volatile compounds from plant foliage, flowers and fruits into the atmosphere and from roots into the soil (Dudareva *et al.* 2006). Volatile compounds are lipophilic liquids of low molecular weight with high vapour pressure that can cross membranes freely and are released into the atmosphere or soil in the absence of a diffusion barrier (Pichersky *et al.* 2006). They function as semio-chemicals when they convey information between organisms. Semiochemicals that are emitted by plants in response to herbivore damage are also termed herbivore-induced plant volatiles (HIPVs) (Dicke & Sabelis 1988). Emitted HIPVs can directly affect herbivores negatively due to toxic, repellent and deterring properties (Lou & Baldwin 2003). Plant-damaging herbivores can also be affected indirectly by HIPVs which attract natural enemies of the attacking herbivores (Mumm & Dicke 2010). However, HIPVs are available to all of the arthropod community and therefore herbivores may also exploit them to their own benefit by locating suitable host plants (Kalberer *et al.* 2001). The induction of HIPVs occurs not only in response to herbivore feeding on plant parts above ground but also following physical plant damage, deposition of eggs, or from insects feeding on plant roots, attracting the respective natural enemies that use eggs, larvae or root feeders as prey or hosts (Hilker & Meiners 2006, Turlings & Ton 2006). Furthermore, HIPV emission takes place not only at the site of damage but also from other, uninfested plant parts (Turlings & Tumlinson 1992). There is evidence that HIPVs such as methyl jasmonate (MeJA), methyl salicylate (MeSA), ethylene (ET) and green leaf volatiles (GLVs) also act as plant-to-plant signals (Farmer 2001). These volatiles activate jasmonic acid- (JA), salicylic acid-(SA) and ET-dependent defence reactions in other parts of the same plant or in neighbouring undamaged plants, priming them for impending attack by boosting the production of endogenous aromatic and terpenoid volatile compounds that enhance the induced defence response of the plants (Frost *et al.* 2008, Heil & Ton 2008). However, an interaction in which plants increase the fitness of their neighbours without improving their own fitness may result in disadvantage for the emitting plant. The neighbouring plants are effectively 'eavesdropping' on their neighbour's volatile emissions, triggering their own defence responses in advance of damage to their own tissues occurring; a benefit derived from damage to their neighbour (Heil & Karban 2010). Plants emit complex blends of HIPVs, often species-specific, comprising hundreds of compounds (Krips *et al.* 2001). These include, for example, terpenoids, GLVs, phenylpropanoids and benzoids, with MeSA being the most studied (Mumm & Dicke 2010). The compounds usually first released are GLVs and these are therefore also called wound signals (Hatanaka 1993), while other volatiles such as MeJA and MeSA follow later and are still detectable from plants after hours or days (Hilker & Meiners 2006, Schaub *et al.* 2010). MeSA also acts as a plant-to-plant signalling volatile, leading to systemic acquired resistance (Heil & Ton 2008).

12.3 The use of synthetic HIPVs in pest management

Although there are no published studies of the use of HIPVs in organic crops, scope for their use can be gauged from the large number of studies undertaken in non-organic systems. These have demonstrated that the use of synthetic HIPVs can increase the diversity or density of natural enemies within crops. These HIPVs have been field-tested in many different fruit and vegetable crops in various countries by placing controlled-release dispensers into the crop (James 2005, Yu *et al.* 2008, Lee 2010, Orre *et al.* 2010) or by spray-applying the HIPVs to the crop foliage (Simpson *et al.* 2011a). Methyl salicylate has been one of the most studied compounds. Table 12.2 lists field studies that tested a range of

Table 12.2 Summary of field studies that led to significantly higher attraction to HIPVs deployed in (non-organic) arable field crops and increased the abundance of beneficial insects.

Natural enemies Order/Family/ Species	HIPV	Crop	Reference
Neuroptera			
Chrysopidae *Chrysopa nigricornis*	Methyl salicylate	Hops, grapes	James (2003a,b), James & Price (2004)
Chrysopa oculata	Methyl salicylate	Grapes	James (2006)
Chrysoperla carnea	2-phenylethanol	Soyabean	Zhu & Park (2005)
Hemerobiidae *Hemerobius* spp.	Methyl salicylate	Grapes	James & Price (2004)
Coleoptera			
Coccinellidae *Stethorus punctum picipes*	cis-3-Hexenyl acetate, methyl salicylate, cis-3-hexen-1-ol, benzaldehyde	Hops, grapes	James (2003b), James & Price (2004), James (2005)
Coccinella septempunctata	Methyl salicylate, cis-3-hexenyl acetate	Soyabean, cotton	Zhu & Park (2005), Yu et al. (2008)
Hemiptera			
Anthocoridae *Orius tristicolor*	cis-3-Hexenyl acetate, methyl salicylate, octyl aldehyde, benzaldehyde	Hops, grapes	James (2003b), James & Price (2004), James (2005, 2012)
Orius similis	Methyl salicylate, 3,7-dimethyl-1,3,6-octatriene, nonanal, cis-3-hexenyl acetate, nonanal + cis-3-hexen-1-ol	Cotton	Yu et al. (2008)
Geocoridae *Geocoris pallens*	Methyl salicylate, trans-2-hexen-1-al, indole	Hops, grapes, open field	James (2003b), James & Price (2004), James (2005)
Miridae *Deraeocoris brevis*	cis-3-hexenyl acetate, methyl salicylate	Hops, grapes	James (2003b), James & Price (2004)
Deraeocoris punctulatus	Octanal	Cotton	Yu et al. (2008)

Table 12.2 (*cont'd*)

Natural enemies Order/Family/ Species	HIPV	Crop	Reference
Diptera			
Syrphidae	Methyl salicylate, 2-phenylethanol	Hops, grapes, open field, soyabean	James (2003b), James & Price (2004), James (2005), Zhu & Park (2005)
Paragus quadrifasciatus	di-Methyl octatriene, nonanal + *cis*-3-hexen-1-ol, octanal	Cotton	Yu *et al.* (2008)
Tachinidae	Benzaldehyde	Hops	James (2005)
Empididae	Methyl salicylate	Grapes, hops	James & Price (2004), James (2012)
Sarcophagidae	Methyl salicylate, benzaldehyde, nonanal, geraniol, *cis*-jasmone	Grapes, hops	James & Price (2004), James (2005)
Chloropidae *Thaumatomyia glabra*	Methyl anthranilate	Open field	James (2005), Landolt *et al.* (2000)
Agromyzidae	Methyl salicylate	Open field	James (2005)
Hymenoptera			
Ichneumonidae *Diadegma semiclausum*	Methyl salicylate	Turnip	Orre *et al.* 2010
Braconidae	Methyl salicylate, *cis*-3-hexenyl acetate, geraniol, methyl anthranilate, methyl jasmonate, *cis*-jasmone, *cis*-3-hexen-1-ol	Grapes	James & Price (2004), James (2005)
Macrocentrus linearis	*cis*-3-Hexen-1-ol, 3,7-dimethyl-1,3, 6-octatriene	Cotton	Yu *et al.* (2008)
Eulophidae *Cerenius menes*	Methyl anthranilate	Field with vegetables and fruit trees	Murai *et al.* (2000)
Mymaridae *Anagrus daanei*	Farnesene, octyl aldehyde, *cis*-3-hexenyl acetate	Hops	James & Price (2004)

(*cont'd*)

Table 12.2 (cont'd)

Natural enemies Order/Family/ Species	HIPV	Crop	Reference
Anagrus spp.	Methyl salicylate, methyl jasmonate, *cis*-3-hexenyl acetate	Grapes	James & Grasswitz (2005), James (2012)
Encyrtidae *Metaphycus* spp.	Methyl salicylate, methyl jasmonate, *cis*-3-hexenyl acetate	Grapes	James & Grasswitz (2005)
Micro-Hymenoptera (no further identification determined in this study)	Methyl salicylate, indole, *cis*-3-hexen-1-ol, *cis*-3-hexenyl acetate	Open field, hops	James (2005, 2012)
Araneae Micryphantidae *Erigonidium graminicolum*	Methyl salicylate, *cis*-3-hexenyl acetate	Cotton	Yu *et al.* (2008)

HIPVs which led to attraction and an increase in abundance of natural enemies. Studies by James & Price (2004) and Simpson *et al.* (2011b) not only showed increased abundance of natural enemies but also reduced pest numbers and crop damage These HIPVs can be used at certain times of the year or when the monitoring of pest numbers indicates that control measures are required.

HIPVs have also been tested in conjunction with habitat manipulation, a concept termed 'attract and reward', to provide food and shelter for the attracted natural enemies in case prey or hosts are low in number (Simpson *et al.* 2011b, 2011c) (Fig. 12.1). In 'attract and reward', a HIPV is deployed in the crop, with the aim of attracting natural enemies. This can be in the form of a controlled-release dispenser or applied directly to the crop as a diluted spray. The reward component consists of a flowering plant species which is intended to provide floral foods and shelter. This concept has been tested in Australia in crops of wine grapes, broccoli and sweetcorn and, in New Zealand, in brassica crops. It was anticipated that both components might work synergistically compared with either 'attract' or 'reward' alone. However, these field experiments provided only modest evidence of a synergistic 'attract and reward' effect, with only one Scelionidae species being increased in abundance near MeSA-treated broccoli plants with reward compared with either MeSA-treated broccoli plants or reward alone (Simpson *et al.* 2011b). A follow-up field experiment by Simpson *et al.* (2011c) and a New Zealand study that also tested this concept in a brassica crop (Orre *et al.* 2010) did not find any synergistic effects. Simpson *et al.* (2011b) concluded that this could have been the result of several signalling cues such as odours from the spray-applied synthetic HIPVs, host volatiles, crop plant volatiles, flower volatiles and flower colour being present simultaneously in the crop systems. For example, arthro-

'Attract and Reward'

Figure 12.1 Field evaluation of 'attract and reward': vines have been treated with exogenous herbivore-induced plant volatiles to attract parasitoids and other natural enemies, while plots of flowering buckwheat provide nectar rewards to enhance the impact on pests (photograph by M. Simpson).

pods may have responded more to host, flower volatiles or colour than to the applied HIPVs. The relatively short distance between treatment plots in these replicated experiments may also have led to a diminution of the effects of individual treatments. However, despite true synergy not being found in 'attract and reward', Simpson *et al.* (2011b) concluded that this concept merits further work, as many natural enemies responded significantly either to the HIPVs or the reward plant and no negative effects on natural enemies were identified from the combined approach. Furthermore, HIPVs could be used early in the season while the reward is being established, and once the reward plants are flowering, HIPV deployment may no longer be necessary.

12.4 Arthropod pest management strategies used in organic farming

Organic farmers aim to keep pest damage under economic thresholds by using a variety of practices in place of synthetic chemical pesticides (Shrivastava *et al.* 2010). The pest management strategies employed in organic systems have been shown to successfully encourage biodiversity (Krauss *et al.* 2011) and this can enhance ecosystem services such as biological control and pollination (Sandhu *et al.* 2010). Management strategies that can positively contribute to biological control include crop rotations, and increasing crop diversity and non-crop habitats for natural enemies. Biodiversity on the farm can also be increased by integrating cover crops (Bugg & Waddington 1994, Altieri 1999). Many field and laboratory studies have tested different herbaceous flowering plant species and these have been shown to benefit natural enemy longevity and fecundity by supplying key ecological resources such as food and shelter (see Thomas *et al.* 1991, Begum *et al.* 2006, van Rijn *et al.* 2006, Olson & Wäckers 2007, Géneau *et al.* 2012). Some cover crops offer advantages other than supporting natural enemies. For example, legume plant species are beneficial for fixing atmospheric nitrogen and thereby improving soil nutrient status (Sanginga *et al.*, 1992, Hartwig & Ammon 2002, Anugroho *et al.* 2010). Further methods for preventing pest outbreaks can include mulching, sticky traps and pheromone-scented lures.

A contrasting series of approaches to suppress pests is based on silicon compounds such as potassium silicate and calcium silicate, which are allowable in organic farming because they are naturally occurring rather than synthetic in origin. They are used to strengthen plant roots and stems, to protect plant leaves and fruit from pathogens, and to constitute a phytolith barrier against insect feeding. Silicon is also the key component of the bio-dynamic preparation '501', which consists of finely ground quartz (silica) placed in a cowhorn and buried in the soil over the summer before being diluted in water and applied to the crop. Although biodynamic preparations such as 501 appear to be the antithesis of rationally based management and have generally not been fully tested under randomized, replicated conditions, there is a strong body of scientific evidence that other silicon com-pounds such as potassium silicate can markedly affect plant defences (Gomes *et al.* 2005, Kvedaras & Keeping 2007). A link to HIPV production has also been suggested in research with cucumbers that were treated with potassium silicate before being attacked by lepidop-teran larvae (Kvedaras *et al.* 2010). Volatiles from silicon-treated plants were more attractive to predatory beetles in Y-tube olfactometer bioassays. Furthermore, when the plants were placed in the field bearing known numbers of pest eggs, more of the eggs were removed (indicating predator activity) from silicon-treated plants than from control plants. This suggests that silicon plays a more profound role in plant defences than was previously recognized and can directly enhance HIPV-based induced resistance in plants attacked by pests, possibly by acting as a signal in inducing systemic chemical defences in plants (Kvedaras *et al.* 2010).

More generally, the use of biological control with natural enemies is an important pest management strategy within organic farming systems. Eilenberg *et al.* (2001) categorize biological control into classical biological control, inoculation biological control, inunda-tive biological control, and conservation biological control. The last-listed of these forms of biological control can include the establishment of flowering ground cover along field edges or within crops, cover cropping, mulching, and diversified cropping systems (Landis *et al.* 2000). Conservation biological control has therefore been a readily accepted part of organic agriculture as it coincides with the principle of increasing biodiversity and incor-porates practices already used.

To date, conservation biological control is the best-studied biological control approach in terms of a potentially synergistic effect with HIPVs, through, for example, the previously-described 'attract and reward' approach (Gurr & Kvedaras 2010). Although results to date look encouraging (Simpson *et al.* 2011b, 2011c), concerns have been expressed about the viability and long-term prospects for 'attract and reward'. Kaplan (2012) warns that the full implications of employing the vast number of HIPV compounds to augment biological control are as yet unknown. The major gaps in current knowledge listed by Kaplan (2012) include whether pest problems are transferred to untreated areas depleted of natural enemies, whether natural enemies lose their responsiveness to volatiles over time, the identity of compound(s) and the release rate(s) most effective, and effects on non-target species (notably pollinators).

A number of the other pest management strategies used in organic farming systems make the use of HIPVs viable, though not necessarily in the form of exogenous applications. These include the use of companion plants, trap cropping, intercropping and cultivar selection (Shrivastava *et al.* 2010). Within these systems, both the qualitative and quantitative characteristics of HIPVs depend upon the type of herbivore, the plant genotype, and various abiotic factors (Turlings *et al.* 1993, Takabayashi *et al.* 1994, Mumm & Dicke 2010). Mumm & Dicke (2010) provide a comprehensive review of what is known about how biotic and abiotic factors affect HIPV emissions. Selection of traits pertaining to yield may have resulted in the loss of the ability to produce certain HIPVs in some modern crop varieties. A notable example is North American maize varieties that were unable to produce (E)-β-caryophyllene, which attracts entomopathogenic nematodes; however, this trait was successfully reinstated in work by Degenhardt *et al.* (2009).

Approaches such as companion planting, trap cropping and intercropping, as used in the 'push–pull' technique, which do not depend on the application of exogenous HIPVs, are more compatible with organic certification standards than are the use of spray-on or dispenser-delivered technologies that generally use synthetic versions of HIPVs because of their low cost and ready availability from chemical suppliers. An indication of the potential uptake of these non-exogenous methods is evident in the success of a 'push–pull' strategy used in sub-Saharan Africa and based on the volatiles produced by a companion plant, desmodium (*Desmodium uncinatum* Jacq.), added to maize (to repel gravid stem-borer moths) and those produced by a second plant species, Napier grass (*Pennisetum purpureum* Schumach), sown as a strip bordering the maize to serve as a trap plant for ovipositing moths (Khan *et al.* 2012). Several studies of the success and impact of the 'push–pull' strategy have been published. Midega *et al.* (2009) showed that the larval parasitoids *Cotesia sesamiae* (Cameron) and *Cotesia flavipes* (Cameron) and the pupal parasitoid *Dentichasmias busseolae* Heinrich were more active under 'push–pull' conditions. Since its inception in 1999, this 'push–pull' method, employing endogenously produced plant chemical cues, has been adopted by nearly 50,000 farmers in sub-Saharan Africa (Khan *et al.* 2012).

12.5 Potential for extending chemical cue use in organic systems

The many field studies which concluded that HIPVs can contribute to improving biological pest control have not explicitly addressed their potential use in organic farming. Nevertheless, some encouragement can be taken from other pest control methods, based on similar chemical ecology principles, already being used in organic agriculture. These include pheromone traps for pest monitoring and disrupting mating behaviour or acting as repellents (Zehnder *et al.* 2007). Furthermore, two HIPV-based products that are allowable in organic farming are currently being marketed in the USA. They are PredaLure®, containing wintergreen essential oil, of which methyl salicylate is a major constituent, and Benallure®, containing 2-phenylethanol, both sold in the form of slow-release dispensers. A remaining hurdle to the wider use of HIPV-based strategies in organic systems is to gain approval for the use of HIPVs that are sprayed directly onto the crop. Nevertheless, there are some indicators that HIPVs sprays may gain approval. Regulations regarding which insecticidal products organic farmers are permitted to use differ between countries. However, there are some general guidelines such as that the insecticidal products must be derived from natural sources rather than being synthetically manufactured. Permitted products that are sprayed onto crops, for example, include botanical extracts such as neem oil, insect pathogens (e.g. *Bacillus thuringiensis*) and fungal derivatives.

Such products are permitted in organic farming but are not without their risks. For example, neem can adversely affect non-target organisms such as arthropod natural enemies (Schmutterer 1997). However, Charleston *et al.* (2006) found no negative effects on parasitoids when neem extract was applied to cabbage plants. The authors also found that damage by *Plutella xylostella* L. (Lepidoptera: Plutellidae) was significantly less in neem-treated plants. Furthermore, their study identified a signalling function by the extract made from the syringe tree *Melia azedarach* L. (Meliaceae), which was applied to the cabbage plants and resulted in induced plant volatile emission which attracted the parasitoid *Cotesia plutellae* (Kurdjumov) (Hymenoptera: Braconidae).

Genetically modified plants are being developed with enhanced or restored ability to produce HIPVs (Kappers *et al.* 2005, Schnee *et al.* 2006, Degenhardt *et al.* 2009, Kos *et al.* 2009), quite possibly compensating for resistance traits lost through plant breeding. While genetically modified plants have no place within organic systems, it is quite likely they will become part of the arsenal for conventional farmers who employ integrated pest management techniques (Kos *et al.* 2009). Therefore, crop variety selection will remain an important consideration in organic farming. An interesting research avenue is to explore the volatiles produced by the 'landrace' and 'heirloom' crop varieties favoured by some organic growers. It is possible that these retain high levels of HIPV production that have been lost from many modern varieties as a result of selection for easily measured agronomic traits such as yield.

In the future, many HIPVs may be developed into permissible organic agriculture pesticides by using plant-derived compounds. Field experiments would then have to be conducted with these plant-derived compounds in various crops to ascertain their efficacy in attracting natural enemies and any side-effects, as they may differ from their synthetic counterparts. Potentially, HIPVs or HIPV blends from natural sources will be less potent than the synthetic versions, but this may prove to be advantageous. Sobby *et al.* (2012) recently reported that maize plants were more attractive to parasitoids when HIPV emissions were lower. They surmised that either the HIPVs were repelling the parasitoids or the actual attractant was being obscured by the dominant sesquiterpenes.

There appears to be scope to maximize the levels of endogenous HIPV production by pest-attacked plants through the use of silicon compounds, which are likely to be allowable in most organic certification schemes. In addition, most studies to date have looked at using synthetic HIPVs as part of a conservation biological control strategy (Kaplan 2012). Investigation is warranted into the potential for using HIPVs beyond that of enhancing conservation biological control. One particularly exciting opportunity is to use HIPVs to enhance the success of classical, inundative and inoculation biological control strategies by controlling the rate of dispersal of biological control agents from release sites (Heimpel & Asplen 2011, Kaplan 2012). If agents disperse too rapidly from the point of release, they can be subject to Allee effects (undercrowding), particularly at the leading edge of the dispersing population. More practically, in the case of the release of an expensive parasitoid (e.g. *Trichogramma*) in a particular field crop or other site, the grower will want the wasps to remain in the area rather than dispersing to areas where biological control is not required. In contrast, the slow dispersal of agents can be disadvantageous when the aim is to establish a new agent widely and to gain control of the target species over a large area.

Inbreeding can also disadvantage the longer-term viability of a poorly dispersed agent population. Under each of these opposite scenarios, chemical cues could be used to regulate the movement of parasitoids and other agents, speeding or slowing dispersal as conditions require.

12.6 Conclusions

Pest management is a major aspect in organic farming and remains a challenging task. The use of HIPVs as chemical cues for arthropod pest management appears to be set to make a significant contribution to future organic agriculture systems. HIPVs have been tested widely in a range of crop species and on several continents, with promising results. What is lacking so far is comprehensive testing under organic production conditions. The key reason for this is that most of the currently used HIPVs would not be allowable under organic registration systems. At present, only two commercially available HIPV-based products are allowable, and only in slow-release dispenser form. Certification authorities are likely to be less inclined to allow the direct application of synthetic versions of natural HIPVs onto crops, but plant-derived compounds such as oil of wintergreen would be allowable.

Despite such constraints to the use of exogenous chemical cues, endogenously produced cues could be exploited more widely in ways that are compatible with organic agriculture. Silicon compounds may allow plants to mount a stronger induced resistance response when subject to attack by pests. Research into 'landrace' and 'heirloom' crop varieties may uncover higher levels of endogenous HIPV production compared with the crop varieties used in modern agriculture. Finally, the use of 'attract and reward', 'push–pull' and related plant diversity approaches constitutes a promising branch of chemical ecology that could maximize the impact of beneficial parasitoids and other natural enemies for pest management in organic agriculture.

References

Altieri, M. (1999) The ecological role of biodiversity in agroecosystems. *Agriculture, Ecosystems and Environment* **74**: 19–31.

Anugroho, F., Kitou, M., Kinjo, K. and Kobashigawa, N. (2010) Growth and nutrient accumulation of winged bean and velvet bean as cover crops in a subtropical region. *Plant Production Science* **13**: 360–6.

Begum, M., Gurr, G.M., Wratten, S.D., Hedburg, P.R. and Nicol, H.I. (2006) Using selective food plants to maximize biological control of vineyard pests. *Journal of Applied Ecology* **43**: 547–54.

Bugg, R.L. and Waddington, C. (1994) Using cover crops to manage arthropod pests of orchards: a review. *Agriculture, Ecosystems and Environment* **50**: 11–28.

CAC (Codex Alimentarius Commission) (2001) *Guidelines for the Production, Processing, Labelling and Marketing of Organically Produced Foods*. First revision. Joint Food and Agriculture Organisation (FAO) and World Health Organisation (WHO) Food Standards Program, Rome [available at http://www.codexalimentarius.org/standards/list-of-standards/en/, accessed 25/6/2012].

Charleston, D.S., Kfir, R., Dicke, M. and Vet, L.E.M. (2006) Impact of botanical extracts derived from *Melia azedarach* and *Azadirachta indica* on populations of *Plutella xylostella* and its natural enemies: a field test of laboratory findings. *Biological Control* **39**: 105–14.

Council of the European Union (2007) Council regulation (EC) No 834/2007 of 28 June 2007 on organic production and labelling of organic products and repealing regulation (EEC) No 2092/91. *Official Journal of the European Union* **189**: 1–23.

Degenhardt, J., Hiltpold, I., Köllner, T.G., Frey, M., Gierl, A., Gershenzon, J., Hibbard, B.E., Ellersieck, M.R. and Turlings, T.C.J. (2009) Restoring a maize root signal that attracts insect-killing nematodes to control a major pest. *Proceedings of the National Academy of Sciences USA* **106**: 13213–18.

Dicke, M. (2009) Behavioural and community ecology of plants that cry for help. *Plant, Cell and Environment* **32**: 654–65.

Dicke, M. and Sabelis, M.W. (1988) How plants obtain predatory mites as bodyguards. *Netherlands Journal of Zoology* **38**: 148–65.

Dudareva, N., Negre, F., Nagegowda, D.A. and Orlova, I. (2006) Plant volatiles: recent advances and future perspectives. *Critical Reviews in Plant Sciences* **25**: 417–40.

Eilenberg, J., Hajek, A. and Lomer, C. (2001) Suggestions for unifying the terminology in biological control. *BioControl* **46**: 387–400.

El-Hage Scialabba, N. and Müller-Lindenlauf, M. (2010) Organic agriculture and climate change. *Renewable Agriculture and Food Systems* **25**: 158–69

Farmer, E.E. (2001) Surface-to-air signals. *Nature* **411**: 854–6.

Forschungsinstitut für Biologischen Landbau (FiBL) (2012) *Organic Farming Statistics* [available at http://www.fibl.org/en/themen/themen-statistiken.html, accessed 18/6/2012].

Frost, C.J., Mescher, M.C., Carlson, J.E. and De Moraes, C.M. (2008) Plant defense priming against herbivores: getting ready for a different battle. *Plant Physiology* **146**: 818–24.

Géneau, C.E., Wäckers, F.L., Luka, H., Daniel, C. and Balmer, O. (2012) Selective flowers to enhance biological control of cabbage pests by parasitoids. *Basic and Applied Ecology* **13**: 85–93.

Gomes, F.B., de Moraes, J.C., dos Santos, C.D. and Gaussian, M.M. (2005) Resistance induction in wheat plants by silicon and aphids. *Scientia Agricola* **62**: 547–51.

Gurr, G.M. and Kvedaras, O.L. (2010) Synergizing biological control: scope for sterile insect technique, induced plant defences and cultural techniques to enhance natural enemy impact. *Biological Control* **52**: 198–207.

Hartwig, N.L. and Ammon, H.U. (2002) Cover crops and living mulches. *Weed Science* **50**: 688–99.

Hatanaka, A. (1993) The biogeneration of green odor by green leaves. *Phytochemistry* **34**: 1201–18.

Heil, M. and Karban, R. (2010) Explaining evolution of plant communication by airborne signals. *Trends in Ecology and Evolution* **25**: 137–44.

Heil, M. and Ton, J. (2008) Long-distance signalling in plant defence. *Trends in Plant Science* **13**: 264–72.

Heimpel, G.E. and Asplen, M.K. (2011) A 'Goldilocks' hypothesis for dispersal of biological control agents. *BioControl* **56**: 441–50.

Hilker, M. and Meiners, T. (2006) Early herbivore alert: insect eggs induce plant defense. *Journal of Chemical Ecology* **32**: 1379–97.

IFOAM (International Federation of Organic Agriculture Movements) (2009a) Definition of organic agriculture [available at http://www.ifoam.org/growing_organic/definitions/doa/index.html, accessed 18/06/2012].

IFOAM (International Federation of Organic Agriculture Movements) (2009b) Principles of organic agriculture [available at http://www.ifoam.org/about_ifoam/principles/index.html, accessed 18/06/2012].

James, D.G. (2003a) Field evaluation of herbivore-induced plant volatiles as attractants for beneficial insects: methyl salicylate and the green lacewing, *Chrysopa nigricornis. Journal of Chemical Ecology* **29**: 1601–9.

James, D.G. (2003b) Synthetic herbivore-induced plant volatiles as field attractants for beneficial insects. *Environmental Entomology* **32**: 977–82.

James, D.G. (2005) Further field evaluation of synthetic herbivore-induced plant volatiles as attractants for beneficial insects. *Journal of Chemical Ecology* **31**: 481–95.

James, D.G. (2006) Methyl salicylate is a field attractant for the goldeneyed lacewing, *Chrysopa oculata*. *Biocontrol Science and Technology* **16**: 107–10.

James, D.G. (2012) Grape and hop plants sprayed with botanical oil pesticides containing herbivore-induced plant volatiles attract insect predators and parasitoids. *Environmental Entomology* in press 3.

James, D.G. and Grasswitz, T.R. (2005) Synthetic herbivore-induced plant volatiles increase field captures of parasitic wasps. *BioControl* **50**: 871–80.

James, D.G. and Price, T.S. (2004) Field-testing of methyl salicylate for recruitment and retention of beneficial insects in grapes and hops. *Journal of Chemical Ecology* **30**: 1613–28.

Kalberer, N.M., Turlings, T.C.J. and Rahier, M. (2001) Attraction of a leaf beetle (*Oreina cacaliae*) to damaged host plants. *Journal of Chemical Ecology* **27**: 647–61.

Kaplan, I. (2012) Attracting carnivorous arthropods with plant volatiles: the future of biocontrol or playing with fire? *Biological Control* **60**: 77–89.

Kappers, I.F., Aharoni, A., van Harpen, T.W.J.M., Luckerhoff, L.L., Dicke, M. and Bouwmeester, H.J. (2005) Genetic engineering of terpenoid metabolism attracts bodyguards to *Arabidopsis*. *Science* **309**: 2070–2.

Khan, Z.R, Midega, C.A.O., Pittchar, J., Bruce, T.J.A. and Pickett, J.A. (2012) 'Push–pull' revisited: the process of successful deployment of a chemical ecology-based pest management tool. In: Gurr, G.M., Wratten, S.D., Snyder, W.E. and Read, D.M.Y. (eds) *Biodiversity and Insect Pests: Key Issues for Sustainable Management*. John Wiley & Sons, Chichester, UK, pp. 259–75.

Kos, M., van Loon, J.J.A., Dicke, M. and Vet, E.M. (2009) Transgenic plants as vital components of integrated pest management. *Trends in Biotechnology* **27**: 621–6.

Krauss, J., Gallenberger, I. and Steffan-Dewenter, I. (2011) Decreased functional diversity and biological pest control in conventional compared to organic crop fields. *PLoS ONE* **6**(5): e19502.

Krips, O.E., Willems, P.E.L., Gols, R., Posthumus, M.A., Gort, G. and Dicke, M. (2001) Comparison of cultivars of ornamental crop *Gerbera jamesonii* on production of spider mite-induced volatiles, and their attractiveness to the predator *Phytoseiulus persimilis*. *Journal of Chemical Ecology* **27**: 1355–72.

Kvedaras, O.L. and Keeping, M.G. (2007) Silicon impedes stalk penetration by the borer *Eldana saccharina* in sugarcane. *Entomologia Experimentalis et Applicata* **125**: 103–10.

Kvedaras, O.L., An, M., Choi, Y.S. and Gurr, G.M. (2010) Silicon enhances natural enemy attraction to pest-infested plants through induced plant defences. *Bulletin of Entomological Research* **100**: 367–71.

Landis, D.A., Wratten, S.D. and Gurr, G.M. (2000) Habitat management to conserve natural enemies of arthropod pests in agriculture. *Annual Review of Entomology* **45**: 175–201.

Landolt, P.J., Wixson, T., Remke, L.J., Lewis, R.R. and Zack, R.S. (2000) Methyl anthranilate attracts males of *Thaumatomyia glabra* (Meigen) (Diptera: Chloropidae). *Journal of the Kansas Entomological Society* **73**: 189–94.

Lee, J.C. (2010) Effect of methyl salicylate-based lures on beneficial and pest arthropods in strawberry. *Environmental Entomology* **39**: 653–60.

Lou, Y.G. and Baldwin, I.T. (2003) *Manduca sexta* recognition and resistance among allopolyploid *Nicotiana* host plants. *Proceedings of the National Academy of Sciences USA* **100**: 14581–6.

Midega, C.A.O., Khan, Z.R., van den Berg, J., Ogol, C.K.P.O., Bruce, T.J. and Pickett, J.A. (2009) Non-target effects of the 'push–pull' habitat management strategy: parasitoid activity and soil fauna abundance. *Crop Protection* **28**: 1045–51.

Mumm, R. and Dicke, M. (2010) Variation in natural plant products and the attraction of bodyguards involved in direct plant defense. *Canadian Journal of Zoology* **88**: 628–67.

Murai, T., Imai, T. and Maekawa, M. (2000) Methyl anthranilate as an attractant for two thrips species and the thrip parasitoid *Ceranisus menes*. *Journal of Chemical Ecology* **26**: 2557–65.

Olson, D.M. and Wäckers, F.L. (2007) Management of field margins to maximize multiple ecological services. *Journal of Applied Ecology* **44**: 13–21.

Orre, G.U.S., Wratten, S.D., Jonsson, M. and Hale, R.J. (2010) Effects of an herbivore-induced plant volatile on arthropods from three trophic levels in brassicas. *Biological Control* **53**: 62–7.

Pichersky, E., Noel, J.P. and Dudareva, N. (2006) Biosynthesis of plant volatiles: nature's diversity and ingenuity. *Science* **311**: 808–11.

Pieterse, C.M.J. and Dicke, M. (2007) Plant interactions with microbes and insects: from molecular mechanisms to ecology. *Trends in Plant Science* **12**: 564–9.

Sandhu, H.S., Wratten, S.D. and Cullen, R. (2010) Organic agriculture and ecosystem services. *Environmental Science and Policy* **13**: 1–7.

Sanginga, N., Mulongoy, K. and Swift, M.J. (1992) Contribution of soil organisms to the sustainability and productivity of cropping systems in the tropics. *Agriculture, Ecosystems and Environment* **41**: 135–52.

Schaub, A., Blande, J.D., Graus, M., Oksanen, E., Holopainen, J.K. and Hansel, A. (2010) Real-time monitoring of herbivore-induced volatile emissions in the field. *Physiologia Plantarum* **138**: 123–33.

Schmutterer H. (1997) Side-effects of neem (*Azadirachta indica*) products on insect pathogens and natural enemies of spider mites and insects. *Journal of Applied Entomology* **121**: 121–8.

Schnee, C., Kollner, T.G., Held, M., Turlings, T.C.J., Gershenzon, J. and Dergenhardt, J. (2006) A maize terpene synthase contributes to a volatile defense signal that attracts natural enemies of maize herbivores. *Proceedings of the National Academy of Sciences USA* **103**: 1129–34.

Shrivastava, G., Rogers, M., Wszelaki, A., Panthee, D.R. and Chen, F. (2010) Plant volatile-based insect pest management in organic farming. *Critical Reviews in Plant Sciences* **29**: 123–33.

Simpson, M., Gurr, G.M., Simmons, A.T., Wratten, S.D., James, D.G., Leeson, G. and Nicol, H. (2011a) Insect attraction to synthetic herbivore-insect plant volatile treated field crops. *Agricultural and Forest Entomology* **13**: 45–57.

Simpson, M., Gurr, G.M., Simmons, A.T., Wratten, S.D., James, D.G., Leeson, G., Nicol, H. and Orre, G.U.S. (2011b) Attract and reward: combining chemical ecology and habitat manipulation to enhance biological control in field crops. *Journal of Applied Ecology* **48**: 580–90.

Simpson, M., Gurr, G.M., Simmons, A.T., Wratten, S.D., James, D.G., Leeson, G., Nicol, H. and Orre, G.U.S. (2011c) Field evaluation of the 'attract and reward' biological control approach in vineyards. *Annals of Applied Biology* **159**: 69–78.

Sobby, I.S., Erb, M., Sarhan, A.A., El-Husseini, M.M., Mandour, N.S. and Turlings, T.C.J. (2012) Less is more: treatment with BTH and laminarin reduces herbivore-induced volatile emissions in maize but increases parasitoid attraction. *Journal of Chemical Ecology* **38**: 348–60.

Takabayashi, J., Dicke, M. and Posthumus, M.A. (1994) Volatile herbivore-induced terpenoids in plant–mite interactions: variation caused by biotic and abiotic factors. *Journal of Chemical Ecology* **20**: 1329–54.

Thomas, M.B., Wratten, S.D. and Sotherton, N.W. (1991) Creation of 'island' habitats in farmland to manipulate populations of beneficial arthropods: predator densities and emigration. *Journal of Applied Ecology* **28**: 906–17.

Turlings, T.C.J. and Ton, J. (2006) Exploiting scents of distress: the prospect of manipulating herbivore-induced plant odours to enhance the control of agricultural pests. *Current Opinion in Plant Biology* **9**: 421–7.

Turlings, T.C.J. and Tumlinson, J.H. (1992) Systemic release of chemical signals by herbivore-injured corn. *Proceedings of the National Academy of Sciences USA* **89**: 8399–402.

Turlings, T.C.J., Wäckers, F.L., Vet, L.E.M., Lewis, W.J. and Tumlinson, J.H. (1993) Learning of host-finding cues by hymenopterous parasitoids. In: Papaj, D.R. and Lewis, W.J. (eds) *Insect Learning*. Chapman & Hall, New York, pp. 51–78.

van Rijn, P.C.J., Kooijman, J. and Wäckers, F.L. (2006) The impact of floral resources on syrphid performance and cabbage aphid biological control. *IOBC/WPRS Bulletin* **29**: 149–52.

Yu, H., Zhang, Y., Wu, K., Xi, W.G. and Yu, Y.G. (2008) Field-testing of synthetic herbivore-induced plant volatiles as attractants for beneficial insects. *Environmental Entomology* **37**: 1410–15.

Yussefi, M. and Willer, H. (2003) *The World of Organic Agriculture: Statistics and Future Prospects.* Tholey-Theley: International Federation of Organic Agriculture Movements.

Zehnder, G., Gurr, G.M., Kuehne, S., Wade, M., Wratten, S.D. and Wyss, E. (2007) Arthropod pest management in organic crops. *Annual Review of Entomology* **50**: 52–87.

Zhu, J. and Park, K.C. (2005) Methyl salicylate, a soybean aphid-induced plant volatile attractive to the predator *Coccinella septempunctata. Journal of Chemical Ecology* **31**: 1733–46.

13

Application of chemical cues in arthropod pest management for forest trees

Timothy D. Paine

Department of Entomology, University of California, Riverside, USA

Abstract

Many natural enemies of insect herbivores in forest systems use semiochemicals to locate suitable hosts or prey in a complex environment. These chemical signals may be kairomones from their host or prey insects, constitutive host plant volatiles, odours produced by fungal or microbial symbionts associated with the herbivore host or prey, or volatiles induced in the plant host by herbivore oviposition or consumption of host tissues. However, the use of semiochemicals as a management tactic for insect herbivores in forest systems faces a number of very significant challenges. Many of the critical pests are episodic in the nature of their damage, so it is often difficult to predict when the tools will be needed. In addition, forest management objectives may limit pest control options. Forests that are grown on a very long rotational harvest cycle, of decades to centuries, may have small annual returns which may limit the implementation of high-cost tactics. Other forms of value for the forests, for example watershed protection, scenic amenity, wilderness habitat, wildlife or pasturage, are difficult to determine over the life of the forest and may also limit the willingness to undertake costly management options. Many forest systems are spread over large areas with limited accessibility, which also limits implementation. The scale of forest systems, as well as the long periods between outbreaks, are major factors in reducing opportunities for implementation of chemical cue-based enhancement of natural enemy activity in many forest environments.

Chemical Ecology of Insect Parasitoids, First Edition. Eric Wajnberg and Stefano Colazza.
© 2013 John Wiley & Sons, Ltd. Published 2013 by John Wiley & Sons, Ltd.

13.1 Forest insect herbivores and natural enemy host/prey finding

The orientation of natural enemies to potential hosts or prey is thought to follow a sequence of behaviours (Rutledge 1996). As with many herbivores, the initial behavioural step would be orientation to an environment that would contain suitable hosts or prey. That would assume orientation to the environment containing the plant host using visual or chemical stimuli as long-range cues. Once in the appropriate environment and on the correct host plant, the searching natural enemies must locate the insect host or prey using a combination of visual, tactile and odour signals as short-range cues. The source of odours that could be used as cues by predators or parasitoids in searching for suitable prey or hosts include plant volatiles (produced with or without damage caused by insect feeding), kairomones produced by the insect host or prey, general fermentation volatiles, or volatiles produced by microorganisms that are specifically associated with the insect host or prey (metabolic volatiles associated with bacteria and fungi transmitted by the insect; see Adams & Six 2008, Boone *et al.* 2008, 2009).

The range of semiochemical stimuli used by either predators or parasitoids can be documented across a taxonomic range of potential prey or hosts. For example, members of a diverse guild of predators are associated with mast scales on Mediterranean pines. Many of these predators exploit the sex pheromones of their prey to locate feeding and oviposition sites. Mendel *et al.* (1995, 2004), for example, demonstrated that anthocorid and hemerobiid predators respond to sex pheromones as kairomones for locating their prey. Adults and larval coccinelid and dasytid beetles also respond to the sex pheromones of their scale hosts: the adult females respond at a distance to locate scale colonies, and the larvae respond at a short range to find prey (Branco *et al.* 2006a, 2006b, 2011).

In addition, conifer sawflies' natural enemies show a wide variety of responses to semiochemical cues. At one end of the spectrum, the generalist tachinid parasitoid, *Drino inconspicua* Meigen (Diptera: Tachinidae), responds to movements of its host, the European spruce sawfly *Gilpinia hercyniae* (Hartig) (Hymenoptera: Diprionidae), but not to volatiles from the herbivore or plant host (Dippel & Hilker 1998). However, foraging *Dahlbominus fuscipennis* (Hymenoptera: Eulophidae) are attracted to the host habitat of *G. hercyniae* by odours from needles fallen from the tree but not to volatiles from the sawfly pupae (Rostás *et al.* 1998). Ovipositor probing was stimulated in *Olesicampe monticola* (Hedwig) (Hymenoptera: Ichneumonidae) in response to fresh frass from larvae of the larch sawfly, *Cephalcia lariciphila* Wachtl (Hymenoptera: Pamphiliidae) (Longhurst & Baker 1981). Parasitoids can also respond to the mating pheromones produced by their hosts. The eulophid egg parasitoids *Chrysonotomyia ruforum* Krausse (Hymenoptera, Eulophidae) and *Dipriocampe diprioni* Ferr. (Hymenoptera, Eulophidae) showed arrestment behaviour when exposed to sex pheromone components of their hosts *Diprion pini* L. (Hymenoptera, Diprionidae) and *Neodiprion sertifer* Geoffroy (Hymenoptera Diprionidae) (Hilker *et al.* 2000). In addition to responding to odours produced by their insect hosts, some parasitoids can respond to changes in specific semiochemical volatile profiles that are elicited in the host plant in response to sawfly oviposition (Hilker *et al.* 2002, 2005, Mumm *et al.* 2005, Mumm & Hilker 2006, Koepke *et al.* 2008, 2010).

Siricid wood wasp larvae excavate galleries in the inner bark and wood of conifer hosts. The larvae are associated with symbiotic fungi that colonize the host tissues. Volatiles from the symbionts associated with *Sirex noctilio* F. (Hymenoptera: Siricidae) are attractive to ovipositing ichneumonid and ibaliid wasp parasitoids (Madden 1968). The volatiles from fungal symbiont cultures 3–4 months old are the most attractive, suggesting that females are responsive to cues associated with host larvae that are at the most suitable developmental stage (Spradbery 1970, 1974). *Ibalia leucospoides* (Hochenwarth) (Hymenoptera: Ibaliidae) females can determine the density of wood wasp larvae in a patch of host log at a distance, and will choose patches with higher numbers of larvae in preference to patches with fewer larvae (Martinez Andres *et al.* 2006, Fischbein *et al.* 2012).

Parasitoids of lepidopteran herbivores may respond to a variety of semiochemicals associated with their herbivore hosts. *Tetrastichus servadeii* Domenichini (Hymenoptera: Eulophidae) and *Ooencyrtus pityocampae* Mercet (Hymenoptera: Encyrtidae) are egg parasitoids of *Thaumetopoea pityocampa* (Den. et Schiff.) (Lepidoptera: Thaumetopoeidae), the pine processionary moth. They respond to different semiochemical cues associated with their host (Battisti 1989). *O. pityocampae* is attracted to the sex pheromone of the moth, while *T. servadeii* is not attracted to the pheromone stimuli, but appears to be attracted to host plant cues.

Bark beetles (Coleoptera: Curculionidae: Scolytinae) colonize the inner bark of host trees. Many species utilize aggregation pheromones that result in mass colonization, which overwhelms the host defences of the trees (Paine *et al.* 1997). The arrival of their natural enemies may be mediated by semiochemicals, including host volatiles and pheromones. There is an evolutionary advantage for the predators and parasitoids to respond to semiochemical stimuli that are associated with the appropriate developmental stage of their host. Natural enemies that are associated with attacking adults or early developmental stages often respond to the aggregation pheromones of the beetles (Tommeras 1985, Grégoire *et al.* 1992, Pettersson 2001a, 2001b, Pettersson & Boland 2003, Hulcr *et al.* 2005, 2006).

The type of semiochemical may determine its potential utility as an approach for management of different guilds of natural enemies, as well as the potential risks. For example, response to the kairomonal signal of host sex or aggregation pheromones would be an important stimulus that would enable predators to arrive at a plant host at the same time as the insect prey. Similarly, egg parasitoids would have an advantage if they responded to host sex pheromones, as long as oviposition followed very soon after mating. However, use of the sex or aggregation pheromones to attract natural enemies could also increase the risk of increasing or concentrating pest densities and increasing damage to the plant host. The release of artificial sources of host pheromones, for example in mating disruption programmes (Yamanaka & Liebhold 2009, Tobin *et al.* 2011, Suckling *et al.* 2012), could result in confusion or disorientation of the natural enemies such that host finding is disrupted. If used in conjunction with traps, for example in mass trapping or trap tree programmes (Lindgren & Borden 1993, El-Sayed *et al.* 2009, Bednarz *et al.* 2011), the release of host pheromones could remove natural enemies from the environment and reduce potential control (Aukema *et al.* 2000).

Release of plant volatiles associated with insect feeding, or volatiles associated with microbial or fungal symbionts, could be valuable for attracting or arresting predators and parasitoids of host larvae and adults if responses to these stimuli can be demonstrated. Use of these materials does not carry the same risk of concentrating the herbivore host as could occur with aggregation or sex pheromones. However, there is still a risk of confusion or disruption in orientation to herbivore feeding sites. An alternative that has been used in a number of systems that could qualify as a chemical cue in the broadest sense is the application of food supplements (Tassan *et al.* 1979; see Simpson, Read & Gurr, Chapter 12, this volume). Food sprays that provide nutrition for adult natural enemies function to arrest the adults in the target environment and can increase the fecundity of females that feed on these supplements. Both of these effects of added nutrition have the potential to enhance the impact of natural enemies.

13.2 Introduction to forest systems

Forest systems present unique challenges for the use of chemical cues for improved success of biological control agents. Conventional agricultural systems are either planted annually or, in the case of perennial cropping systems, are subject to annual cultural practices (e.g. pruning and sanitation). In both cases, the crops are typically uniform in spacing, are planted at a specific density, are of uniform age, and have a uniform genotype. Consequently, many of the herbivores feeding on the plants are annual pests that can be the focus of conventional integrated pest management practices implemented every year. Because the community of herbivores and the temporal pattern in their colonization of cropping systems is often well established and reasonably predictable in an annual cycle, it is also reasonable to exploit natural enemies as a management tactic. The herbivore will either overwinter in or around established agricultural environments or move into the environments in a seasonal pattern. Natural enemies will also be resident in and around the cropping system or also move into the crops as the pest populations arrive. The annual population cycles of pests and natural enemies may facilitate both the timing of semiochemical releases and the predictability of the response by target natural enemies.

Forest systems may differ fundamentally from annual or perennial agriculture. Unmanaged stands of forest trees can extend over large areas with limited access. These forests are communities with high species diversity, high genetic diversity, and high spatial complexity. The communities may be dominated by a small number of species that characterize a particular forest type. Although there will be a dominant size class of individuals of these species, there will also frequently be a variety of age classes and a diversity in genotypes. In addition, there may also be a very large number of subdominant species as well as understorey shrubs and forbs. The high species diversity is also reflected in high spatial complexity in layers of vegetation from the forest floor through the canopy.

The harvest of commercial agricultural crops typically occurs on an annual cycle. Although there are examples of selective cutting of individual mature trees, forest harvest cycles are typically on the order of decades to centuries. The value of trees may be significant, but on these time scales, the annual return across a landscape may be relatively small. Consequently, the cost of implementing management tactics may severely restrict the potential options. Moreover, the value of these trees in unmanaged forests extends beyond the commercial value of the timber. They may include important non-monetary elements, including habitat conservation, water resources, recreation and aesthetic quality. These alternative values may have a significant impact on the suitability of management options that can be applied to the stands.

Managed forests, including plantations and seed orchards, often have higher economic value and, potentially, easier access. These forests may be highly variable in species diversity, age structure and spatial complexity. The range in managed forests can extend from minimal management on native forest lands or plantations of native trees on reforested lands to plantations of non-native species on agricultural or previously non-forested lands. Plantations that originate from a defined or selected seed source may have very narrow genetic diversity in the dominant species. Similarly, the plantations tend to be relatively uniform in age, have limited species diversity, and are planted at relatively uniform densities. The lack of community diversity may increase the risk of insect outbreaks and affect the potential for using chemical cues for enhancing the effectiveness of natural enemies.

Urban forests and parks present a third forest type and a different management challenge for enhancing the impact of natural enemies. Urban forest communities are often highly diverse mixtures of native and non-native tree species. Within neighbourhoods, single species may be of uniform age, but across a community the street trees of urban forests may be a mosaic of species and ages. However, these represent the most intensively managed systems with the highest economic value. That is, while the trees have limited commercial value because they are rarely harvested for timber, they add very significantly to the quality of life in an urban environment and add greatly to property values. The opportunities for use of a diversity of chemical cues could be the greatest in these systems.

Many herbivores of forest trees are present at low levels. These can be described as perennial pests that feed on trees each year. Furthermore, many of the most critical forest insect pests are episodic. Environmental conditions (e.g. drought or environmental change associated with global climate change) may become suitable, or forest conditions (e.g. tree susceptibility or suitability associated with tree maturity) may become optimal for rapid increases in pest populations. Unlike perennial pests, outbreaks of these episodic pests are very difficult to predict. Finally, there are increasing numbers of introduced herbivores that have become established in susceptible forest communities (e.g. emerald ash borer, red turpentine beetle, gold spotted oak borer, gypsy moth) and are now causing great devastation in managed and unmanaged forests (Coleman *et al.* 2011, Lu *et al.* 2011, Pugh *et al.* 2011, Couture *et al.* 2012). These are of particular importance because the potential use of semiochemicals for enhancing natural enemy activity may be limited by whether or not biological control agents have accompanied the herbivore or been introduced as part of a classical biological control programme.

The different forest types, level of management, value of the resources, and the different outbreak potential of forest insect pests provide constraints on the possible use of semiochemicals to reduce herbivore damage. In addition, the community of natural enemies, including the diversity and relative abundance of generalist and specialist species, must be considered in assessing the potential for manipulation. Consequently, potential targets and opportunities must be evaluated differently in the three different forest systems.

The potential for use of chemical cues for enhancing biological control or natural enemy success in forest environments will be a function of the target pest species, the community of natural enemies, the relative amenability to manipulation, the ease of application, and the type of forest system involved. The likelihood of implementation, the cost, and the probability of success must be related back to management goals for the forest, the value and type of resource to be protected, and the intensity of management in the forest. The risk of damage and the potential benefits that could accrue must be assessed against the cost over the life of the forest or the duration of the rotation cycle. Unfortunately, uncertainty in any of these areas could limit implementation of this or any other pest management tactic. However, there are some example systems that can be evaluated.

Table 13.1 Insect herbivore types found in North American forest systems and semiochemicals used by their natural enemies.

Insect herbivore	Natural enemy	Semiochemical cue
Bark beetle	Predator	Host sex or aggregation pheromones as kairomones
		Host volatiles
	Parasitoid	Microbial associate volatiles
		Host volatiles
Gypsy moth	Predator	Volatiles from damaged foliage for host location
	Parasitoid	Volatiles from damaged foliage for host location
		Host sex pheromones as kairomones
Eucalyptus longhorned borer	Parasitoid	Stress volatiles of the tree as cues for adult beetles and natural enemies

13.3 Examples from North America

Table 13.1 gives a summary of the insect herbivore types discussed below, along with the semiochemicals used by their natural enemies.

13.3.1 Native bark beetles in plantation and unmanaged forests

Bark beetles (Coleoptera: Curculionidae: Scolytinae) are capable of causing high levels of mortality in both plantation and unmanaged conifer forests in many parts of the world. The population dynamics of the diverse community of native bark beetles in North American forests has been studied extensively in an effort to understand the causes of mortality and to develop sustainable management approaches. They can kill large numbers of trees in both managed and unmanaged forests, so the management approaches that are applicable are determined, in part, by the forest type and the value of the resource.

Natural enemies have been identified as important causes of beetle mortality (Stephen & Dahlsten 1974, 1976, Linit & Stephen 1983, Reeve 1997). Forest entomologists have observed that predators arrive on host trees at the same time as their bark beetle hosts. The predators feed on both arriving adult bark beetles but also lay eggs on trees as they become infested. It was subsequently determined that the predators have a kairomonal response to the bark beetle sex or aggregation pheromones (Vite & Williams 1970, Dixon & Payne 1979, Bedard *et al.* 1980, Billings & Cameron 1984, Mizell *et al.* 1984, Herms *et al.* 1991, Raffa & Dahlsten 1995, Salom *et al.* 1995, Sullivan *et al.* 2000, Erbilgin & Raffa 2001a, 2001b, Zhou *et al.* 2001, Dahlsten *et al.* 2004, Miller *et al.* 2005, Shepherd *et al.* 2005, Costa & Reeve 2011, Hofstetter *et al.* 2012).

The arrival of parasitoids must be timed to coincide with the development of the appropriate developmental stage of the host insect (Pettersson & Boland 2003). Although there are some notable exceptions (Camors & Payne 1972), a response to bark beetle pheromones

Figure 13.1 *Dendroctonus ponderosae* Hopkins (Coleoptera: Curculionidae: Scolytinae) colonize many *Pinus* species in pheromone-mediated mass attacks. Two species of symbiotic fungi carried in a maxillary mycangium are inoculated into the inner bark tissues and grow to line the pupal chambers. The fungi are critical for larval and teneral adult nutrition. Volatiles produced by the mycangial fungi are used as cues by parasitoids to locate suitable host insect life stages.

would not be too early for larval or pupal parasitoids and the production of pheromones would be completed by the time immature stages were present. Moreover, many bark beetles are associated with a community of fungal symbionts that colonize inner bark and wood tissues of the tree following inoculation by the beetles (Fig. 13.1). Many parasitoids will respond to volatiles associated with these fungi (Sullivan *et al.* 1997, 2003, Sullivan & Berisford 2004, Adams & Six 2008, Boone *et al.* 2008, 2009). Finally, adult parasitoids typically require supplemental sources of food to achieve maximum fecundity and may thus respond to food odours (Stephen & Browne 2000, Vanlaerhoven *et al.* 2005).

Although both predators and parasitoids will respond to semiochemicals, there are potential problems in implementing a pest management strategy based on such chemical compounds. As noted previously, release of aggregation or sex pheromones to attract predators could attract bark beetles as well as natural enemies. It is possible to manipulate the components of the pheromone lure to minimize the on-target effects (Aukema & Raffa 2005). The release of volatiles associated with the symbiotic fungi present less risk of attracting bark beetles, but release of the semiochemicals and the response of the parasitoids must be closely timed to the phenology of host development in order to be effective. The cost of materials and application technology and potential for benefit across the forest rotation cycle or other management objectives of the forest may significantly limit the potential for implementation.

13.3.2 Introduced defoliator in urban and unmanaged forests

The gypsy moth, *Lymantria dispar* L. (Lepidoptera: Lymantriida), was introduced into North America in 1868 (Forbush & Fernald 1896) and has now become widely distributed in urban and unmanaged deciduous forests (Fig. 13.2). A community of introduced natural enemies of the herbivore, including 12 species of parasitoids (Hajek 2007), has become established in the environment and exerts some degree of control on the pest population (Reardon 1981, Lee & Pemberton 2010). Unfortunately, some of these introduced parasi-

Figure 13.2 Feeding by larvae of *Lymantria dispar* on plant host tissues causes the release of volatiles that can be used by predators or parasitoids in searching for suitable prey or hosts.

toids (e.g. the dipteran parasitoid *Compsilura concinnata* (Meigen)) are generalist species that attack a wide range of lymantriid and saturniid moths (Boettner *et al.* 2000, Elkinton & Boettner 2012).

The impact of the parasitoid community on the population dynamics of the gypsy moth has been documented (Gould *et al.* 1992a, 1992b). However, there is little information available to document how the natural enemies respond to behavioural stimuli in their environment. Semiochemicals may be used at very close range to ensure male mating success (Danci *et al.* 2011, Hrabar *et al.* 2012). However, it is unlikely that these compounds would have any utility for increasing the rate of parasitism on a landscape scale. Consequently, the lack of research on host finding in the parasitoid and predator communities limits the prospect of using chemical cues to enhance biological control of gypsy moth in these forest systems.

13.3.3 Introduced wood borer in plantation and urban environments

Eucalyptus spp. are native to Australia and New Guinea but have been widely planted around the world in plantations for fibre production and also in urban environments (Doughty 2000). The cerambycid borer, *Phoracantha semipunctata* F. (Coleoptera: Cerambycidae), has been introduced into all of these environments over the last century and has caused very significant tree mortality (Paine *et al.* 2011). Males and females are attracted to stressed or damaged trees in response to host volatiles (Hanks *et al.* 1998, Paine *et al.* 2000). Males search for females on the bark surface, mate, and then guard the females until oviposition is complete (Hanks *et al.* 1996a, 1996b).

Biological control of the borer has been achieved by the introduction of the encyrtid egg parasitoid, *Avetianella longoi* Siscaro (Hymenoptera, Encyrtidae) (Hanks *et al.* 1995). The parasitoid has a fitness advantage if it arrives at the host tree very soon after the eggs are deposited (Fig. 13.3) and uses eggs that are less than 60 hours old (Luhring *et al.* 2000). Although it has not yet been studied, it is a reasonable hypothesis that the parasitoid is using very similar semiochemical cues as those used by adult beetles searching for suitable host trees, specifically volatiles released by stressed or damaged host trees, to locate the habitat occupied by ovipositing beetles.

Figure 13.3 The egg parasitoid *Avetianella longoi*, arrives at beetle egg masses within hours of oviposition. Both sexes of the beetle host *Phoracantha semipunctata* are attracted to volatiles released by stressed *Eucalyptus* trees. Shortly after mating, females lay batches of spindle-shaped pale-yellow eggs under exfoliating bark or in tight crevices. The synchronous arrival suggests that the beetles and the parasitoids use the same plant volatile cues.

If the parasitoids are attracted to a specific combination of host volatiles, then it might be possible to release those chemical signals to facilitate aggregation of the natural enemies. However, if the beetles and the parasitoids are using the same stimuli, then there is a significant risk that releasing the volatiles will also result in an aggregation of ovipositing female borers. Unless there is a very large population of parasitoids in the local environment, the result could be increased tree mortality instead of tree protection.

13.4 Conclusions

There are significant opportunities to use chemical signals derived from host plants or host insect sources to enhance biological control by natural enemies in diverse agricultural systems. The opportunities are much more limited in forest environments. The forest systems, whether they are unmanaged, plantation or urban, have many different values that extend well beyond their production of fuel and fibre. Even if the primary value is derived from harvesting the timber, the harvest cycles may be on the order of decades to centuries. Consequently, the annual value of the wood produced may be low compared with agricultural systems and the total value is not realized until the forest reaches harvest maturity. This places an economic limit on the return on the costs associated with any management strategy. Other forms of value for the forests, for example watershed protection, scenic amenity, wilderness habitat, wildlife or pasturage, are subject to changes in societal priorities and are much more difficult to determine. The ambiguity of these values over the life of a forest may also limit the willingness to undertake costly management options. Finally, the episodic nature of many insect pests in forest systems and the extensive size of the resource make it difficult to implement management tactics. The logistics of scale with their associated costs, as well as the long periods between outbreaks, are major factors in reducing opportunities for the implementation of chemical cue-based enhancement of natural enemy activity in many forest environments.

References

Adams, A.S. and Six, D.L. (2008) Detection of host habitat by parasitoids using cues associated with mycangial fungi of the mountain pine beetle, *Dendroctonus ponderosae*. *Canadian Entomologist* **140**: 124–7.

Aukema, B.H. and Raffa, K.F. (2005) Selective manipulation of predators using pheromones: responses to frontalin and ipsdienol pheromone components of bark beetles in the Great Lakes region. *Agricultural and Forest Entomology* **7**: 193–200.

Aukema, B.H., Dahlsten, D.L. and Raffa, K.F. (2000) Exploiting behavioral disparities among predators and prey to selectively remove pests: maximizing the ratio of bark beetles to predators removed during semiochemically based trap-out. *Environmental Entomology* **29**: 651–60.

Battisti, A. (1989) Field studies on the behavior of two egg parasitoids of the pine processionary moth *Thaumetopoes pityocampa*. *Entomophaga* **34**: 29–38.

Bedard, W.D., Wood, D.L., Tilden, P.E., Lindahl, K.Q., Silverstein, R.M. and Rodin, J.O. (1980) Field responses of the western pine beetle (Coleoptera: Scolytidae) and one of its predators to host-produced and beetle-produced compounds. *Journal of Chemical Ecology* **6**: 625–41.

Bednarz, B., Kacprzyk, M. and Cebrat, R. (2011) The influence of rich odours on bark beetles infestation of trap-trees in spruce (*Picea abies* L. Karst) stands. *Sylwan* **155**: 179–87.

Billings, R.F. and Cameron, R.S. (1984) Kairomonal responses of Coleoptera, *Monochamus titillator* (Cerambycidae), *Thanasimus dubius* (Cleridae), and *Temnochila virescens* (Trogositidae), to behavioral chemicals of the southern pine beetle (Coleoptera: Scolytidae). *Environmental Entomology* **13**: 1542–8.

Boettner, G.H., Elkinton, J.S. and Boettner, C.J. (2000) Effects of a biological control introduction on three nontarget native species of saturniid moths. *Conservation Biology* **14**: 1798–806.

Boone, C.K., Six, D.L., Krauth, S.J. and Raffa K.F. (2009) Assemblage of Hymenoptera arriving at logs colonized by *Ips pini* (Coleoptera: Curculionidae: Scolytinae) and its microbial symbionts in western Montana. *Canadian Entomologist* **141**: 172–99.

Boone, C.K., Six, D.L., Zheng, Y. and Raffa, K.F. (2008) Parasitoids and dipteran predators exploit volatiles from microbial symbionts to locate bark beetles. *Environmental Entomology* **37**: 150–61.

Branco, M., Franco, J.C., Dunkelblum, E., Assael, F., Protasov, A., Ofer, D. and Mendel, Z. (2006a) A common mode of attraction of larvae and adults on insect predators to the sex pheromone of their prey (Hemiptera: Matsucoccidae). *Bulletin of Entomological Research* **96**: 179–85.

Branco, M., Lettere, M., Franco, J.C., Binazzi, A. and Jactel, H. (2006b) Kairomonal response of predators to three pine bast scale sex pheromones. *Journal of Chemical Ecology* **32**: 1577–86.

Branco, M., van Halder, I., Franco, J.C., Constantin, R. and Jactel, H. (2011) Prey sex pheromone as kairomone for a new group of predators (Coleoptera: Dasytidae, *Aplocnemus* spp.) of pine bast scales. *Bulletin of Entomological Research* **101**: 667–74.

Camors, F.B. and Payne, T.L. (1972) Response of *Heydenia unica* (Hymenoptera: Pteromalidae) to *Dendroctonus frontalis* (Coleoptera: Scolytidae) pheromones and a host-tree terpene. *Annals of the Entomological Society of America* **65**: 31–3.

Coleman, T.W., Grulke, N.E., Daly, M., Godinez, C., Schilling, S.L., Riggan, P.J. and Seybold, S. (2011) Coast live oak, *Quercus agrifolia*, susceptibility and response to goldspotted oak borer, *Agrilus auroguttatus*, injury in southern California. *Forest Ecology and Management* **261**: 1852–65.

Costa, A. and Reeve, J.D. (2011) Upwind flight response of the bark beetle predator *Thanasimus dubius* towards olfactory and visual cues in a wind tunnel. *Agricultural and Forest Entomology* **13**: 283–90.

Couture, J.J., Meehan, T.D. and Lindroth, R.L. (2012) Atmospheric change alters foliar quality of host trees and performance of two outbreak insect species. *Oecologia* **168**: 863–76.

Dahlsten, D.L., Six, D.L., Rowney, D.L., Lawson, A.B., Erbilgin, N. and Raffa, K.F. (2004) Attraction of *Ips pini* (Coleoptera: Scolytinae) and its predators to natural attractants and synthetic semiochemicals in northern California: implications for population monitoring. *Environmental Entomology* **33**: 1554–61.

Danci, A., Inducil, C., Schaefer, P.W. and Gries, G. (2011) Early detection of prospective mates by males of the parasitoid wasp *Pimpla disparis* Viereck (Hymenoptera: Ichneumonidae). *Environmental Entomology* **40**: 405–11.

Dippel, C. and Hilker, M. (1998) Effects of physical and chemical signals on host foraging behavior of *Drino inconspicua* (Diptera: Tachinidae), a generalist parasitoid. *Environmental Entomology* **27**: 682–7.

Dixon, W.N. and Payne, T.L. (1979) Aggregation of *Thanasimus dubius* on trees under mass-attack by the southern pine beetle. *Environmental Entomology* **8**: 178–81.

Doughty, R.W. (2000) *The Eucalyptus: A Natural and Commercial History of the Gum Tree*. The Johns Hopkins University Press, Baltimore, MD.

Elkinton, J.S. and Boettner, G.H. (2012) Benefits and harm caused by the introduced generalist tachinid, *Compsilura concinnata*, in North America. *BioControl* **57**: 277–88.

El-Sayed, A.M., Suckling, D.M., Byers, J.A., Jang, E.B. and Wearing, C.H. (2009) Potential of 'lure and kill' in long-term pest management and eradication of invasive species. *Journal of Economic Entomology* **102**: 815–35.

Erbilgin, N. and Raffa, K.F. (2001a) Kairomonal range of generalist predators in specialized habitats: responses to multiple phloeophagous species emitting pheromones vs. host odors. *Entomologia Experimentalis et Applicata* **99**: 205–10.

Erbilgin, N. and Raffa, K.F. (2001b) Modulation of predator attraction to pheromones of two prey species by stereochemistry of plant volatiles. *Oecologia* **127**: 444–53.

Fischbein, D., Bettinelli, J., Bernstein, C. and Corley, J.C. (2012) Patch choice from a distance and use of habitat information during foraging by the parasitoid *Ibalia leucospoides*. *Ecological Entomology* **37**: 161–8.

Forbush, E.H. and Fernald, C.H. (1896) *The Gypsy Moth*. Wright and Potter Printing Co., Boston, MA.

Gould, J.R., Elkinton, J.S. and van Driesche, R.G. (1992a) Assessment of potential methods of measuring parasitism by *Brachymeria intermedia* (Nees) (Hymenoptera: Chalcididae) of pupae of the gypsy moth. *Environmental Entomology* **21**: 394–400.

Gould, J.R., Elkinton, J.S. and van Driesche, R.G. (1992b) Suitability of approaches for measuring parasitoid impact on *Lymantria dispar* (Lepidoptera: Lymantriidae) populations. *Environmental Entomology* **21**: 1035–45.

Grégoire, J.C., Couillien, D., Drumont, A., Meyer, H. and Francke, W. (1992) Semiochemicals and the management of *Rhizophagus grandis* Gyll (Col, Rhizophagidae) for the biocontrol of *Dendroctonus micans* Kug (Col, Scolytidae). *Zeitschrift fur Angewandte Entomologie* **14**: 110–12.

Hajek, A.E. (2007) Classical biological control of gypsy moth: introduction of the entomopathogenic fungus *Entomophaga maimaiga* into North America. In: Vincent, C., Goettel, M. and Lazarovits, G. (eds) *Biological Control: International Case Studies*. CABI Publications, Wallingford, UK, pp. 53–62.

Hanks, L.M., Gould, J.R., Paine, T.D., Millar, J.G. and Wang, Q. (1995) Biology and host relations of *Avetianella longoi* (Hymenoptera: Encyrtidae), an egg parasitoid of the eucalyptus longhorned borer (Coleoptera: Cerambycidae). *Annals of the Entomological Society of America* **88**: 666–71.

Hanks, L.M., Millar, J.G. and Paine, T.D. (1996a) Body size influences mating success of the eucalyptus longhorned borer (Coleoptera: Cerambycidae). *Journal of Insect Behavior* **9**: 369–82.

Hanks, L.M., Millar, J.G. and Paine, T.D. (1996b) Mating behavior of the eucalyptus longhorned borer (Coleoptera: Cerambycidae) and the significance of long 'horns'. *Journal of Insect Behavior* **9**: 383–93.

Hanks, L.M., Millar, J.G. and Paine, T.D. (1998) Dispersal of the eucalyptus longhorned borer (Coleoptera: Cerambycidae) in urban landscapes. *Environmental Entomology* **27**: 1418–24.

Herms, D.A., Haack, R.A. and Ayres, B.D. (1991) Variation in semiochemical-mediated prey–predator interaction: *Ips pini* (Scolytidae) and *Thanasimus dubius* (Cleridae). *Journal of Chemical Ecology* **17**: 1705–14.

Hilker, M., Blaske, V., Kobs, C. and Dippel, C. (2000) Kairomonal effects of sawfly sex pheromones on egg parasitoids. *Journal of Chemical Ecology* **26**: 2591–601.

Hilker, M., Kobs, C., Varma, M. and Schrank, K. (2002) Insect egg deposition induces *Pinus sylvestris* to attract egg parasitoids. *Journal of Experimental Biology* **205**: 455–61.

Hilker, M., Stein, C., Schroder, R., Varama, M. and Mumm, R. (2005) Insect egg deposition induces defence responses in *Pinus sylvestris*: characterisation of the elicitor. *Journal of Experimental Biology* **208**: 1849–54.

Hofstetter, R.W., Gaylord, M.L., Martinson, S. and Wagner, M.R. (2012) Attraction to monoterpenes and beetle-produced compounds by syntopic *Ips* and *Dendroctonus* bark beetles and their predators. *Agricultural and Forest Entomology* **14**: 207–15.

Hrabar, M., Danci, A., Schaefer, P.W. and Gries, G. (2012) In the nick of time: males of the parasitoid wasp *Pimpla disparis* respond to semiochemicals from emerging mates. *Journal of Chemical Ecology* **38**: 253–261.

Hulcr, J., Pollet, M., Ubik, K. and Vrkoc, J. (2005) Exploitation of kairomones and synomones by *Medetera* spp. (Diptera: Dolichopodidae), predators of spruce bark beetles. *European Journal of Entomology* **102**: 655–62.

Hulcr, J., Ubik, K. and Vrkoc, J. (2006) The role of semiochemicals in tritrophic interactions between the spruce bark beetle *Ips typographus*, its predators and infested spruce. *Journal of Applied Entomology* **130**: 275–83.

Koepke, D., Beyaert, I., Gershenzon, J., Hilker, M. and Schmidt, A. (2010) Species-specific responses of pine sesquiterpene synthases to sawfly oviposition. *Phytochemistry* **71**: 909–17.

Koepke, D., Schroeder, R., Fischer, H.M., Gershenzon, J., Hilker, M. and Schmidt, A. (2008) Does egg deposition by herbivorous pine sawflies affect transcription of sesquiterpene synthases in pine? *Planta* **228**: 427–38.

Lee, J.-H. and Pemberton, R.W. (2010) Parasitoid complex of the Asian gypsy moth (*Lymantria dispar*) (Lepidoptera: Lymantriidae) in Primorye Territory, Russia Far East. *Biocontrol Science and Technology* **20**: 197–211.

Lindgren, B.S. and Borden, J.S. (1993) Displacement and aggregation of mountain pine beetles, *Dendroctonus ponderosae* (Coleoptera: Scolytidae), in response to their antiaggregation and aggregation pheromones. *Canadian Journal of Forest Research* **23**: 286–90.

Linit, M.J. and Stephen, F.M. (1983) Parasite and predator component of within-tree southern pine beetle (Coleoptera: Scolytidae) mortality. *Canadian Entomologist* **115**: 679–88.

Longhurst, C. and Baker, R. (1981) Host location in *Olesicampe monticola*, a parasite of larvae of larch sawfly *Cephalcia lariciphila*. *Journal of Chemical Ecology* **7**: 203–8.

Lu, M., Wingfield, M.J., Gillette, N. and Sun, J.-H. (2011) Do novel genotypes drive the success of an invasive bark beetle-fungus complex? Implications for potential reinvasion. *Ecology* **92**: 2013–19.

Luhring, K.A., Paine, T.D., Millar, J.G. and Hanks, L.M. (2000) Suitability of the eggs of two species of eucalyptus longhorned borers (*Phoracantha recurva* and *Phoracantha semipunctata*) as hosts for the encyrtid parasitoid *Avetianella longoi*. *Biological Control* **19**: 95–104.

Madden, J.L. (1968) Behavioral responses of parasites to symbiotic fungus associated with *Sirex noctilio* F. *Nature* **218**: 189–90.

Martinez Andres, S., Fernandez-Arhex, V. and Corley, J.C. (2006) Chemical information from the fungus *Amylostereum areolatum* and host-foraging behaviour in the parasitoid *Ibalia leucospoides*. *Physiological Entomology* **31**: 336–40.

Mendel, Z., Assael, F. and Dunkelblum, E. (2004) Kairomonal attraction of predatory bugs (Heteroptera: Anthocoridae) and brown lacewings (Neuroptera: Hemerobiidae) to sex pheromones of *Matsucoccus* species (Hemiptera: Matsucoccidae). *Biological Control* **30**: 134–40.

Mendel, Z., Zegelman, L., Hassner, A., Assael, F., Harel, M., Tam, S. and Dunkelblum, E. (1995) Outdoor attractancy of males of *Matsucoccus josephi* (Homoptera: Matsucoccidae) and *Elatophilus hebraicus* (Hemiptera: Anthocoridae) to synthetic female sex-pheromone of *Matsucoccus josephi*. *Journal of Chemical Ecology* 21: 331–41.

Miller, D.R., Borden, J.H. and Lindgren, B.S. (2005) Dose-dependent pheromone responses of *Ips pini*, *Orthotomicus latidens* (Coleoptera: Scolytidae), and associates in stands of lodgepole pine. *Environmental Entomology* 34: 591–7.

Mizell, R.F., Frazier, J.L. and Nebeker, T.E. (1984) Response of the clerid predator *Thanasimus dubius* (F) to bark beetle pheromones and tree volatiles in a wind-tunnel. *Journal of Chemical Ecology* 10: 177–87.

Mumm, R. and Hilker, M. (2006) Direct and indirect chemical defence of pine against folivorous insects. *Trends in Plant Science* 11: 351–8.

Mumm, R., Tiemann, T., Varama, M. and Hilker, M. (2005) Choosy egg parasitoids: specificity of oviposition-induced pine volatiles exploited by an egg parasitoid of pine sawflies. *Entomologia Experimentalis et Applicata* 115: 217–25.

Paine, T.D., Hanks, L.M., Millar, J.G. and Paine, E.O. (2000) Attractiveness and suitability of host tree species for colonization and survival of *Phoracantha semipunctata* (Coleoptera: Cerambycidae). *Canadian Entomologist* 132: 907–13.

Paine, T.D., Raffa, K.F. and Harrington, T.C. (1997) Interactions among scolytid bark beetles, their associated fungi, and live host conifers. *Annual Review of Entomology* 42: 179–206.

Paine, T.D., Steinbauer, M.J. and Lawson, S.A. (2011) Native and exotic pests of eucalyptus: a world-wide perspective. *Annual Review of Entomology* 56: 181–201.

Pettersson, E.M. (2001a) Volatiles from potential hosts of *Rhopalicus tutela* a bark beetle parasitoid. *Journal of Chemical Ecology* 27: 2219–31.

Pettersson, E.M. (2001b) Volatile attractants for three Pteromalid parasitoids attacking concealed spruce bark beetles. *Chemoecology* 11: 89–95.

Pettersson, E.M. and Boland, W. (2003) Potential parasitoid attractants, volatile composition through-out a bark beetle attack. *Chemoecology* 13: 27–37.

Pugh, S.A., Liebhold, A.M. and Morin, R.S. (2011) Changes in ash tree demography associated with emerald ash borer invasion, indicated by regional forest inventory data from the Great Lake States. *Canadian Journal of Forest Research* 41: 2165–75.

Raffa, K.F. and Dahlsten, D.L. (1995) Differential responses among natural enemies and prey to bark beetle pheromones. *Oecologia* 102: 17–23.

Reardon, R.C. (1981) Rearing, evaluation and attempting to establish exotic species and redistribut-ing established species. In: Doane, C.C. and McManus, M.L. (eds) *The Gypsy Moth: Research Toward Integrated Pest Management*. USDA Technical Bulletin 1584, pp. 348–51.

Reeve, J.D. (1997) Predation and bark beetle dynamics. *Oecologia* 112: 48–54.

Rostás, M., Dippel, C. and Hilker, M. (1998) Infochemicals influencing the host foraging behaviour of *Dahlbominus fuscipennis*, a pupal parasitoid of the European spruce sawfly (*Gilpinia hercyniae*). *Entomologia Experimentalis et Applicata* 86: 221–7.

Rutledge, C.E. (1996) A survey of identified kairomones and synomones used by insect parasitoids to locate and accept their hosts. *Chemoecology* 7: 121–31.

Salom, S.M., Grosman, D.M., McClellan, Q.C. and Payne, T.L. (1995) Effect of an inhibitor-based suppression tactic on abundance and distribution of the southern pine-beetle (Coleoptera: Sco-lytidae) and its natural enemies. *Journal of Economic Entomology* 88: 1703–16.

Shepherd, W.P., Sullivan, B.T., Goyer, R.A. and Klepzig, K.D. (2005) Electrophysiological and olfac-tometer responses of two histerid predators to three pine bark beetle pheromones. *Journal of Chemical Ecology* 31: 1101–10.

Spradbery, J.P. (1970) Host finding by *Rhyssa persuasoria* (L), an ichneumonid parasite of siricid woodwasps. *Animal Behaviour* 18: 103–14.

Spradbery, J.P. (1974) Responses of *Ibalia* species (Hymenoptera: Ibaliidae) to fungal symbionts of siricid woodwasp hosts. *Journal of Entomology Series A – Physiology and Behaviour* **48**: 217–22.

Stephen, F.M. and Browne, L.E. (2000) Application of Eliminade™ parasitoid food to boles and crowns of pines (Pinaceae) infested with *Dendroctonus frontalis* (Coleoptera: Scolytidae). *Canadian Entomologist* **132**: 983–5.

Stephen, F.M. and Dahlsten, D.L. (1974) Natural enemies and inset associates of mountain pine beetle, *Dendrocyonus ponderosae* (Coleoptera: Scolytidae), in sugar pine. *Canadian Entomologist* **106**: 1211–17.

Stephen, F.M. and Dahlsten, D.L. (1976) Arrival sequence of arthropod complex following attack by *Dendroctonus brevicomis* (Coleoptera: Scolytidae) in ponderosa pine. *Canadian Entomologist* **108**: 283–304.

Suckling, D.M, Tobin, P.C, McCullough, D.G. and Hermes, D.A. (2012) Combining tactics to exploit Allee effects for eradication of alien pest populations. *Journal of Economic Entomology* **105**: 1–13.

Sullivan, B.T. and Berisford, C.W. (2004) Semiochemicals from fungal associates of bark beetles may mediate host location behavior of parasitoids. *Journal of Chemical Ecology* **30**: 703–17.

Sullivan, B.T., Berisford, C.W. and Dalusky, M.J. (1997) Field response of southern pine beetle parasitoids to some natural attractants. *Journal of Chemical Ecology* **23**: 837–56.

Sullivan, B.T., Dalusky, M.J. and Berisford, C.W. (2003) Interspecific variation in host-finding cues if parasitoids if the southern pine beetle (Coleoptera: Scolytidae). *Journal of Entomological Science* **38**: 631–43.

Sullivan, B.T., Pettersson, E.M., Seltmann, K.C. and Berisford, C.W. (2000) Attraction of the bark beetle parasitoid *Roptrocerus xylophagorum* (Hymenoptera: Pteromalidae) to host-associated olfactory cues. *Environmental Entomology* **29**: 1138–51.

Tassan, R.L., Hagen, K.S. and Sawall, E.F. (1979) Influence of field food sprays on the egg-production rate of *Chrysopa carnea* (Neroptera: Chrysopidae). *Environmental Entomology* **8**: 81–5.

Tobin, P.C., Berec, L. and Liebhold, A.M. (2011) Exploiting Allee effects for managing biological invasions. *Ecology Letters* **14**: 615–24.

Tommeras, B.A. (1985) Specialization of the olfactory receptor-cells in the bark beetle *Ips typographus* and its predator *Thanasimus formicarius* to bark beetle pheromones and host tree volatiles. *Journal of Comparative Physiology A – Sensory Neural and Behavioral Physiology* **157**: 335–41.

Vanlaerhoven, S.L., Stephen, F.M. and Browne, L.E. (2005) Adult parasitoids of the southern pine beetle, *Dendroctonus frontalis* Zimmermann (Coleoptera: Scolytidae), feed on artificial diet on pine bolts, pine canopy foliage and understory hardwood foliage. *Biocontrol Science and Technology* **15**: 243–54.

Vite, J.P. and Williams, D.L. (1970) *Thanasimus dubius* prey perception. *Journal of Insect Physiology* **16**: 223–39.

Yamanaka, T. and Liebhold, A.M. (2009) Spatially implicit approaches to understand the manipulation of mating success for insect invasion management. *Population Ecology* **51**: 427–44.

Zhou, J.L., Ross, D.W. and Niwa, C.G. (2001) Kairomonal response of *Thanasimus undatulus*, *Enocleris sphegeus* (Coleoptera: Cleridae), and *Temnochila chlorodia* (Coleoptera: Trogositidae) to bark beetle semiochemicals in eastern Oregon. *Environmental Entomology* **30**: 993–8.

Index

Note: Page numbers in *italic* refer to figures; those in **bold** refer to tables.

Chemical Ecology of Insect Parasitoids, First Edition. Eric Wajnberg and Stefano Colazza.
© 2013 John Wiley & Sons, Ltd. Published 2013 by John Wiley & Sons, Ltd.

Printed and bound by CPI Group (UK) Ltd, Croydon, CR0 4YY

16/04/2025

14658462-0003